D0146879

Logic, Sets & Numbers

THIRD EDITION

Roethel / Weinstein / Foley

Logic, Sets & Numbers

THIRD EDITION

Louis F. Roethel
Nassau Community College

Abraham Weinstein
Nassau Community College

Robert G. Foley
Nassau Community College

PWS Publishing Company
Boston

PWS PUBLISHING COMPANY
20 Park Plaza, Boston, MA 02116-4324

Mathematics Editor: Peter W. Fairchild

Signing Representative: Ray Coleman

Production: Editing, Design & Production, Inc.

Technical Illustrator: Art By Ayxa

 ™

International Thomson Publishing
The trademark ITP is used under license

Printed and bound in the United States of America.

9 10 -- 95

Library of Congress Cataloging in Publication Data

Roethel, Louis F.
  Logic, sets & numbers.

  Includes index.
  1. Mathematics—1961-   . I. Weinstein, Abraham.
II. Foley, Robert G.   III. Title
QA39.2.R64   1983   510   83-6737
ISBN 0-534-02687-7

PREFACE

This textbook is designed to provide the nonscience student with a meaningful, up-to-date introductory mathematics course for which a knowledge of arithmetic is the only prerequisite. It is designed to foster understanding and facility with some of the basic concepts of mathematics.

It is our hope that this book will help the reader to accomplish what we believe to be the major goal in the study of mathematics, namely, the development of skill and insight into the solving of problems—both mathematics problems and problems that are encountered in everyday living. We have tried to present the material with a balance of intuition and mathematical rigor, with rigor being sacrificed where we believe that its use will lead to unnecessary difficulty.

Set theory is a fundamental concept used in all mathematics. It is introduced on an intuitive basis and then developed as a formal mathematical structure. Although it is likely that many students have had some previous experience with set theory, no prior knowledge is assumed. A wide variety of methods of treating and applying the theory of sets is presented.

Chapters 2 and 3 introduce the study of symbolic logic. We believe that many concepts discussed in the introductory chapter will enable the reader to understand symbolic logic more easily. The methods developed in symbolic logic will then be applied to the solutions of problems in later chapters and will also reinforce the concepts of set theory.

The relationship between symbolic logic and set theory is more firmly established in Chapter 4. Here it is shown how these two areas of mathematics complement one another.

Chapter 5 is an introduction to the fundamental principles of probability, an essential tool of most experimental and statistical research. This is followed by a chapter devoted to a discussion of high-speed electronic computers, how they work and what we must do in order to use them to solve problems. Chapter 7 is an investigation of our number system, the fundamental operations ($+$, $\times$, $-$, $\div$), and some of the basic properties of the system. The final chapter is devoted to some of the basic concepts of descriptive statistics.

While symbolic logic and the elements of set theory are the unifying thread that provides continuity to the material, they do not bind the chapters inseparably. With the exception of Chapters 3 and 4, the chapters are relatively independent of each other. Individual topics can be selected to provide a variety of courses geared to the needs and interests of the reader.

We would like to take this opportunity to thank the members of the Mathematics Department of Nassau Community College for their valuable comments, which assisted us in the preparation of this manuscript. Special thanks to James J. Corbett, Radford University, for his helpful suggestions and proofreading.

Louis F. Roethel
Abraham Weinstein
Robert G. Foley

CONTENTS

CHAPTER 3:   Arguments and Proofs   127

CHAPTER 7: Number Systems 285

CHAPTER 8: Statistics 329

Logic, Sets & Numbers

THIRD EDITION

Roethel / Weinstein / Foley

1

SET THEORY

The theory of sets, which was developed by the mathematicians Boole and Cantor in the nineteenth century, has now become a fundamental tool of mathematics. This concept, which brought clarity and organization to many difficult areas of mathematics, is today being used in the study of almost all mathematics.

Basic Concepts

Set and *element* are the undefined terms in our study. We cannot give a precise mathematical definition of these ideas. Instead we depend on an intuitive understanding of their meanings. This is similar to the terms point and line that are used in the study of geometry. We cannot define these terms but we know what they mean. If we try to define or explain what we mean when we use the terms point and line we can see the difficulty involved. Intuitively, a *set* is a well-defined collection of objects or ideas. These objects or ideas are called the *elements* or members of the set. They can be anything: numbers, people, furniture, cars, etc.

The following examples will be used to illustrate the set concept:

(a) the set of letters *a*, *e*, *i*, *o*, and *u*

(b) the set of letters of the alphabet

(c) the set of states Alaska, Maine, and Florida

(d) the set of numbers *1*, *2*, *5*, and *9*

(e) the set of odd whole positive numbers

(f)   the set of students in your school

(g)   the set of Presidents Eisenhower, Kennedy, and Johnson

(h)   the set of number 1, a chair, and an automobile

(i)   the set of whole numbers

(j)   the set of days of the week that begin with the letter A

All of the examples can be written in a more concise form by use of the following notation. Capital letters such as A, B, and C will be used to name sets so that it will not be necessary to list or describe the elements each time we refer to the sets. Lowercase letters such as x, y, and z will be used to represent the elements of the set. In representing sets that list the elements we will separate the elements by commas and enclose them in braces, "{ }". In representing sets that describe their properties we will use the symbol x to denote an arbitrary element of the set, and the symbol "|", which is read "such that." Words such as *good, tall, handsome,* and *pretty* should not be used in describing a set since these terms are not well defined.

The preceding examples can now be written:

$A = \{$a, e, i, o, u$\}$

$B = \{x \mid x$ is a letter of the alphabet$\}$

$C = \{$Alaska, Maine, Florida$\}$

$D = \{1, 2, 5, 9\}$

$E = \{x \mid x$ is an odd whole positive number$\}$

$F = \{x \mid x$ is a student in your school$\}$

$G = \{$President Eisenhower, President Kennedy, President Johnson$\}$

$H = \{1,$ chair, automobile$\}$

$I = \{x \mid x$ is a whole number$\}$

$J = \{x \mid x$ is a day of the week that begins with the letter A$\}$

A set is said to be *finite* if it contains some fixed number of elements; if it does not, we say that the set is *infinite*. Sets E and I are examples of infinite sets, while the other examples are finite sets.

A set and an element are related in one of two ways. Either the element belongs to the set or the element does not belong to the set. The symbol "$\in$" is used to indicate the "belongs to" relation and "$\notin$" means "does not belong to." For example,

$$A = \{\text{a, e, i, o, u}\}; \quad \text{then a} \in A, \text{b} \notin A, u \in A, \text{and } t \notin A$$

$$E = \{1, 3, 5, \ldots\}; \quad \text{then } 3 \in E, 6 \notin E, 13 \in E, \text{and } 100 \notin E$$

The method of representing sets by listing the elements that belong to the set is called the *roster method* (examples a, c, d, and g). The method of representing sets by describing the properties that the elements must possess in order to belong to the set is called the *descriptive method* (examples b, e, f, i, and j).

Consider the set $A = \{a, e, i, o, u\}$, which is read, "A is the set of elements a, e, i, o, and u." This set could also have been written in the descriptive method, $A = \{x \mid x$ is a vowel$\}$, which is read, "A is the set of elements such that each element is a vowel."

When the number of elements is small it is relatively easy to use the roster method. Even when the number of elements is large we may still be able to represent the set in roster form by using three dots in the list to indicate that the sequence continues infinitely. For example, $\{1, 2, 3, \ldots\}$ represents the set of natural numbers. If the three dots are placed in middle of a sequence, such as $\{1, 3, 5, \ldots, 99\}$, it means that the sequence continues up to and including the last term, in this case 99. There are instances, however, when we find it *necessary* to express the set in descriptive form. For example, "The set of real numbers between zero and one" can be expressed in words or even symbolically, $\{x \mid 0 < x < 1, x \in R\}$, but it cannot be expressed in roster form.

A third method which can be used to represent sets is graphing. The set $\{1, 2, 3, 4\}$ can be graphed on the number line as and sets of ordered pairs such as $\{(1, 2), (0, 3)\}$ can be graphed on the Cartesian plane.

It will be beneficial to become familiar with all three methods of representing sets so that you will be able to read and understand them regardless of which one is used. You will also be able to communicate your own ideas in the most effective way possible.

Set Relations

Sets can be related to one another in a number of ways. They can be completely different, have some elements in common, be exactly the same, have the same number of elements, or one set can be part of another. To discuss these relations we use the following definitions:

Definition 1

Equal Sets: Two sets A and B are equal ($A = B$) if and only if every element of set A is an element of set B and every element of set B is an element of set A. That is, the two sets have exactly the same elements. If two sets A and B are not equal we write $A \neq B$.

Example 1

Let $A = \{1,\ 2,\ 3,\ 4\}$ and $B = \{3,\ 4,\ 1,\ 2\}$; then $A = B$. That is, $\{1,\ 2,\ 3,\ 4\} = \{3, 4, 1, 2\}$.

The order of the elements in a set is unimportant. If the order is changed, the set is still the same.

Example 2

Let $A = \{g,\ e,\ o,\ r\}$ and $B = \{x \mid x$ is a letter in President Washington's first name$\}$ Since his first name is George and all these letters make up set A, we have $A = B$.

An element in a set is never repeated. In this example set A is formed from the letters of the name George. The letters g and e are not repeated in constructing the set $A = \{g,\ e,\ o,\ r\}$.

Example 3

If $A = \{a,\ e,\ i,\ o,\ u\}$ and $B = \{x \mid x$ is a letter of the alphabet$\}$, then $A \neq B$ since there are other letters, such as b, in the alphabet which are elements of set B but not elements of set A.

Definition 2

Equivalent Sets: Two sets A and B are equivalent ($A \simeq B$) if and only if for each element in A there is exactly one element in B and for each element in B there is exactly one element in A. If two sets A and B are not equivalent, we write $A \not\simeq B$.

Example 1

Let $A = \{1,\ 2,\ 3\}$ and $B = \{a,\ b,\ c\}$; then $A \simeq B$. The finite sets A and B both contain three elements. Finite sets are equivalent when they have the same number of elements.

Example 2

Let $A = \{1, 3, 5, 7, \ldots\}$ and $B = \{2, 4, 6, 8, \ldots\}$; then $A \simeq B$. The elements of the infinite sets A and B can be paired off so that for each element in A there is an element in B and conversely. We can illustrate this as follows:

$$A = \{1, 3, 5, 7, 9, \ldots\}$$

$$B = \{2, 4, 6, 8, 10, \ldots\}$$

When discussing infinite sets, we do not talk about the number of elements in the set.

Definition 3

Subset: Set A is a subset of set B if and only if each element of set A is an element of set B. That is, if $x \in A$, then $x \in B$. Set A is contained in set B.

   To indicate that A is a subset of B we write $A \subseteq B$. This is read "A is contained in B" or "A is a subset of B." The negation of this idea, or "A is not a subset of B" is written $A \nsubseteq B$. From this definition we see that every set is a subset of itself; that is, for any set A, we have $A \subseteq A$. In this case we call it the *improper* subset.

Definition 4

Proper Subset: Set A is a proper subset of set B if and only if set A is contained in set B and not equal to B. That is, if $A \subseteq B$ and $A \neq B$, then A is a proper subset of B. This is symbolized $A \subset B$.

Example 1

Given $A = \{1, 2, 3, 4, 5\}$, $B = \{2, 4\}$, and $C = \{2, 4, 5, 6\}$, we have $B \subset A$, B is a proper subset of A; $B \subset C$, B is also a proper subset of C; $A \nsubseteq C$ and $C \nsubseteq A$, that is, A is not a subset of C and C is not a subset of A.

Example 2

Given $A = \{1, 2, 3, 4, \ldots\}$, $B = \{1, 4, 9, 16, \ldots\}$, and $C = \{x \mid x$ is a whole positive number squared$\}$, we have $B \subseteq C$ because every element in set B is an element in set C. Also, $C \subseteq B$ because every element in set C is an element in set B. Thus we conclude $B = C$, B and C are subsets of one another; $B \subset A$, B is a proper subset of A; C is also a proper subset of A since $B = C$; $A \nsubseteq B$, that is, A is not a subset of B.

Definition 5

Disjoint Sets: Two sets A and B are disjoint if and only if no element of set A is an element of set B and no element of set B is an element of set A; that is, sets A and B have no elements in common.

Example 1

Given $A = \{1, 3, 5, 7\}$ and $B = \{2, 4, 6, 8\}$, then A and B are disjoint since A and B have no elements in common.

Example 2

Given $A = \{2, 3, 5, 7\}$ and $B = \{1, 3, 6, 8\}$, then A and B are not disjoint since $3 \in A$ and $3 \in B$.

Before we continue our study of set relations it is necessary to discuss two important sets.

Definition 6

Universal Set: The universal set, denoted by "U," is the set containing all the elements under discussion. All the sets under consideration are subsets of the universal set U.

Example 1

If we were discussing the sets formed from the letters in people's names, our universal set would then be all the letters of the alphabet, $U = \{a, b, c, d, \ldots, z\}$

Example 2

If we were considering a study to determine ages of college students, our universal set U would be the set of all college students in the world.

The opposite of the universal set is the null set.

Definition 7

Null Set: The null or empty set, denoted by "$\varnothing$" or "$\{\ \}$," is the set containing no elements. The null set is a subset of every set.

Example 1

The set of past female presidents of the United States.

Example 2

The set of all odd numbers that are exactly divisible by 2.

These examples illustrate sets which are well defined but which contain no elements. Since the null set contains no elements, it is defined to be a subset of every set. That is, for any set A, we have $\emptyset \subset A$.

We see that different sets contain a various number of subsets.

Definition 8

Power Set: By the power set of a given set A, we mean the set of all possible subsets of set A. If set A contains n elements, then there are 2^n subsets of A.

Example 1

Find the power set of $A = \{1, 2\}$.

Since the set contains 2 elements, there are $2^2 = 4$ subsets of set A. They are $\emptyset$, $\{1\}$, $\{2\}$, and $\{1, 2\}$, and the power set of A is $\{\emptyset, \{1\}, \{2\}, \{1, 2\}\}$. Note that each element of the power set is itself a set. That is, the power set is a set of sets.

Example 2

Find the power set of $B = \{a, b, c\}$.

Since the set contains 3 elements, there are $2^3 = 8$ subsets of set B.

$$\{\emptyset, \{a\}, \{b\}, \{c\}, \{a, b\}, \{a, c\}, \{b, c\}, \{a, b, c\}\}$$

Applications of the Power Set

There are many applications of the power set idea—the one we will explore is coalitions in voting bodies.

A *voting body* is any entity that decides policy by vote. Let us assume that an executive committee of an organization consists of five members: Art, Betty, Connie, Dave, and Ed. If Art, Betty, and Dave vote the same way on an issue, and assuming majority rule, they would form a *winning coalition*; Connie and Ed would form a *losing coalition*.

The committee could split into many different coalitions. One likely split might be men versus women. Art, Dave, and Ed would form a winning coalition; Betty and Connie would form a losing coalition. Another split might be everyone against Ed. This produces a four-person winning coalition and a one-person losing coalition.

In this voting body of five members, how many possible coalitions are there? How many that win and how many that lose? To answer these questions it is helpful to represent each committee member by the first letter of his or her name. The committee is then a set.

$$C = \{a, b, c, d, e\}$$

Every coalition is a subset of *C*. All possible coalitions are listed below.

$\{a\}$	$\{a, b\}$	$\{b, d\}$	$\{a, b, c\}$	$\{a, c, d\}$	$\{a, b, c, d\}$
$\{b\}$	$\{a, c\}$	$\{b, e\}$	$\{a, b, d\}$	$\{a, c, e\}$	$\{a, c, d, e\}$ $\varnothing$
$\{c\}$	$\{a, d\}$	$\{c, d\}$	$\{a, b, e\}$	$\{a, d, e\}$	$\{a, b, d, e\}$ C
$\{d\}$	$\{a, e\}$	$\{c, e\}$	$\{b, c, d\}$	$\{b, d, e\}$	$\{a, b, c, e\}$
$\{e\}$	$\{b, c\}$	$\{d, e\}$	$\{b, c, e\}$	$\{c, d, e\}$	$\{b, c, d, e\}$

There are several interesting features of this listing:

(a) The set *C* itself a coalition (subset) indicating the situation when all members agree.

(b) The empty set is a losing coalition (subset) in *C*.

(c) There are 32 possible coalitions—half winning and half losing.

(d) If *A* is a proper subset of *B* then *A* is a less powerful coalition than *B*.

The collection of all possible coalitions or subsets is called the power set of the voting body set.

With the five-member executive committee and majority rule there was always a winning coalition on every vote. However, if the committee consists of four, six or, indeed, any even number of members, then blocking coalitions are possible under majority rule. A blocking coalition is one that neither wins nor loses. For example, if $S = \{a, b, c, d\}$, then $\{a, b\}$, $\{a, c\}$, $\{a, d\}$, $\{b, c\}$, $\{b, d\}$, and $\{c, d\}$ are blocking coalitions. The situation can be resolved by giving individual members different voting power. For example, assume $T = \{a, b, c, d\}$ represents the executive committee of a club consisting of the president (a), vice-president (b), the secretary (c), and the treasurer (d). The members might be given different votes, such as $a = 3$, $b = 2$, $c = 1$, and $d = 1$.

The voting power of a coalition will then be the sum of the voting powers of its members. We will use the letter m to represent the "measure" of the set. Thus,

$$
\begin{aligned}
m\,\{a, b, c, d\} &= 7 & \qquad m\,\{a, c\} &= 4 \\
m\,\{a, b, c\} &= 6 & m\,\{a, d\} &= 4 \\
m\,\{b, c, d\} &= 4 & m\,\{c, d\} &= 2 \\
m\,\{a, b, d\} &= 6 & m\,\{a\} &= 3 \\
m\,\{a, c, d\} &= 6 & m\,\{b\} &= 2 \\
m\,\{a, b\} &= 5 & m\,\{c\} &= 1 \\
m\,\{b, c\} &= 3 & m\,\{d\} &= 1 \\
m\,\{b, d\} &= 3 & m\,\{\ \} &= 0
\end{aligned}
$$

Since the highest voting power (measure) of any coalition is seven, the winning coalitions are those with a measure of four or more. Those with a voting power of three or fewer are losers. No blocking coalitions are possible when the highest voting power is an odd number.

Exercises 1.1

For problems 1–4 use the following:

(a) The presidents of the U.S.

(b) The real numbers between 7 and 100.

(c) All short people.

(d) $\{2, 4, 6, \ldots, 100\}$

(e) All old people.

(f)   All teenagers.

(g) $\{1, 3, 8, \ldots\}$

1. Which of the above are well-defined sets?

2. Which of the above are finite sets? Infinite sets?

3. Which of the above is/are equal to the set $\{1, 3, 5, \ldots, 99\}$?

4. Which of the above is/are equivalent to $\{1, 3, 5, \ldots, 99\}$?

5. Name 3 methods of representing sets.

6. List the elements in the set of letters in the word Mississippi.

7. Describe in set notation the days of the week Tuesday and Thursday.

8. Graph the following sets:
   (a) $\{1, 2, 3\}$ on the number line.
   (b) $\{(1, 2), (3, 4)\}$ on the Cartesian plane.

9. If $U = \{1, 2, 3, 4\}$, $A' = \{1, 3\}$, $B = \{2, 4\}$, list the elements in (a) A (b) $A \cap B$ (c) $A \cup B$ (d) $A - B$.

10. How many proper subsets does a set containing 6 elements have?

11. (T, F) George Washington is a member of the set of presidents of the U.S.

12. (T, F) $\{8\} \in \{2, 4, 6, \ldots, 100\}$.

13. (T, F) $8 \notin \{2, 4, 6, \ldots, 100\}$.

14. (T. F) All sets that are equivalent are equal.

15. (T. F) The sets $\{1, 2, 3\}$ and $\{a, b, c\}$ are disjoint.

16. (T, F) $\varnothing \subseteq \{1, 3, 8, \ldots\}$.

In problems 17 and 18, a voting body consists of $S = \{a, b, c\}$ with equal voting power.

17. Name the winning, losing, and blocking coalitions using majority rule.

18. If the voting power is changed to $a = 4$, $b = 3$, and $c = 1$, name the winning, losing and blocking coalitions for: (a) majority rule (b) 2/3 majority rule.

Venn Diagrams

In order to better understand set relations and set operations, which will be discussed in the next section, we now introduce Venn diagrams. A Venn diagram is the pictorial representation of sets. Rectangles usually represent the universal set while circles or any other convenient figures represent the subsets of the universal set. The elements of a set are thought to be contained within the circle representing that set.

Example 1

Describe the following Venn diagram in terms of A and U.

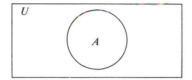

(a)   Set A is a proper subset of the universal set U.

   or

(b)   If $x \in A$, then $x \in U$ and there are elements x such that $x \in U$ and $x \notin A$.

Example 2

Describe the following Venn diagram in terms of A, B, and U.

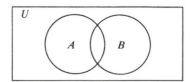

(a)   Sets A and B are neither disjoint nor subsets of one another. They do not make up the entire universal set.

   or

(b)   There are elements x such that $x \in A$ and $x \in B$. There are elements x such that $x \in A$ and $x \notin B$. There are elements x such that $x \in B$ and $x \notin A$. There are elements x such that $x \notin A$ and $x \notin B$.

Example 3

Describe the following Venn diagram in terms of A, B, and U.

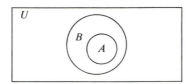

(a)   Set A is a proper subset of B. Set B does not make up the entire universal set.

or

(b)   If $x \in A$, then $x \in B$. There are elements x such that $x \in B$ and $x \notin A$. There are elements x such that $x \notin B$.

Example 4

Describe the following Venn diagram in terms of A, B, C, and U.

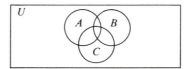

(a)   Sets A and B are neither disjoint nor subsets of one another. Sets A and C are neither disjoint nor subsets of one another. Sets B and C are neither disjoint nor subsets of one another. Sets A, B, and C have elements in common but they do not make up the entire universal set.

or

(b)   There are elements x such that $x \in A$, $x \in B$, and $x \in C$; elements x such that $x \in A$, $x \in B$, and $x \notin C$; elements x such that $x \in A$, $x \in C$, and $x \notin B$; elements x such that $x \in B$, $x \in C$, and $x \notin A$; elements x such that $x \in A$, $x \notin B$, and $x \notin C$; elements x such that $x \in B$, $x \notin A$, and $x \notin C$; elements x such that $x \in C$, $x \notin A$, and $x \notin B$; and finally elements x such that $x \notin A$, $x \notin B$, and $x \notin C$.

Example 5

Describe the following Venn diagram in terms of A, B, C, and U.

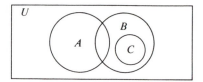

(a) Sets A and C are disjoint. Set C is a proper subset of B. Sets A and B are not disjoint nor is A a subset of B. Sets A, B, and C do not make up the entire universal set.

or

(b) If $x \in C$, then $x \in B$. There are elements x such that $x \in B$ and $x \notin C$. There are no x such that $x \in A$ and $x \in C$. There are x such that $x \in A$ and $x \in B$. There are x such that $x \in A$ and $x \notin B$. There are x such that $x \in B$ and $x \notin A$. Finally, there are x such that $x \notin A$ and $x \notin B$.

In the preceding examples we were given the diagram and asked to describe how the sets are related. Now suppose we are given the description of the set relations and asked to construct the Venn diagram.

Example 6

Construct the Venn diagram which satisfies the following conditions. If $x \in C$, then $x \in A$. There are x such that $x \in A$ and $x \notin C$. There are no x such that $x \in A$ and $x \in B$. There are x such that $x \notin A$ and $x \notin B$.

The first statement tells us that every element of C is an element of A or that C is a subset of A. The second states that there are elements in A which are not in C. From the first and second statements we see that C is a proper subset of A. The third tells us that sets A and B are disjoint. Finally, the last tells us that sets A and B do not make up the entire universal set. This leads to the following diagram.

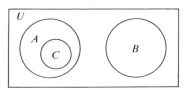

Set Operations

The three basic operations of sets are complementation, union, and intersection. We shall use Venn diagrams to illustrate definitions and examples.

Definition 9

Complement: The complement of a set A is the set of elements in the universal set which are not in set A; that is, the set of all elements in the universal set outside of A. We denote the complement of A by "A'".

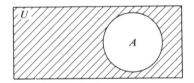

The shaded area of the above figure represents the complement of A. If an element does not belong to set A, then it must belong to the complement of A. Symbolically we write $A' = \{x \mid x \in U \text{ and } x \notin A\}$.

Another method of showing the relationship between a set A and its complement is by use of a *membership table*. According to the definition of complement, for any element x in the universal set U, we have either $x \in A$ or $x \notin A$. That is, for any element $x \in U$,

$$\text{if } x \in A, \quad \text{then } x \notin A'$$

$$\text{if } x \notin A, \quad \text{then } x \in A'$$

Expressing their relationships by a membership table, we have

Table 1

A	A'
$\in$	$\notin$
$\notin$	$\in$

Example 1

Let $U = \{1, 2, 3, 4, 5, 6, 7, 8\}$ and $A = \{1, 3, 6\}$; then $A' = \{2, 4, 5, 7, 8\}$.

Example 2

Let $U = \{x \mid x$ is a student in your school$\}$ and $A = \{x \mid x$ is a math student in your school$\}$; then $A' = \{x \mid x$ is a student in your school who is not taking math$\}$.

When we unite the elements of two sets to form a new set, we are performing the following operation.

Definition 10

Union: The union of two sets A and B, written $A \cup B$, is the set of elements that belong to A or to B or to both. That is, it is the set made up by combining all the elements of set A with all the elements of set B.

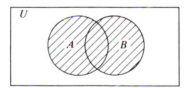

The shaded area of the preceding figure represents the union of sets A and B. If an element belongs to either of the sets, it will belong to their union. Symbolically we write $A \cup B = \{x \mid x \in A \text{ or } x \in B\}$. According to the definition of the union of two sets, $x \in (A \cup B)$ if and only if $x \in A$ or $x \in B$. That is,

$$\text{if } x \in A \text{ and } x \in B, \quad \text{then } x \in (A \cup B)$$

$$\text{if } x \in A \text{ and } x \notin B, \quad \text{then } x \in (A \cup B)$$

$$\text{if } x \notin A \text{ and } x \in B, \quad \text{then } x \in (A \cup B)$$

$$\text{if } x \notin A \text{ and } x \notin B, \quad \text{then } x \notin (A \cup B)$$

Expressing their relationships by a membership table, we have

Table 2

A	B	$A \cup B$
$\in$	$\in$	$\in$
$\in$	$\notin$	$\in$
$\notin$	$\in$	$\in$
$\notin$	$\notin$	$\notin$

Example 1

Let $A = \{1, 3, 4, 6\}$ and $B = \{2, 3, 5, 9\}$; then $A \cup B = \{1, 2, 3, 4, 5, 6, 9\}$.

Example 2

Let $A = \{x \mid x$ is a male student in your school$\}$ and $B = \{x \mid x$ is a female student in your school$\}$; then $A \cup B = \{x \mid x$ is a student in your school$\}$.

The last of the three basic set operations is the one that considers the elements common to two sets.

Definition 11

Intersection: The intersection of two sets A and B, written $A \cap B$, is the set of all elements which belong to both A and B; that is, the set made up of the elements common to A and B.

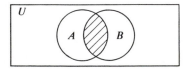

The shaded area of the above figure represents the intersection of sets A and B. If an element belongs to both A and B, it will belong to their intersection. Symbolically we write $A \cap B = \{x \mid x \in A \text{ and } x \in B\}$. According to the definition of the intersection of two sets, $x \in (A \cap B)$ if and only if $x \in A$ and $x \in B$. That is,

$$\text{if } x \in A \text{ and } x \in B, \quad \text{then } x \in (A \cap B)$$

$$\text{if } x \in A \text{ and } x \notin B, \quad \text{then } x \notin (A \cap B)$$

$$\text{if } x \notin A \text{ and } x \in B, \quad \text{then } x \notin (A \cap B)$$

$$\text{if } x \notin A \text{ and } x \notin B, \quad \text{then } x \notin (A \cap B)$$

Expressing their relationships by a membership table, we have

Table 3

A	B	$A \cap B$
$\in$	$\in$	$\in$
$\in$	$\notin$	$\notin$
$\notin$	$\in$	$\notin$
$\notin$	$\notin$	$\notin$

Example 1

Let $A = \{1, 2, 3, 5, 6\}$ and $B = \{2, 3, 4, 6, 8\}$; then $A \cap B = \{2, 3, 6\}$.

Example 2

Let $A = \{1, 2, 3\}$ and $B = \{4, 5, 6\}$; then $A \cap B = \varnothing$. This is true since A and B have no elements in common. Their intersection is the null set.

A fourth set operation can be defined, but it is merely a combination of the operations of intersection and complementation.

Definition 12

Difference: The difference of set A minus set B, written $A - B$, is the set of elements which belong to A but which do not belong to B. That is, it is the set of elements common to A and B'. (*NOTE:* $A - B = A \cap B'$).

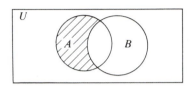

The shaded area of the above figure represents the difference of set A minus set B. It is also the intersection of set A with the complement of B. Symbolically we write $A - B = \{x \mid x \in A \text{ and } x \notin B\}$. According to this definition of the difference of two sets, $x \in (A - B)$ if and only if $x \in A$ and $x \notin B$. That is,

if $x \in A$ and $x \in B$,   then $x \notin (A - B)$

if $x \in A$ and $x \notin B$,   then $x \in (A - B)$

if $x \notin A$ and $x \in B$,   then $x \notin (A - B)$

if $x \notin A$ and $x \notin B$,   then $x \notin (A - B)$

Expressing their relationships by a membership table, we have

Table 4

A	B	$A - B$
$\in$	$\in$	$\notin$
$\in$	$\notin$	$\in$
$\notin$	$\in$	$\notin$
$\notin$	$\notin$	$\notin$

Example 1

Let $A = \{1, 2, 3, 4, 5, 6\}$ and $B = \{3, 5, 6, 7, 9\}$; then $A - B = \{1, 2, 4\}$ and $B - A = \{7, 9\}$.

Example 2

Let $A = \{1, 2, 3\}$ and $B = \{1, 2, 3, 4, 5\}$; then $A - B = \emptyset$ and $B - A = \{4, 5\}$. In this example $A - B = \emptyset$ since $A \subset B$ and, therefore, A and B' would have no elements in common.

No doubt many readers have already observed that the number of rows in the membership table is found by the formula 2^n, where n represents the number of sets involved in the statement. This is the same formula we used previously to determine the number of subsets that a set has; only then, "n" represented the number of elements in the set.

Exercises 1.2

1. Describe the following Venn diagram in terms of A, B, C, and U.

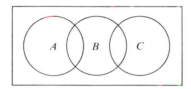

2. Construct the Venn diagram which satisfies the following conditions:
   If $x \in A$, then $x \in B$. There are x such that $x \in B$ and $x \notin A$.
   There are x such that $x \in C$ and $x \in B$.
   There are x such that $x \notin C$ and $x \in B$.
   There are no x such that $x \in A$ and $x \in C$.
   There are x such that $x \notin A$, $x \notin B$, and $x \notin C$.

3. Draw a Venn diagram to represent each of the following:
   (a) Two sets with at least one element in common.
   (b) Two sets that have nothing in common.
   (c) Three sets.
   (d) Two disjoint sets.
   (e) $A \cap B = \emptyset$.

4. Using Venn diagrams represent:
   (a) The complement of set A (b) $A \cup B$
   (c) $(A \cup B)'$ (d) $A \cap B$ (e) $A - B$

5. If $U = \{1, 2, 3, 4, 5, 6, 7, 8, 9\}$, $A = \{1, 2, 3\}$, $B = \{3, 4, 5, 6\}$. Use the roster method to list the elements in question 4(a)–(e).

6. Express the following relationships using membership tables:
   (a) $A \cup B'$ (b) $A \cap B'$ (c) $(A \cup B)'$

Rules Governing Set Operations

Having defined the set operations of complement, union, and intersection and the ideas of null and universal set, we can now list some important rules governing sets.

Let A be an arbitrary subset of U.

(a) $U' = \varnothing$ and $\varnothing' = U$. The null set and the universal set are complements of each other.

(b) $A \cup A' = U$ and $A \cap A' = \varnothing$. The union of any set and its complement equals the universal set, and the intersection of any set and its complement equals the null set. A set and its complement are disjoint.

(c) $A \cup U = U$ and $A \cap U = A$. The union of any set with the universal set equals the universal set and the intersection of any set with the universal set equals itself.

(d) $A \cup \varnothing = A$ and $A \cap \varnothing = \varnothing$. The union of any set with the null set is equal to itself and the intersection of any set with the null set equals the null set.

These rules follow directly from the definitions.

The Application of Operations to Sets

We can apply the operations of complement, union, and intersection to a number of sets to form a more complex set.

Example 1

Let $U = \{1, 2, 3, 4, 5, 6, 7, 8, 9\}$, $A = \{2, 3, 5, 6\}$, $B = \{3, 4, 6, 7\}$, and then let $C = \{5, 6, 7, 8\}$. Find (a) $(A \cap B)'$, (b) $A \cup (B' \cap C)$, and (c) $(A \cup B)' \cap (A \cup C')$.

(a) To find $(A \cap B)'$ we first find $A \cap B$: $A \cap B = \{3, 6\}$. Now we find the complement of $A \cap B$. We have $(A \cap B)' = \{1, 2, 4, 5, 7, 8, 9\}$.

(b) To find $A \cup (B' \cap C)$ we first find the complement of B: $B' = \{1, 2, 5, 8, 9\}$. We then find the intersection of B' and C: $B' \cap C = \{5, 8\}$. Finally we unite A with $B' \cap C$: $A \cup (B' \cap C) = \{2, 3, 5, 6, 8\}$.

(c) To find $(A \cup B)' \cap (A \cup C')$ we first find $A \cup B$: $A \cup B = \{2, 3, 4, 5, 6, 7\}$.
Then we find its complement: $(A \cup B)' = \{1, 8, 9\}$. Next we find the comple-
ment of C: $C' = \{1, 2, 3, 4, 9\}$. We unite it with A: $A \cup C' = \{1, 2, 3, 4, 5, 6, 9\}$.
Finally we intersect $(A \cup B)'$ and $A \cup C'$: $(A \cup B)' \cap (A \cup C') = \{1, 9\}$.

These complex set relations can be represented by Venn diagrams.
When drawing these set diagrams we shall use the following figures.

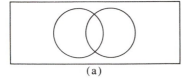
(a)

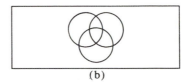
(b)

Diagram a is used when two sets are involved and b is used for three
sets. The relative positions of the circles have been selected so that all the
possible types of intersections will be represented. If there are four or more
sets involved, figures other than circles are required.

Example 2

Draw a Venn diagram and shade in the set $(A \cup B)'$.

First we shade in the set $A \cup B$ with horizontal lines.

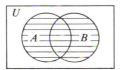

Then we shade the complement $(A \cup B)'$ with vertical lines.

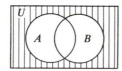

Example 3

Draw a Venn diagram and shade in the set $A \cap (B \cup C)$.

First we shade in set A with horizontal lines, and then set $B \cup C$ with vertical lines.

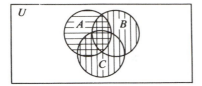 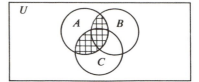

The intersection of A and $B \cup C$ is the region where the horizontal and vertical lines intersect.

This technique is convenient when the set in question is relatively simple. However, if the set notation is more complex, then this method can become confusing. Let us examine the Venn diagram for two sets. We see that when there are two sets, the diagram has four distinct regions, which can be arbitrarily numbered as follows:

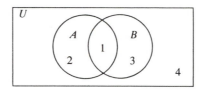

Even though the numbering is arbitrary, the above option, and the one for the three-circle diagram that will follow, is recommended because it will enable the reader to systematically prepare the corresponding membership table. Each of the four regions can be described by one of the four rows in the following membership table. Using the above numbering system then, the first column will start with two $\in$'s followed by two $\notin$'s. The second column will then alternate $\in$ and $\notin$.

	Set	
Region	A	B
1	$\in$	$\in$
2	$\in$	$\notin$
3	$\notin$	$\in$
4	$\notin$	$\notin$

In the above case we have $U = \{1, 2, 3, 4\}$, $A = \{1, 2\}$, and $B = \{1, 3\}$. The universal set U, which contains sets A and B, is divided into four mutually disjoint subsets. These sets can be described in terms of sets A and B as follows:

$$\{1\} = A \cap B = \{x \mid x \in A \text{ and } x \in B\}$$

$$\{2\} = A \cap B' = \{x \mid x \in A \text{ and } x \notin B\}$$

$$\{3\} = A' \cap B = \{x \mid x \notin A \text{ and } x \in B\}$$

$$\{4\} = A' \cap B' = \{x \mid x \notin A \text{ and } x \notin B\}$$

Now using these concepts, we can shade in the region of a Venn diagram that corresponds to some given set.

Example 4

Construct the Venn diagram and shade the set $[(A \cap B') \cup (B \cap A)]'$. Verify the result by the use of a membership table.

Since there are two sets, A and B, we use the following figure:

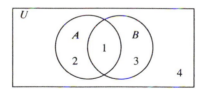

We have $U = \{1, 2, 3, 4\}$, $A = \{1, 2\}$, and $B = \{1, 3\}$. First we find B' and then intersect it with A: $B' = \{2, 4\}$ and $A \cap B' = \{2\}$. Next we find $B \cap A$: $B \cap A = \{1\}$. Now we unite $A \cap B'$ and $B \cap A$: $(A \cap B') \cup (B \cap A) = \{1, 2\}$. Finally, we find the complement of it: $[(A \cap B') \cup (B \cap A)]' = \{3, 4\}$. We shade in the appropriate region, $\{3, 4\}$, below:

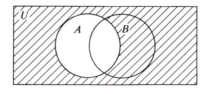

If we apply the membership tables that were used to define complement, union, intersection, and difference (Tables 1, 2, 3, and 4), we can construct a membership table for any set.

We can now construct the Venn diagram for $[(A \cap B') \cup (B \cap A)]'$ by use of a membership table:

Region	A	B	$[(A$	$\cap$	$B')$	$\cup$	$(B$	$\cap$	$A)]$	$'$
1	∈	∈	∈	∉	∉	∈	∈	∈	∈	∉
2	∈	∉	∈	∈	∈	∈	∉	∉	∈	∉
3	∉	∈	∉	∉	∉	∉	∈	∉	∉	∈
4	∉	∉	∉	∉	∈	∉	∉	∉	∉	∈
Steps	1	1	2	4	3	8	5	7	6	9

Step 1 Since there are two sets A and B, we construct a membership table with four rows and then fill in the membership values for A and B.

Step 2 Fill in the membership values for A.

Step 3 Fill in the membership values for B'.

Step 4 Fill in the membership values for $A \cap B'$.

Step 5 Fill in the membership values for B.

Step 6 Fill in the membership values for A.

Step 7 Fill in the membership values for $B \cap A$.

Step 8 Fill in the membership values for $(A \cap B') \cup (B \cap A)$.

Step 9 Fill in the membership values for $[(A \cap B') \cup (B \cap A)]'$.

From the table we see that an element belongs to set $[(A \cap B') \cup (B \cap A)]'$ when it belongs to B and not to A (row 3) and also when it does not belong to either A or B (row 4). The rows of the membership table which have a " ∈ result" correspond to the regions of the Venn diagram that are to be shaded, {3, 4}.

Now we examine the Venn diagram for three sets. We see that the diagram has eight distinct regions which can be arbitrarily numbered as follows:

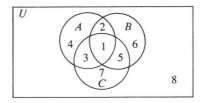

Each of these eight regions can be described by one of the eight rows in the following membership table. Notice that the first column will contain four ∈'s, followed by four ∉'s; the second column will have two ∈'s and two ∉'s and alternate; and the third column will have one ∈ and one $\notin$ alternating. This pattern will appear only if we use the suggested numbering system for the regions.

Region	Set		
	A	B	C
1	$\in$	$\in$	$\in$
2	$\in$	$\in$	$\notin$
3	$\in$	$\notin$	$\in$
4	$\in$	$\notin$	$\notin$
5	$\notin$	$\in$	$\in$
6	$\notin$	$\in$	$\notin$
7	$\notin$	$\notin$	$\in$
8	$\notin$	$\notin$	$\notin$

We have $U = \{1, 2, 3, 4, 5, 6, 7, 8\}$, $A = \{1, 2, 3, 4\}$, $B = \{1, 2, 5, 6\}$, and $C = \{1, 3, 5, 7\}$. The universal set U, which contains sets A, B, and C, is divided into eight mutually disjoint subsets. These disjoint sets can be described in terms of sets A, B, and C as follows:

$$\{1\} = A \cap B \cap C = \{x \mid x \in A \text{ and } x \in B \text{ and } x \in C\}$$

$$\{2\} = A \cap B \cap C' = \{x \mid x \in A \text{ and } x \in B \text{ and } x \notin C\}$$

$$\{3\} = A \cap B' \cap C = \{x \mid x \in A \text{ and } x \notin B \text{ and } x \in C\}$$

$$\{4\} = A \cap B' \cap C' = \{x \mid x \in A \text{ and } x \notin B \text{ and } x \notin C\}$$

$$\{5\} = A' \cap B \cap C = \{x \mid x \notin A \text{ and } x \in B \text{ and } x \in C\}$$

$$\{6\} = A' \cap B \cap C' = \{x \mid x \notin A \text{ and } x \in B \text{ and } x \notin C\}$$

$$\{7\} = A' \cap B' \cap C = \{x \mid x \notin A \text{ and } x \notin B \text{ and } x \in C\}$$

$$\{8\} = A' \cap B' \cap C' = \{x \mid x \notin A \text{ and } x \notin B \text{ and } x \notin C\}$$

Using this information, we can shade in the region of a Venn diagram that corresponds to some given set.

Example 5

Construct the Venn diagram and shade in the set $[(A \cap B) - C] \cup [(B \cap C) \cup A]$. Verify the result by the use of a membership table.

Since there are three sets involved we use the following diagram:

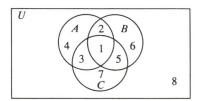

We have $U = \{1, 2, 3, 4, 5, 6, 7, 8\}$, $A = \{1, 2, 3, 4\}$, $B = \{1, 2, 5, 6\}$, and $C = \{1, 3, 5, 7\}$. First we find $A \cap B$: $A \cap B = \{1, 2\}$. Next we find $(A \cap B) - C$: $(A \cap B) - C = \{2\}$. Now we find $B \cap C$ and unite it with A: $B \cap C = \{1, 5\}$ and $(B \cap C) \cup A = \{1, 2, 3, 4, 5\}$. Finally we unite $(A \cap B) - C$ with $(B \cap C) \cup A$: $[(A \cap B) - C] \cup [(B \cap C) \cup A] = \{1, 2, 3, 4, 5\}$. We shade in the appropriate region, $\{1, 2, 3, 4, 5\}$:

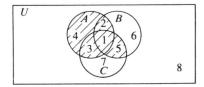

We can now construct the Venn diagram for $[(A \cap B) - C] \cup [(B \cap C) \cup A]$ by use of a membership table:

Regions	A	B	C	[(A	$\cap$	B)	$-$	C]	$\cup$	[(B	$\cap$	C)	$\cup$	A]
1	∈	∈	∈	∈	∈	∈	∉	∈	∈	∈	∈	∈	∈	∈
2	∈	∈	∉	∈	∈	∈	∈	∉	∈	∈	∉	∉	∈	∈
3	∈	∉	∈	∈	∉	∉	∉	∈	∈	∉	∉	∈	∈	∈
4	∈	∉	∉	∈	∉	∉	∉	∉	∈	∉	∉	∉	∈	∈
5	∉	∈	∈	∉	∉	∈	∉	∈	∈	∈	∈	∈	∈	∉
6	∉	∈	∉	∉	∉	∈	∉	∉	∉	∈	∉	∉	∉	∉
7	∉	∉	∈	∉	∉	∉	∉	∈	∉	∉	∈	∉	∉	∉
8	∉	∉	∉	∉	∉	∉	∉	∉	∉	∉	∉	∉	∉	∉
Steps	1	1	1	2	4	3	6	5	12	7	9	8	11	10

The rows of the membership table which have the membership value $\in$ correspond to the region of the Venn diagram that is shaded, $\{1, 2, 3, 4, 5\}$.

Example 6

Determine if the following statement is true: $(A \cap B)' = A' \cap B'$.

We shall solve this example by the use of Venn diagrams. If the shaded region of
one set corresponds precisely with the shaded region of the other set, the two sets
are equal.

Construct a Venn diagram for $(A \cap B)'$:

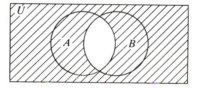

Next construct a Venn diagram for $A' \cap B'$:

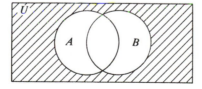

Finally, by comparing the two Venn diagrams, we see that the shaded region for
$(A \cap B)'$ is different than that of $A' \cap B'$. Hence the sets $(A \cap B)'$ and $A' \cap B'$
do not contain precisely the same elements. Therefore, $(A \cap B)' \neq A' \cap B'$.

We could also have solved this problem by membership tables since two
sets are equal if and only if they have precisely the same membership
values:

A	B	$A \cap B$	$(A \cap B)'$	A	B	A'	B'	$A' \cap B'$
∈	∈	∈	∉	∈	∈	∉	∉	∉
∈	∉	∉	∈	∈	∉	∉	∈	∉
∉	∈	∉	∈	∉	∈	∈	∉	∉
∉	∉	∉	∈	∉	∉	∈	∈	∈

Comparing the entries in the columns of $(A \cap B)'$ and $A' \cap B'$ we see that they do not have the same membership values. Therefore, $(A \cap B)' \neq A' \cap B'$.

Example 7

In a certain school, there are 63 students in the senior class: 23 are studying mathematics, 24 are studying English, 26 are studying history, 10 are studying both mathematics and English, 11 are studying both mathematics and history, 9 are studying both English and history, and 6 are studying all three subjects.

(a)   How many are taking only mathematics?

(b)   How many are not taking any of these subjects?

This problem can be solved by means of a Venn diagram. Let each of three circles represent a set of students taking each of the three subjects. The numbers that we will now place in the different regions of the diagram represent how many students satisfy the conditions of that section. Recalling the suggested numbering system for a three-set (circle) diagram (it is drawn again for ready reference), we place a 6 in region I because that section represents students taking all three subjects. Since there are 10 taking both mathematics *and* English and not history (regions I and II combined) we subtract the 6 already in region I from the 10. This leaves 4 for region II alone. In this same manner we get 3 taking English and history but not mathematics (region V) and 5 taking mathematics and history but not English (region III).

To answer part (a), we consider that 23 students are taking mathematics (combined regions I, II, III, IV). Since region IV represents those students taking only mathematics, we need only find the sum of the numbers in regions I, II, and III, (which is 15) and subtract it from 23. Answer 8.

In this same way we find that 11 students are taking just English (region VI) and 12 are taking just history (region VII). To answer part (b), we see that there is a total of 49 students taking at least one of the three subjects (the sum of regions I, II, III, IV, V, VI, VII). Subtracting 49 from 63 leaves 14 students not enrolled in any of the three subjects (region VIII).

REGION NUMBERS

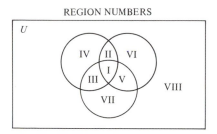

STUDENT PROBLEM

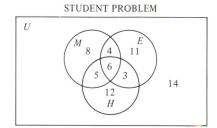

We have seen that a Venn diagram can be used in several ways to picture and clarify various types of problems. In some cases the numbers in the diagram represent the actual elements in these regions; in others the numbers represent the region number itself; and in still others the number represents how many elements are in that region. Be careful not to use two different methods within the same diagram.

Sets and Open Sentences

Let us consider the following sentence: It is a country in Europe. We cannot tell whether this sentence is true or false until more information is given. Such a sentence is called an *open sentence*.

Suppose the " It " can be replaced by an element of the set {United States, Canada, China, France}. If " It " is replaced by France then the resulting sentence " France is a country in Europe," is a true sentence. If " It " is replaced by any other element of the given set, for example, " Canada is a country in Europe," then the sentence is false.

In the sentence " It is a country in Europe," " It " is a symbol which holds the place for the elements of a given set of countries. Such a symbol is called a *variable*.

The set from which the replacements for the variable are selected is called the *replacement set*. In the above example, {United States, Canada, China, France} is the replacement set. The set of elements which replace the variable in an open sentence and form true statements is called the *solution set*. In the above example, " France " is the solution set of the given open sentence " It is a country in Europe."

In mathematics, equations and inequalities such as:

$$x + 6 = 8 \quad \text{(a number plus 6 equals 8)}$$
$$3y - 1 = 16 \quad \text{(3 times a number less 1 equals 16)}$$
$$x > 3 \quad \text{(a number greater than 3)}$$
$$3 + y < 6 \quad \text{(3 plus a number is less than 6)}$$

are often called *open sentences*. In these sentences, x and y represent numbers and these letters are called *variables*. These sentences can be either true or false depending on what numbers from a given replacement set are used in place of the variable. The solution set contains all the numbers of the replacement set that are solutions to the sentence.

For example, using the replacement set $\{0, 1, 2, 3\}$, find the solution set for the open sentence $x + 6 = 8$.

In order to solve the above example replace the variable x in the open sentence $x + 6 = 8$ by each member of the replacement set.

$$\text{If} \quad x = 0, \quad \text{then } 0 + 6 = 8, \text{ which is false}$$
$$x = 1, \quad \text{then } 1 + 6 = 8, \text{ which is false}$$
$$x = 2, \quad \text{then } 2 + 6 = 8, \text{ which is true}$$
$$x = 3, \quad \text{then } 3 + 6 = 8, \text{ which is false}$$

The solution set is $\{2\}$ because the open sentence forms a true statement.

Example 1

Find the solution set for the open sentence $16 - x > 11$. The replacement set is $\{0, 1, 2, 3, 4, 5, 6, 7, 8, 9, 10\}$.

Solution

$$\text{If} \quad x = 0, \quad \text{then } 16 - 0 > 11; \quad 16 > 11 \text{ which is true}$$
$$x = 1, \quad \text{then } 16 - 1 > 11; \quad 15 > 11 \text{ which is true}$$
$$x = 2, \quad \text{then } 16 - 2 > 11; \quad 14 > 11 \text{ which is true}$$
$$x = 3, \quad \text{then } 16 - 3 > 11; \quad 13 > 11 \text{ which is true}$$
$$x = 4, \quad \text{then } 16 - 4 > 11; \quad 12 > 11 \text{ which is true}$$
$$x = 5, \quad \text{then } 16 - 5 > 11; \quad 11 > 11 \text{ which is false}$$
$$x = 6, \quad \text{then } 16 - 6 > 11; \quad 10 > 11 \text{ which is false}$$
$$x = 7, \quad \text{then } 16 - 7 > 11; \quad 9 > 11 \text{ which is false}$$
$$x = 8, \quad \text{then } 16 - 8 > 11; \quad 8 > 11 \text{ which is false}$$
$$x = 9, \quad \text{then } 16 - 9 > 11; \quad 7 > 11 \text{ which is false}$$
$$x = 10, \quad \text{then } 16 - 10 > 11; \quad 6 > 11 \text{ which is false}$$

The solution set is $\{0, 1, 2, 3, 4\}$ because the open sentence forms a true statement.

Example 2

Find the solution set $x < 7$ and $x > 3$. The replacement set is {0, 1, 2, 3, 4, 5, 6, 7, 8, 9, 10}.

Solution

Open Sentence	Solution Set
$x > 7$	$A = \{0, 1, 2, 3, 4, 5, 6\}$
$x > 3$	$B = \{4, 5, 6, 7, 8, 9, 10\}$
$x < 7$ and $x > 3$	$A \cap B = \{4, 5, 6\}$

Example 3

Find the solution set $x < 3$ or $x > 7$. The replacement set is {0, 1, 2, 3, 4, 5, 6, 7, 8, 9, 10}.

Solution

Open Sentence	Solution Set
$x < 3$	$A = \{0, 1, 2\}$
$x > 7$	$B = \{8, 9, 10\}$
$x < 3$ or $x > 7$	$A \cup B = \{0, 1, 2, 8, 9, 10\}$

Example 4

Find the solution set $x \leq 4$. The replacement set is {0, 1, 2, 3, 4, 5, 6, 7, 8, 9, 10}.

Solution

The statement $x \leq 4$ means $x < 4$ or $x = 4$.

Open Sentence	Solution Set
$x < 4$	$A = \{0, 1, 2, 3\}$
$x = 4$	$B = \{4\}$
$x \leq 4$	$A \cup B = \{0, 1, 2, 3, 4\}$

We have seen that when the solution set is represented by an inequality symbol such as "$\leq$" (see Example 4) it actually means "less than or equal to." This means that we must union the answer we get from the "less than" part with the answer we get from the "equal to" part. We must also union the results that we get from two separate statements if they are joined by the word "or," as in Example 3. On the other hand, the word "and" between two separate statements means that we must intersect the solution sets we get from each part.

If the replacement set is changed to the set of real numbers, the roster method of representing the solution set is inadequate. Using the descriptive method the solution to Example 2 could be written more concisely as, $\{x \mid 3 < x < 7\}$. This is called a compound inequality. Notice that $x > 3$ (x is greater than 3) means the same thing as $3 < x$ (3 is less than x).

So long as the number on the extreme left is less than or equal to the number on the extreme right, the symbol $<$ stands for "less than" and is interpreted "less than," but if you are reading it from right to left, it would be interpreted "greater than." A "less than" compound inequality, in proper numerical order, is read with an AND between the two parts and means to intersect the two separate solution sets. For example, $5 < x < 7$ represents the intersection of the set "x is less than 7" with the set "x is greater than 5" (i.e., all numbers between 5 and 7, not including 5 and 7).

You will find it helpful to begin the reading of a compound inequality in the middle, go right, state the proper connective (and then go back to the middle and go left (remembering to reverse the meaning of the symbol).

The compound inequality $3 < x < 7$ can also be represented graphically on a number line as $\overset{3}{\circ}\!\!-\!\!-\!\!-\!\!-\!\!\overset{7}{\circ}$. We use a hollow point to start and end the graph because the numbers 3 and 7 are not in the solution set. If we started the graph at 4 with a solid point and ended it at 6 with a solid point we would have left out all the numbers between 3 and 4 as well as the numbers between 6 and 7 (an infinite number). We can also see that $\overset{3}{\circ}\!\!-\!\!-\!\!-\!\!-\!\!\overset{7}{\circ}$ is the intersection or overlapping part of the two separate graphs $\overset{3}{\circ}\!\!-\!\!-\!\!\rightarrow$ and $\leftarrow\!\!-\!\!-\!\!-\!\!\overset{7}{\circ}$.

If the inequality symbol has the "equal to" part attached, (i.e., $\leq$ or $\geq$) the graphs have solid end points. Consider the graph of $4 \leq x \leq 8$: $\overset{}{\underset{4}{\bullet}}\!\!-\!\!-\!\!-\!\!\overset{}{\underset{8}{\bullet}}$.

Exercises 1.3

1. $U = \{1, 2, 3, 4, 5, 6, 7, 8, 9\}$, $A = \{3, 5, 7, 9\}$, $B = \{3, 4, 6, 7\}$, and $C = \{1, 2, 3, 4, 5\}$. Draw a Venn diagram for the above (with the elements in the diagram), then find (a) $A \cap B$ (b) $(A \cap B)'$ (c) $(A \cup B) \cap C'$ (d) $(A \cap B)' \cup C$.

2. Same sets and questions as problem #1 but use the roster method.

3. Draw a membership table for question 1(a)–(d).

4. Twenty-five people went to dinner. There were two choices of vegetables: peas and carrots. Three people ordered both. Ten had only carrots. Eight had neither peas nor carrots. How many had only peas?

5. There were 60 people at a party. Thirty were Irish, 30 were Polish, and 35 were German. Seven people were all three, Irish, Polish, and German. Thirteen were

Polish and German and not Irish. Eight were Polish and Irish and not German. Ten were Irish and German and not Polish. (a) How many people are only Irish? (b) How many are neither Irish, Polish, nor German?

6. Find the solution set for each of the open sentences. The replacement set is the set, $\{1, 2, 3, \ldots, 20\}$.

(a) $5x - 2 = 13$ (b) $2x + 4 = 36$ (c) $3x + 7 = 4$
(d) $x - 3 > 15$ (e) $x/3 > 5$ (f) $x \le 4$ and $x \ge 1$
(g) $x < 5$ or $x > 3$ (h) $x - 4 = -4 + x$
(i) $x + 3 = x - 7$ (j) $x + 12 = 12$
(k) $x > 9$ and $x < 14$

7. Using the set of real numbers as a replacement set, graph the following:

(a) $x \le 8$ (b) $x > 4$ (c) $x \le 5$ and $x > 2$
(d) $x > 4$ or $x < 3$ (e) $2 \le x \le 6$

Flow Charting Set Operations*

An alternate method of determining whether an element is or is not a member of a set is flow charting. To use the flow charting method we will denote the parallelogram regions for inputs (given) and outputs (results). In some instances a flow chart will contain a decision box, usually diamond shape, that asks a question. In such situations you must answer the question "yes" or "no" and follow the arrows out of the decision box in the indicated direction. For example:

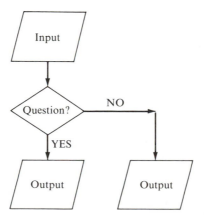

---

* This topic is optional. If it is omitted there will be no loss of continuity.

Now we shall use the flow charting method to reinforce the definitions of the basic operations of complementation, union, intersection, and difference to sets.

Complement.

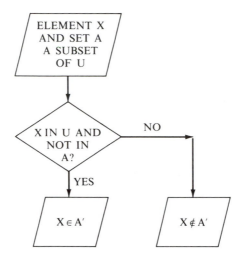

We can see that in this example the question in the decision diamond is phrased in a negative way so that the "yes" exit is down and the "no" exits to the right.

Union.

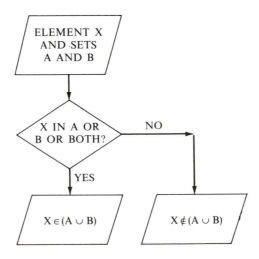

Intersection.

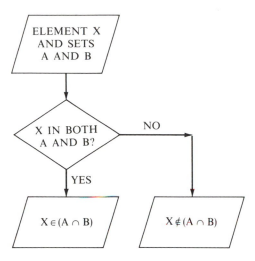

Difference.

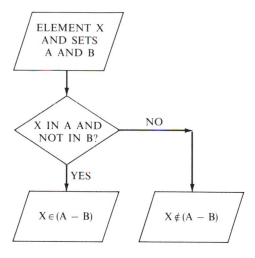

Some flow charts are more involved and have a sequence of rules to process, such as in the next illustration. Rules to process are usually denoted in rectangular boxes.

For example, construct a flow chart that can be used to determine whether an element is a member of the shaded region of the following Venn diagram in terms of *A* and *B*.

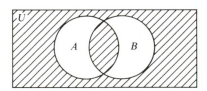

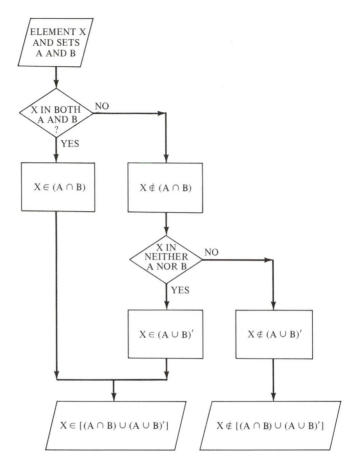

It might be easier to follow the flow of the chart if the question, "x in both A and B?" is interpreted, "Are there any x's that are in both A and B?" The fact that x is an element of $A \cap B$ does not eliminate the possibility that there could be x's that are members of other parts of the universe. So, we must continue to follow the "no" line and pinpoint other regions that contain x.

When we summarize at the bottom of the flowchart (output) we write $X \in$ the *intersection* of the sets that appear in the rectangles that are in a *single* "yes" ($\in$) branch, (see example) and then *union* that result with the result of any other "yes" branch (see Example 1).

When we summarize the "no" ($\notin$) branche(s) we union all the sets that appear in the rectangles that flow directly from a "no" exit from a decision diamond. Even though it would be correct to intersect all the sets that appear in rectangles in a single "no" ($\notin$) branch, as we did with the "yes" ($\in$) sets, it is more inclusive to use the union. Our rationale is this: If an element is outside ($\notin$) set A and is also outside ($\notin$) set B, it must be outside the union. (i.e., If $x \notin A$ and $x \notin B$, then $x \notin (A \cup B)$).

Example 1

Construct a flow chart that can be used to determine whether an element is a member of the shaded region of the following Venn diagram in terms of sets A, B, and C.

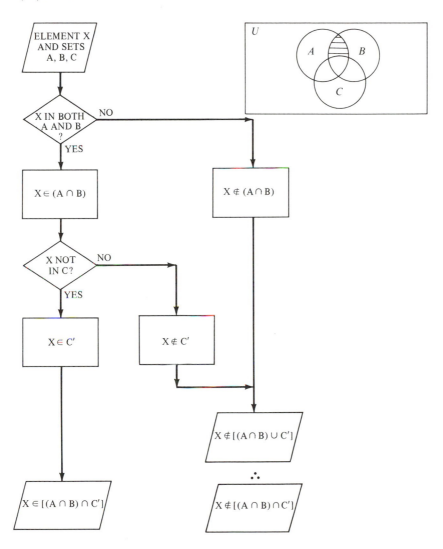

If the total set in the summary of the "no" ($\notin$) branches does not read the same as the total set in the summary of the "yes" ($\in$) branches, we may be able to reconcile the difference between them by changing one or more of the "union" symbols in the "no" ($\notin$) output to intersection symbols. (Clearly, if an element is not in the union of two sets, it is not in the intersection.)

Example 2

Construct a Venn diagram that represents the following flow chart:

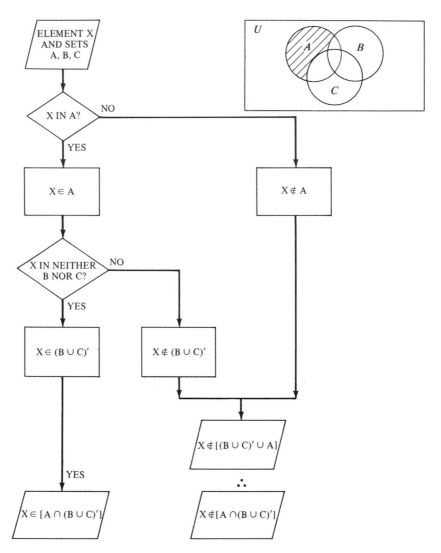

One of the difficulties of constructing a flow chart is finding an appropriate question to ask in the decision diamond. The following systematic (but sometimes tedious) approach will enable us to construct a flow chart to represent any shaded region or combination of regions of a two-circle Venn diagram (16 possible drawings) or a three-circle Venn diagram (256 possible drawings).

Represent the shaded area using set notation (i.e., capital letters for sets, and the symbols for union, intersection, and complement). Any of the four regions of a two-circle drawing can be represented by a statement of the form — $\cap$ —, where one of the blanks will be filled by A, if the region is inside A, or A' if the region is outside A. The other blank is filled by B if it is inside B and B' if it is outside B.

In like manner, any one of the eight regions of a three-circle diagram can be represented by a statement of the form — $\cap$ — $\cap$ —, where one of the blanks is filled in by A or A', one by B or B', and the third by C or C' as outlined above.

Collectively, then, the shaded areas can be represented by the union of these separate regions. For example:

Example 3

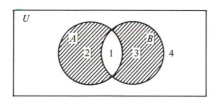

Region 2 is $A \cap B'$
Region 3 is $B \cap A'$
Thus the total shaded area is $(A \cap B') \cup (B \cap A')$

Example 4

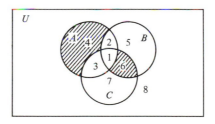

Region 4 is $A \cap B' \cap C'$
Region 6 is $A' \cap B \cap C$
The total shaded area is $(A \cap B' \cap C') \cup (A' \cap B \cap C)$

Example 5

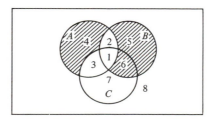

Region 4 is $A \cap B' \cap C'$
Region 5 is $A' \cap B \cap C'$
Region 6 is $A' \cap B \cap C$
The total shaded area is $(A \cap B' \cap C') \cup (A' \cap B \cap C') \cup (A' \cap B \cap C)$

If two shaded regions have a common side, we will be able to represent the shaded area more concisely by combining them. In Example 5 then we can say:

Region 4 = $A \cap B' \cap C'$

Regions 5 & 6 = $A' \cap B$

The shaded area is $(A \cap B' \cap C') \cup (A' \cap B)$

Once the area has been symbolized we can begin the construction of the flow chart. In the first decision diamond we write the question, "Is X in (the first statement in the symbolic representation)?" For instance, in Example 5 we could say, "Is x in $A \cap B' \cap C$?" or in words, "Is x in A and not in either B or C?"

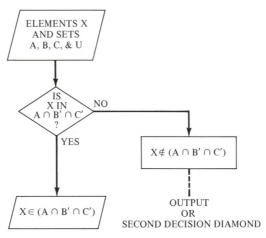

The second question (if necessary) will appear in the second diamond in the "no" branch, the third question in the second "no" branch, etc.

If the translation is not too complicated we may wish to show the "flow" in more detail.

For example:

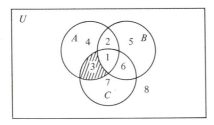

This shaded area can be represented as $A \cap B' \cap C$ and the flow chart could be

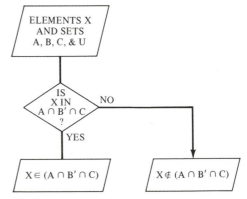

In more detail, the same shaded area can be represented as:

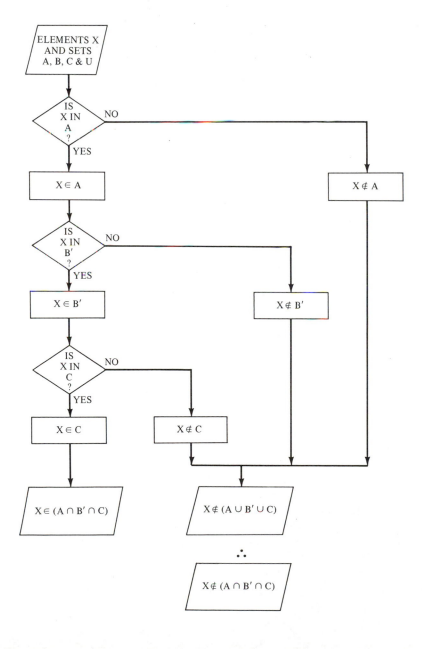

We could even take part of the intersection $A \cap B' \cap C$ and use that as the first question. Consider the following:

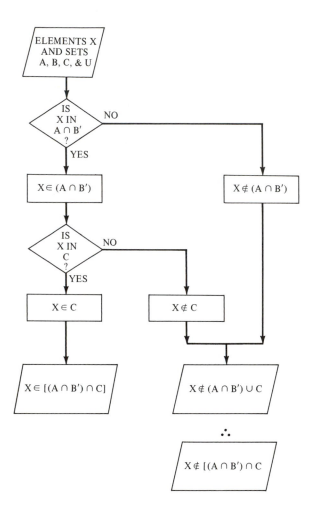

The kind of flow chart you end up with for a particular Venn diagram will depend on your skill in translating the picture into set notation and on the amount of detail that is desired.

The diagram below shows how each region of an eight region (three-circle) Venn diagram can be separated from the others. In this picture regions I, III, V, and VII come "out" from the "yes" branches. If you wish to have region VI exit from a "yes" branch instead of region V you simply change the last decision diamond question from C to C'. If you wish to include both regions V and VI in the "yes" branch, end that branch *before* the decision diamond that now splits them.

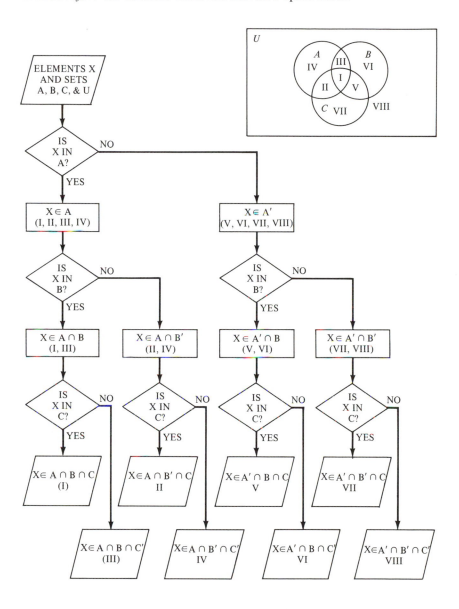

By putting the correct questions in the diamonds and ending the branches at the right points it will be possible to represent any combination of regions by unioning the ends of the "yes" branches (i.e., $x \in \{$The union of the "yes" branches"$\}$). The set that represents the union of the "no" branches will then be the complement of the set that represents the union of the "yes" branches. This means that we may represent the union of the ends of the "no" branches

by $x \in$ {The union of the "no" branches} *or*

by $x \in$ {The union of the "yes" branches}' *or*

by $x \notin$ {The union of the "yes" branches}.

After you have finished the next section on Boolean algebra, you may wish to return to flow charting and do the same problem starting with different questions. Then if your output areas are different you may be able to show that they are equal by applying the laws of Boolean algebra.

Exercises 1.4

1–3 Construct a flow chart that can be used to determine whether an element is a member of the shaded region.

1.

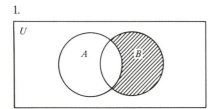

2

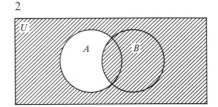

3

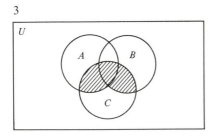

4–6 Construct a flow chart that can be used to determine whether an element is a member of the set symbolized by:

4. $A \cup B'$
5. $(A \cap B) \cup C$
6. $(A \cup B)' \cap C$

Boolean Algebra

So far we have dealt with sets and their operations strictly on an intuitive basis. We have seen that sets follow certain rules, some of which have been demonstrated by Venn diagrams and membership tables. It is important to note that by using Venn diagrams and membership tables we only demonstrated these properties and did not prove them. In this section we shall present an axiomatic development of set theory which will enable us to prove these relationships. An axiomatic development of any system has five major parts:

(a) *Undefined Terms* are terms which are used without specific mathematical definition; their meanings are purely intuitive.

(b) *Relations* are properties that connect two objects in some specified order.

(c) *Axioms* are self-evident and generally accepted principles.

(d) *Definitions* are explanations of what certain terms mean.

(e) *Theorems* are general conclusions which are proved using the undefined terms, definitions, axioms, and already proven theorems.

In our axiomatic development of set theory we have two undefined terms: "set" and "element." The undefined relation which relates elements to sets is that of "belongs to."

Axiomatic Development of Boolean Algebra

There are six axioms which we shall use in our development of Boolean algebra:

Ax$_1$ *Closure Axiom.* For any elements A and B of S, $A \cap B$ and $A \cup B$ exist and are elements of S.

Ax$_2$ *Commutative Axiom.* For any elements A and B of S, $A \cap B = B \cap A$ and $A \cup B = B \cup A$.

Ax$_3$ *Associative Axiom.* For any elements A, B, and C of S, $A \cap (B \cap C) = (A \cap B) \cap C$ and $A \cup (B \cup C) = (A \cup B) \cup C$.

Ax$_4$ *Distributive Axiom.* For any elements A, B, and C of S, $A \cap (B \cup C) = (A \cap B) \cup (A \cap C)$ and $A \cup (B \cap C) = (A \cup B) \cap (A \cup C)$.

Ax$_5$ *Identity Axiom.* Elements $\varnothing$ (null set) and U (universal set) exist such that for any element A of S, we have $A \cap U = A$ and $A \cup \varnothing = A$.

Ax$_6$ *Inverse Axiom.* For any element A of S, there exists an element A' of S called the complement of A, such that $A \cap A' = \varnothing$ and $A \cup A' = U$.

Our basic definitions are those of equality and subset.

Definition 13

Equality of Sets: Two sets A and B are equal if and only if every element of A belongs to B and every element of B belongs to A.

Definition 14

Subset: For any elements A and B of S, we say that A is a subset of B, $A \subset B$, if and only if $A \cap B = A$.

Before we prove theorems of Boolean algebra we shall state a principle that will reduce our work considerably. If we examine the six axioms, we see that when the operations of $\cup$ and $\cap$ and the sets $\varnothing$ and U are interchanged throughout the statement of the axioms we still have exactly the same set of axioms. For example, for the inverse axiom the second statement $A \cup A' = U$ can be obtained from the first statement $A \cap A' = \varnothing$ by replacing the $\cap$ by $\cup$ and $\varnothing$ by U. This leads to the following general conclusion.

The Principle of Duality.   Any theorem (or axiom) of Boolean algebra remains valid if the operations of union ($\cup$) and intersection ($\cap$) and the sets null ($\varnothing$) and universal (U) are interchanged throughout the statement of the theorem (or axiom). For example, the dual of

$$(A \cap \varnothing) \cup (U \cap B) = B$$

is

$$(A \cup U) \cap (\varnothing \cup B) = B$$

This principle is extremely helpful since it enables us to cut our work in half. If we prove a theorem, then its dual is also true.

The following are some of the basic theorems of Boolean algebra. Some will be proven and the remainder will be left as exercises.

Theorem 1

$$A \cap A = A \text{ and } A \cup A = A \text{ (idempotent laws)}$$

Proof

(1)	$A = A \cap U$	A_5 (Identity)
(2)	$A = A \cap (A \cup A')$	1, A_6 (Inverse)
(3)	$A = (A \cap A) \cup (A \cap A')$	2, A_4 (Distributive)
(4)	$A = (A \cap A) \cup \varnothing$	3, A_6 (Inverse)
(5)	$A = A \cap A$	4, A_5 (Identity)
(6)	$A = A \cup A$	5, Principle of duality

Theorem 2

$$A \cap \varnothing = \varnothing \text{ and } A \cup U = U$$

Theorem 3

$$U' = \varnothing \text{ and } \varnothing' = U$$

Theorem 4

$$(A')' = A$$

Proof

(1)	$A' \cap (A')' = \varnothing$	A_6 (Inverse)
(2)	$A = A \cup \varnothing$	A_5 (Identity)
(3)	$A = A \cup [(A') \cap (A')']$	1, 2, Substitution
(4)	$A = [A \cup A'] \cap [A \cup (A')']$	3, A_4 (Distributive)
(5)	$A = U \cap [A \cup (A')']$	4, A_6 (Inverse)
(6)	$U = (A') \cup (A')'$	A_6 (Inverse)
(7)	$A = [A' \cup (A')'] \cap [A \cup (A')']$	5, 6, Substitution

(8) $A = [(A')' \cup A'] \cap [(A')' \cup A]$ 7, A_2 (Commutative)
(9) $A = (A')' \cup (A' \cap A)$ 8, A_4 (Distributive)
(10) $A = (A')' \cup \varnothing$ 9, A_6 (Inverse)
(11) $A = (A')'$ 10, A_5 (Identity)

Theorem 5

$A \cup (A \cap B) = A$ and $A \cap (A \cup B) = A$ (absorption laws)

Theorem 6

$\varnothing$ and U are unique

Proof

In proving that $\varnothing$ is unique we shall use the indirect method. That is, we assume another $\varnothing$ exists, and call it $\varnothing_1$, such that $\varnothing \neq \varnothing_1$.

(1) $\varnothing \neq \varnothing_1$ Premise for the indirect proof
(2) $\varnothing = \varnothing \cup \varnothing_1$ A_5 (Identity)
(3) $\varnothing = \varnothing_1 \cup \varnothing$ 2, A_2 (Commutative)
(4) $\varnothing = \varnothing_1$ 3, A_5 (Identity)
(5) $\varnothing$ is unique   Steps 1 and 4 contradict; the
                          assumed premise $\varnothing \neq \varnothing_1$ is
                          false; therefore, $\varnothing = \varnothing_1$;
                          this shows that there is only
                          one $\varnothing$
(6) U is unique   5, Principle of duality

Theorem 7

For any set A of S, A' is unique.

Theorem 8

$(A \cap B)' = A' \cup B'$ and $(A \cup B)' = A' \cap B'$ (DeMorgan's law)

Hint: to show that $(A \cap B)' = A' \cup B'$, we must prove that

$$(A \cap B) \cap (A' \cup B') = \varnothing \quad \text{and} \quad (A \cap B) \cup (A' \cup B') = U$$

Theorem 9

If $A \subseteq B$ and $B \subseteq C$, then $A \subseteq C$

Proof

(1)	$A \subseteq C$	Given
(2)	$B \subseteq C$	Given
(3)	$A \cap B = A$	1, Definition 14
(4)	$B \cap C = B$	2, Definition 14
(5)	$A \cap (B \cap C) = A$	3, 4, Substitution
(6)	$(A \cap B) \cap C = A$	5, A_3 (Associative)
(7)	$A \cap C = A$	3, 6, Substitution
(8)	$A \subseteq C$	7, Definition 14

Theorem 10

$A \subseteq A$ (every set is a subset of itself)

Theorem 11

$\varnothing \subseteq A$ (the null set is a subset of every set)

Theorem 12

$$\text{If } A \subseteq B \text{ and } B \subseteq A, \text{ then } A = B$$

Theorem 13

$$\text{If } A \subseteq B, \text{ then } B' \subseteq A'$$

Theorem 14

$$\text{If } A \cap B = \varnothing, \text{ then } A \cup B' = B'$$

Proof

(1)	$A \cap B = \varnothing$	Given
(2)	$B' \cup (A \cap B) = (B' \cup A) \cap (B' \cup B)$	A_4 (Distributive)
(3)	$B' \cup (A \cap B) = (B' \cup A) \cap U$	2, A_6 (Inverse)
(4)	$B' \cup (A \cap B) = B' \cup A$	3, A_5 (Identity)
(5)	$B' \cup \varnothing = B' \cup A$	1, 4, Substitution
(6)	$B' = B' \cup A$	5, A_5 (Identity)
(7)	$B' = A \cup B'$	6, A_2 (Commutative)

Theorem 15

$$\text{If } A \cap B = \varnothing, \text{ then } B \cap A' = B$$

Theorem 16

$$\text{If } A \subseteq B, \text{ then } A \cap B' = \varnothing$$

Theorem 17

$$\text{If } A \subseteq B, \text{ then } A' \cup B = U$$

Laws of Boolean Algebra

The following is a list of the basic laws governing set operations. They are extremely useful in establishing the validity of set equalities and in reducing a set to its simplest possible form.

(a) Commutative
$$A \cap B = B \cap A$$
$$A \cup B = B \cup A$$

(b) Associative
$$A \cap (B \cap C) = (A \cap B) \cap C$$
$$A \cup (B \cup C) = (A \cup B) \cup C$$

(c) Distributive
$$A \cap (B \cup C) = (A \cap B) \cup (A \cap C)$$
$$A \cup (B \cap C) = (A \cup B) \cap (A \cup C)$$

(d) Absorption
$$A \cup (A \cap B) = A$$
$$A \cap (A \cup B) = A$$

(e) DeMorgan's
$$(A \cap B)' = A' \cup B'$$
$$(A \cup B)' = A' \cap B'$$

(f) Identity
$$A \cap U = A \; ; A \cup U = U$$
$$A \cup \emptyset = A; A \cap \emptyset = \emptyset$$

(g) Inverse
$$A \cap A' = \emptyset$$
$$A \cup A' = U$$

(h) Idempotent
$$A \cap A = A$$
$$A \cup A = A$$

(i) Complement
$$\emptyset' = U$$
$$U' = \emptyset$$

(j) Double Complement
$$(A')' = A$$

(k) Difference
$$A - B = A \cap B'$$

Example 1

Simplify the following set notation:

(1)	$\{[(B \cup U) \cap (A \cup \emptyset)] - C\} \cup C$	Given
(2)	$[(U \cap A) - C] \cup C$	Identity
(3)	$(A - C) \cup C$	Identity
(4)	$C \cup (A - C)$	Commutative
(5)	$C \cup (A \cap C')$	Difference
(6)	$(C \cup A) \cap (C \cup C')$	Distributive
(7)	$(C \cup A) \cap U$	Inverse
(8)	$C \cup A$	Identity

Example 2

Verify the following equality by using the laws of Boolean algebra:

(1)	$A - (B \cup C) = (A - B) - C$	Given
(2)	$A \cap (B \cup C)' = (A - B) - C$	Difference
(3)	$A \cap (B' \cap C') = (A - B) - C$	DeMorgan's
(4)	$(A \cap B') \cap C' = (A - B) - C$	Associative
(5)	$(A - B) \cap C' = (A - B) - C$	Difference
(6)	$(A - B) - C = (A - B) - C$	Difference

Example 3

Verify the following equality:

(1)	$[A \cup (A \cup C)'] \cap (A \cup C) = A \cap (A \cup B)$	Given
(2)	$[A \cup (A' \cap C')] \cap (A \cup C) = A$	Absorption and DeMorgan's
(3)	$[(A \cup A') \cap (A \cup C')] \cap (A \cup C) = A$	Distributive
(4)	$[U \cap (A \cup C')] \cap (A \cup C) = A$	Inverse
(5)	$(A \cup C') \cap (A \cup C) = A$	Identity
(6)	$A \cup (C' \cap C) = A$	Distributive
(7)	$A \cup \emptyset = A$	Inverse
(8)	$A = A$	Identity

Exercises 1.5

1. (T, F) $(A \cup \emptyset) \cup (U \cap B) = B$
2. (T, F) $A \cap \emptyset = U'$
3. (T, F)$(\emptyset')' = \emptyset$
4–6 Verify the following by using the laws of Boolean algebra.
4. $(A \cup C') \cap (A \cup C) = A$
5. $A = (A \cap A) \cup \emptyset$
6. $A \cap (C \cap \emptyset) - (B \cap U) = A \cap B'$

Chapter Exercises

1. Use the roster method to list the elements in the following sets:

   (a) the set of letters in the name of your school
   (b) the set of the number of dots on a die
   (c) the set of whole numbers between 3 and 10
   (d) the set of all whole numbers between 3 and 4
   (e) the set of even numbers greater than 10 but less than 24

2. Use the descriptive method in writing the following sets:

   (a) the set of numbers *1, 2, 3, 4, 5, 6, 7, 8, 9*
   (b) the set of days *Saturday* and *Sunday*
   (c) the set of letters *b, c, d, f, g, h, j, k, l, m, n, p, q, r, s, t, v, w, x, y, z*
   (d) the set of numbers *1, 3, 5,* and *7*
   (e) the set of months *January, June,* and *July*

3. Let $U = \{1, 2, 3, 4, 5, 6, 7, 8, 9\}$, $A = \{1, 3, 4, 7, 9\}$, $B = \{2, 3, 6, 8\}$, $C = \{1, 2, 3, 4, 5\}$, $D = \{6, 7, 8, 9\}$, and $E = \{1, 4, 9\}$. List the elements in the following sets

   (a) $A \cap C$ (f) $A \cap (B \cup D)$
   (b) $A \cup (B \cap C)$ (g) $B - (C \cup D)$
   (c) $A \cap D'$ (h) $A - (C - D)$
   (d) $(A \cap C) \cup (A \cap D)$ (i) $(A' \cap D) \cup (A \cap C)$
   (e) $(A \cup C)'$ (j) $(A' \cup B)' \cap (B - D)'$

4. Using the sets of problem 3, determine which of the following are true:

   (a) $A - B = A$ (e) $A - D = D - A$
   (b) $A' \cap (A \cup B) = A - B$ (f) $B \cap (C \cap E) = B \cap D$
   (c) $B \cap E = E \cap C$ (g) $(A - C) \cup (A' \cup C) = U$
   (d) $(D - A) \subset B$ (h) $[A \cap (A \cup B)] - B = \varnothing$

5. (a) If $N = \{a, b, c\}$ represents a three-member voting body, list all coalitions and label the winning and losing ones.
   (b) If $M = \{a, b, c, d\}$ represents a four-member voting body, list all coalitions and label the winning and losing ones.

6. Consider the voting body $H = \{a, b, c, d, e\}$ with vote assignments given by the following table:

a	b	c	d	e
2	1	2	4	1

   (a) List all winning coalitions assuming majority rule.

   (b) List all losing coalitions associated with each winning coalition.

   (c) List all blocking coalitions.

   (d) Next assume that a coalition must have two-thirds of all votes to win. List all winning coalitions under this new rule.

   (e) List all blocking coalitions under the two-thirds rule.

7. Consider the voting body $H = \{a, b, c, d, e\}$ with vote assignments given by the following table:

a	b	c	d	e
2	1	2	4	1

Notice that the number assigned to a subset is called the measure of the subset and is written $m(\{a, b\}) = 3$.

Let $M = \{a, b, c\}$, $N = \{d, e\}$, and $P = \{b, c, d, e\}$. Find:

(a) $m(M) =$

(b) $m(M \cup N) =$

(c) $m(M \cup P) =$

(d) $m(M \cap N) =$

(e) $m(\varnothing) =$

(f) $m(M') =$

(g) $m(N) =$

(h) $m(P) =$

(i) $m(N \cup P) =$

(j) $m(N \cap P) =$

(k) $m(H) =$

(l) $m(N') =$

8. Given the following Venn diagrams, shade in the region that represents the following sets:

(a) $(A \cup B) \cap (B - A)$

(b) $(A \cap B') \cup B$

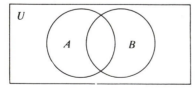

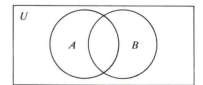

(c) $(A \cap B) \cup (A \cup C')$

(d) $[A \cup (B \cap C)'] \cap (B \cup C)$

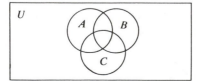

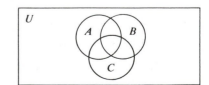

9. Describe each of the following Venn diagrams in terms of sets A, B, C, and U.

(a)   (b)

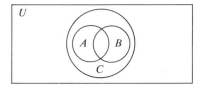   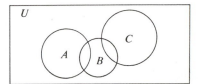

(c)   (d)

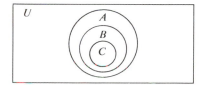   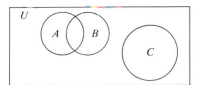

(e)

10. Let A, B, and C be subsets of U. Construct the Venn diagram described by the following conditions:

   (a) Sets A and C are disjoint; A and B have elements in common but they are not a subset of one another; B and C have elements in common but they are not a subset of one another. There are elements that do not belong to either A, B, or C.
   (b) Sets A and B are disjoint. Set C is a proper subset of B. There are elements that do not belong to either A, B, or C.
   (c) Set C is a proper subset of A. Sets A and B have elements in common but they are not a subset of one another. Sets B and C have elements in common but they are not a subset of one another. Finally, there are no elements that are not in either A, B, or C.
   (d) Sets C and A are equal. Set C is a proper subset of B. There are elements that are not in either A, B, or C.

11. Using the region numbers indicated,

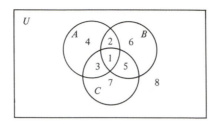

list the region numbers representing the following:

(a) $A \cap B$
(b) $A' \cup B$
(c) $A \cup B$
(d) $A \cup (B \cap C)$
(e) $C' \cap B'$
(f) $(A \cup B \cup C)'$

12. Given the following Venn diagrams, describe the numbered sets below in terms of sets A, B, C, and D.

(a)   (b)

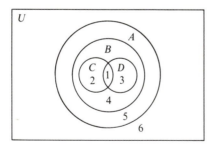   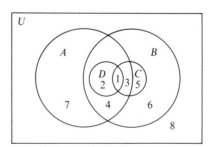

{1} =   {1} =
{2} =   {2} =
{3} =   {3} =
{4} =   {4} =
{5} =   {5} =
{6} =   {6} =
                                               {7} =
                                               {8} =

13. Construct a flow chart that can be used to determine whether an element is a member of the shaded region.

(a)   (b)   (c)

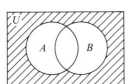   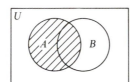   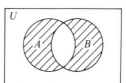

14. Using Venn diagrams or membership tables, determine whether or not the following statements are true.

(a) $A \cap (B - C) = (A \cap B) - C$
(b) $(A \cup B) - (A \cup C) = (B - A) - C$
(c) $(A \cap B) \cup (A \cap C) = A \cup (B \cap C)$
(d) $(A \cup B) - (C - A) = A \cup (B - C)$

15. Using set theory and Venn diagrams, solve the following problems:

(a) Twenty-five boys went on a picnic. There were two activities: baseball and swimming. Fifteen boys went swimming, five participated in both, and three boys did neither. How many boys played only baseball?

(b) In a certain high school, there are 60 students in the senior class. There are 20 students in mathematics, 25 in history, and 28 in science. Four students are taking both math and history, five both history and science, and six both math and science. No student is taking all three courses. How many students are taking both math and science? How many are not taking science but are taking history? How many are not taking both math and history?

(c) Given that $U = 12$ elements, $A = 8$ elements, $B = 6$ elements, and $A \cap B = 5$ elements, find the number of elements in

(1) $(A \cap B)'$ (6) $A' \cup B$
(2) $(A \cup B)'$ (7) $(A \cap B)' - A$
(3) $A \cap B'$ (8) $A' \cup B'$
(4) $B - A$ (9) $A' \cap B'$
(5) $U - (A \cup B)'$ (10) $(A \cap B') - (A - B)$

(d) Given that $U = 50$ elements, $A = 19$ elements, $B = 20$ elements, $C = 19$ elements, $(A \cap C) = 7$ elements, $(B \cap C) = 8$ elements, $(A \cap B) = 9$ elements, and $(A \cap B \cap C) = 5$ elements, find the number of elements in

(1) $A \cup B \cup C$ (6) $A - (B - C)$
(2) $(A \cup B \cup C)'$ (7) $(A - B) - C$
(3) $B - C$ (8) $(B \cap C) \cup A$
(4) $B \cup C'$ (9) $(A' \cap B') \cup C$
(5) $A \cap (B \cup C)'$ (10) $(A - B) \cap B$

16. Find the solution set for each of the open sentences. The replacement set is {0, 1, 2, 3, 4, 5, 6, 7, 8, 9, 10}.

(a) $2x - 1 = 7$
(b) $x - 3 > 5$
(c) $x - 2 < 6$
(d) $\dfrac{x}{4} > 2$
(e) $x \geq 2$ and $x \leq 4$

(f) $x < 5$ or $x > 7$
(g) $x \leq 2$ or $x > 8$
(h) $x + 6 = 6 + x$
(i) $x \geq 2$ and $x < 6$
(j) $x + 12 < 11$

17. Using the laws of Boolean algebra, simplify the following statements:

(a) $[A \cup (B \cap B')] \cap [A \cap (B \cap U)]$
(b) $[(A \cup C) \cap (A \cup C')] \cup [A' \cap (A' \cup B)]$
(c) $\{[(A \cap A) \cup (A \cap B)] - (A \cap B)\} - B$

18. Using the laws of Boolean algebra, determine whether or not the following statements are true.

(a) $(A - B) \cup (A - C) = A - (B - C)$
(b) $(C - A) - (B - A) = C - (A \cup B)$
(c) $A \cap (B - C) = (A \cap B) - C$
(d) $A - (A' - B) = A \cap (A \cup B)$

19. The following are proofs of theorems of Boolean algebra. Fill in the reason for each step and give the number of the previous step used. (*Remember:* you can use only the axioms, definitions, and previously proven theorems.)

(a) Theorem 2. $A \cap \emptyset = \emptyset$.

(1) $A \cap \emptyset = \emptyset \cap A$
(2) $A \cap \emptyset = A \cap A' \cap A$
(3) $A \cap \emptyset = (A' \cap A) \cap A$
(4) $A \cap \emptyset = A' \cap (A \cap A)$
(5) $A \cap \emptyset = A' \cap A$
(6) $A \cap \emptyset = \emptyset$

(b) Theorem 15. If $A \cap B = \emptyset$, then $B \cap A' = B$.

(1) $A \cap B = \emptyset$
(2) $B \cap U = B$
(3) $A' \cup A = U$
(4) $B \cap (A' \cup A) = B$
(5) $(B \cap A') \cup (B \cap A) = B$
(6) $(B \cap A') \cup (A \cap B) = B$
(7) $(B \cap A') \cup \emptyset = B$
(8) $B \cap A' = B$

20. Prove the following theorems of Boolean algebra:

(a) Theorem 3
(b) Theorem 5
(c) Theorem 7
(d) Theorem 8
(e) Theorem 10

(f) Theorem 11
(g) Theorem 12
(h) Theorem 13
(i) Theorem 16
(j) Theorem 17

Review Test

Use the following information for problems 1–6:

$$U = \{1, 2, 3, \ldots, 10\}$$
$$A = \{2, 4, 6, 8, 10\}$$
$$B = \{1, 3, 5, 7, 9\}$$
$$C = \{3, 7\}$$
$$D = \{1, 2, 3, 4\}$$

1. $B \cup C =$ (a) U (b) B (c) C (d) $\varnothing$
   (e) none of these

2. $C \cap D =$ (a) $\varnothing$ (b) B (c) $\{1, 2, 3, 4, 7\}$ (d) $\{3\}$
   (e) none of these

3. $A' \cap U =$ (a) A (b) B (c) U (d) $\varnothing$
   (e) none of these

4. $D' \cap C =$ (a) $\varnothing$ (b) B (c) $\{1, 2, 3, 4, 7\}$ (d) $\{3\}$
   (e) none of these

5. $(A \cap B)' =$ (a) A (b) $\varnothing$ (c) U (d) B'
   (e) none of these

6. Which of the following is true?

   (a) $A \subset B$ (b) $D \subset B$ (c) $B \subset C$ (d) $C \subset B$
   (e) none of these

7. How many *proper* subsets are there for a set containing five elements?

   (a) 24   (b) 25   (c) 31   (d) 32   (e) 10

8. The shaded set below is

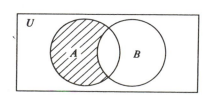

(a) A (b) B' (c) $A \cup B$ (d) $A' \cap B$ (e) $A \cap B'$

9. Consider the voting body $H = \{a, b, c\}$ with vote assignments given by the following table

a	b	c
2	1	3

If $M = \{a, b\}$ and $N = \{a, c\}$ find $m\,(M \cap N)$.

(a) 6   (b) 2   (c) 3   (d) 1

10. The shaded set below is

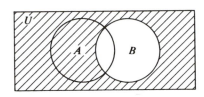

(a) $A' \cap B'$ (b) B' (c) $A' \cup B$ (d) $A \cup B'$
(e) $(A' \cap B)'$

11. The shaded set below is

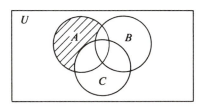

(a) $A \cup (B \cap C)$ (b) $A \cap (B \cap C)'$ (c) $A - (B \cup C)$
(d) $A - (B \cap C)$ (e) none of these

12. The Venn Diagram representing the statement $(A \cap B) - C$ is

(a)   (b)

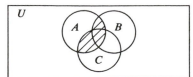 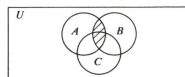

(c)   (d)

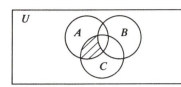   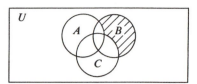

(e)   none of these

13. $(A \cup \varnothing)' \cap U =$ (a) A (b) $\varnothing$ (c) U (d) A'
                                          (e)   none of these

14. $A \cap (T \cup A') =$ (a) A (b) T (c) $A \cap T$ (d) $T \cup A'$
                                (e) $\varnothing$

Use the following information for problems 15–16:

There are 800 students in a school: 600 are enrolled in history, 300 are enrolled in mathematics, and 200 are enrolled in both history and mathematics.

15. The number of students that are enrolled only in history is

(a)   300   (b)   600   (c)   100   (d)   400   (e)   200

16. The number of students that are enrolled in neither history nor mathematics is

(a)   300   (b)   600   (c)   100   (d)   800
(e)   none of these

17. The statement that describes the following Venn diagram is

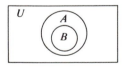

(a) $A \cap B = \varnothing,\ A \subset U,\ B \subset U,\ A \neq \varnothing,$ and $B \neq \varnothing$
(b) $A \subset U,\ B \subset A,\ B \neq U,\ B \neq \varnothing,$ and $A \neq \varnothing$
(c) $A \subset B,\ B \subset U,\ A \neq \varnothing,\ B \neq U,$ and $B \neq \varnothing$
(d) $A \cup B \neq \varnothing,\ A \neq \varnothing,\ B \neq \varnothing,\ B \nsubseteq A,\ A \subset U,$ and $B \subset U.$
(e)   none of these

18. Sets A and C are disjoint. Set C is a proper subset of B. Sets A and B are disjoint. Sets A, B, and C do not make up the entire universal set. The Venn diagram representing these statements is

(a)

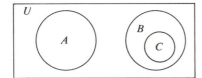

(b)

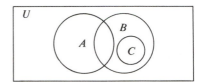

(c)

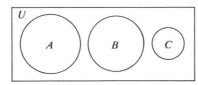

(d)

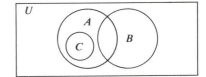

(e)   none of these

19.   We define the *symmetric difference* of two sets A and B, denoted by $A \triangle B$, as follows: $A \triangle B = (A - B) \cup (B - A)$. Which of the following membership tables represents $A \triangle B$?

(a)

A	B	$A \triangle B$
$\in$	$\in$	$\in$
$\in$	$\notin$	$\notin$
$\notin$	$\in$	$\notin$
$\notin$	$\notin$	$\in$

(b)

A	B	$A \triangle B$
$\in$	$\in$	$\notin$
$\in$	$\notin$	$\in$
$\notin$	$\in$	$\in$
$\notin$	$\notin$	$\notin$

(c)

A	B	$A \triangle B$
$\in$	$\in$	$\notin$
$\in$	$\notin$	$\in$
$\notin$	$\in$	$\notin$
$\notin$	$\notin$	$\notin$

(d)

A	B	$A \triangle B$
$\in$	$\in$	$\notin$
$\in$	$\notin$	$\notin$
$\notin$	$\in$	$\in$
$\notin$	$\notin$	$\notin$

(e)   none of these

20. Is the statement $(A \cup B) - (C - A)$ equal to the statement $A \cup (B - C)$?

    (a) yes   (b) no

21. If $(A \cap B) = \{0\}$, then A and B have no elements in common.

    (a) true   (b) false

22. If the replacement set is $\{1, 2, 3, 4, 5, 6\}$, find the solution set of $x < 5$ and $x > 1$.

    (a) $\varnothing$ (b) $\{1, 2, 3, 4, 5\}$ (c) $\{2, 3, 4\}$ (d) $\{2, 3, 4, 5\}$

23. If the replacement set is $\{1, 2, 3, \ldots, 18\}$, find the solution set of $x > 15$ or $x \leq 3$.

    (a) $\{16, 17, 18\}$ (b) $\{1, 2, 3\}$ (c) U (d) $\{1, 2, 3, 16, 17, 18\}$

24. If the replacement set is $\{1, 2, 3, 4, 5, 6, 7, 8, 9\}$, find the solution set of $3 < x$ and $x \leq 6$.

    (a) $\{3, 4, 5\}$ (b) $\{4, 5, 6\}$ (c) $\{3, 4, 5, 6\}$ (d) $\{4, 5\}$

25. If the replacement set is $\{0, 1, 2, 3\}$ find the solution set of $3x + 1 \geq 10$.

    (a) $\{0, 1, 2, 3\}$ (b) $\{0, 1, 2\}$ (c) $\varnothing$ (d) $\{3\}$

26. If $A \subset B$ and $B \subset C$, then $A \subset C$.

    (a) true   (b) false

27. Since $(A \cap \varnothing) \cup [U \cap (B \cap C)] = B \cap C$, we know that $(A \cup U) \cap [\varnothing \cup (B \cup C)] = B \cup C$. This is true by the

    (a) idempotent law   (b) absorption law
    (c) DeMorgan's law   (d) principle of duality
    (e) none of these

28. The statement $A \cup (A \cap B') = A$ is true by the

    (a) idempotent law   (b) DeMorgan's law
    (c) commutative law   (d) absorption law
    (e) none of these

29. The statement $A \cap (B \cup C) = (A \cap B) \cup (A \cap C)$ is true by the

    (a) associative law   (b) commutative law
    (c) distributive law   (d) DeMorgan's law
    (e) none of these

30. The statement $(A - B) \cup (A - B) = A - B$ is true by the

    (a) difference law   (b) idempotent law

    (c) absorption law   (d) associative law

    (e) none of these

2

STATEMENTS AND TRUTH TABLES

"Logic is the science of reasoning." This statement is a very broad definition of the word logic. In this book we will narrow our scope of study. In our study of logic discussion will be limited to the study of deductive reasoning. We will not attempt to discuss how we reason, but we will develop a system based on a specific set of rules that will enable us to examine the results of the deductive reasoning process. The system which we will develop is called Symbolic or Mathematical Logic.

We believe that the study of logic is a very practical undertaking. The reasoning processes which will be developed in these chapters can be applied to problem solving and to verbal and written expression in mathematics as well as other subjects. They will enable the reader to communicate his thoughts more clearly and effectively, and provide guidelines for judging the reasoning of others.

Before beginning the development of our system we will examine several illustrative questions which will be the subject of our future investigations. These questions will be answered in the following chapters.

Consider the following questions.

1   When is the statement *It is not the case that if you read the Times, your vote will be prejudiced in the coming election* a true statement?

2   Is the statement *He is not interested in politics or he would run for office* equivalent to the statement *If he is interested in politics, then he will run for office*?

3   What do we mean when we say that two statements are equivalent?

4   If we agree with the statement *If Mr. Dixon is truly concerned about the welfare of his country, he votes in every election* and we know for a fact that *Mr. Dixon votes in every election* must we then agree with the statement *Mr. Dixon is truly concerned about the welfare of his country*? That is, if we consider the first two statements to be true statements, are we then forced to accept the third statement as true?

The three statements of question 4, when taken together form a structure which is called an argument. Our study of logic will be concerned with examining the structure of arguments in order to determine whether or not they are logical or valid.

From the illustrative questions just presented we see that our study of logic is concerned with examining individual statements or sequences of statements called arguments. Therefore we begin our study with a discussion of simple statements which are the basic units or building blocks of symbolic logic.

Simple Statements

Sentences are usually classified as declarative, interrogative, exclamatory, and imperative. In our study of logic we shall deal only with declarative statements which are either "true" or "false." For example:

1   *It is raining.*

2   *John is at home.*

3   *Some men convicted of crimes are innocent.*

We *will not* deal with the following types of statements.

4   *Did it rain?*

5   *Go home.*

6   *Maybe he is at home.*

Examining the declarative sentences 1, 2 and 3, we see that they can be either true or false. They can not have both truth values simultaneously.

In our discussion we will not be concerned with the actual truth or falsity of the statements involved since truth is very often relative. What we will do is examine a statement for the case when it is true and also examine the same statement for the case when it is false. That is, we will assign both truth values to the statement and examine it for each possibility. Now we have the following definition.

Definition 1

Simple Statement: A simple statement is a declarative sentence containing one idea which is either true or false but not both.

Compound Statements

Some statements are composite, that is, they are made up of two or more simple statements joined together by connectives. The following are examples of compound or composite statements.

1 *He is working hard or he is on vacation* is a compound statement made up of the simple statements *He is working hard* and *He is on vacation.* The simple statements are joined together by the connective *or.*

2 *If it is raining, then John is at home* is a compound statement made up of the simple statements *It is raining* and *John is at home.* The simple statements are joined together by the connective *if . . . , then . . . .*

3 *He is a tall strong athlete* is a compound statement made up of the simple statements *He is tall, He is strong,* and *He is an athlete.* The connective understood in the construction of the compound statement is *and.* That is, *He is tall and he is strong and he is an athlete.*

The truth value of a compound statement is determined by the truth values of its component simple statements and by the connectives used in its construction.

Definition 2

Compound Statement: A compound statement is a statement consisting of two or more simple statements joined by connectives.

Logical Connectives

Before we can begin our study of compound statements, we must examine the connectives that are used in their formation. In the following discussion, lowercase letters of the alphabet such as a, b, c, p, q, r, etc., will be used to represent simple statements. The use of letters to represent simple statements enables us to present compound statements in a clear concise form and helps avoid the possibility of being misled or confused by the wording of the statements. This symbolic representation of statements allows us to concentrate on statement structure and not get confused by the words involved.

Definition 3

Logical Connective: A logical connective is the word or phrase that is used to join simple statements in order to form a compound statement.

There are four basic logical connectives that we will use in our study of symbolic logic. They are conjunction (*and*), disjunction (*or*), conditional (*if* . . . , *then* . . .) and biconditional (*if and only if*). We will now examine each of the connectives in detail.

Conjunction. Two statements joined by the word *and* form a compound statement called the conjunction of the two statements. The symbol " $\wedge$ " is used to represent the word *and*. The conjunction of the statements p and q is denoted symbolically by $p \wedge q$. It is read *p and q*. It can also be read *p but q*. The truth value of the compound statement $p \wedge q$ which is determined by the truth value of p and by the truth value of q has the following property.

Rule 1 (R_1). If p is true and q is true then $p \wedge q$ is true; otherwise $p \wedge q$ is false. That is, the conjunction of two statements is true, if and only if both statements are true. It is false when at least one component statement is false.

Example

Consider the following compound statements.

(a) $1 + 1 = 2$ and $2 \times 2 = 4$ $T \wedge T = T$

(b) $1 + 1 = 2$ and $2 \times 2 = 5$ $T \wedge F = F$

(c) $1 + 1 = 3$ and $2 \times 2 = 4$ $F \wedge T = F$

(d) $1 + 1 = 3$ and $2 \times 2 = 5$ $F \wedge F = F$

The first compound statement is true since both component statements are true. Each of the other compound statements is false, since each has at least one component statement that is false.

Example

If p represents the statement *It is raining* and q the statement *John is at home*, then $p \wedge q$ represents the compound statement *It is raining and John is at home*, or *It is raining but John is at home*. The truth values of the compound statement are given in the following table.

Table 1*

Component Statements				Compound Statement
Actual Weather Conditions	*It is raining* p	*John is at home* q	John's Actual Location	*It is raining and John is at home* $p \wedge q$
Raining	T	T	Home	T
Raining	T	F	Not home	F
Not raining	F	T	Home	F
Not raining	F	F	Not home	F

* Notice that the structure of this table is the same as the membership table for $A \cap B$ in Chapter 1. Just change the $\in$ to T and $\notin$ to F.

Disjunction.   Two statements joined by the word *or* form a compound statement called the disjunction of the two statements. The symbol " $\vee$ " is used to represent the word *or*. The disjunction of the statement p with the statement q is denoted symbolically by $p \vee q$. It is read p *or* q.

Note: The word *or* has two distinct meanings. In the exclusive sense it means one or the other but not both. In the inclusive sense it means one or the other or both. We will use only the inclusive or in the following material.

The truth value of the compound statement $p \lor q$, which is determined by the truth value of p and by the truth value of q, has the following property.

Rule 2 (R_2).   If either p is true, or q is true, or both p and q are true, then $p \lor q$ is true; otherwise $p \lor q$ is false. That is, the disjunction of two statements is true if and only if at least one of the component statements is true. It is false only when both of the component statements are false.

Example

Consider the following compound statements.

(a) $1 + 1 = 2$ or $2 \times 2 = 4$ $T \lor T = T$

(b) $1 + 1 = 2$ or $2 \times 2 = 5$ $T \lor F = T$

(c) $1 + 1 = 3$ or $2 \times 2 = 4$ $F \lor T = T$

(d) $1 + 1 = 3$ or $2 \times 2 = 5$ $F \lor F = F$

The first three compound statements are true since at least one of the component statements is true. The last compound statement is false since both component statements are false. The structure of this table matches the membership table for $A \cup B$.

Example

If p represents the statement *It is raining* and q represents the simple statement *John is at home*, then $p \lor q$ represents the compound statement *It is raining or John is at home*. The truth values of the compound statement are given in the following table.

Table 2

Component Statements				Compound Statement
Actual Weather Conditions	*It is raining* p	*John is at home* q	John's Actual Location	*It is raining or John is at home* $p \lor q$
Raining	T	T	Home	T
Raining	T	F	Not home	T
Not raining	F	T	Home	T
Not raining	F	F	Not home	F

Conditional.   Two statements joined by the phrase if ... , then ...
form a compound statement called a conditional. The symbol "→" is
used to represent the words *if* ... , *then* .... The *if* statement is called
the antecedent and the *then* statement is called the consequent. The
conditional, *if p, then q,* is denoted symbolically by $p \to q$. It is read p
conditional q. It can also be read p *only if q* and p *is sufficient for q*. Statements of the form p *is necessary for q* and p, *if q* are also conditionals.
These particular examples are translated $q \to p$. Note that in these two
cases the part that actually appeared second in the English sentence is
written first in the symbolic translation. The truth value of the conditional,
which is determined by the truth value of the antecedent and by the
truth value of the consequent, has the following property.

Rule 3 (R_3).   The conditional $p \to q$ is true for all truth values of p
and q except when the antecedent p is true and the consequent q is false.

Note: It is important to remember that this rule is a definition of
conditional and that it is by definition that false implies true and false
implies false are both true. A rationale for this definition will be given
later in this chapter.

Example

Consider the following compound statements.

(a)   If $1 + 1 = 2$, then $2 \times 2 = 4$ $T \to T = T$

(b)   If $1 + 1 = 2$, then $2 \times 2 = 5$ $T \to F = F$

(c)   If $1 + 1 = 3$, then $2 \times 2 = 4$ $F \to T = T$

(d)   If $1 + 1 = 3$, then $2 \times 2 = 5$ $F \to F = T$

The second compound statement is false since the antecedent is true and the
consequent is false. The other three compound statements are true.

Example

If p represents the simple statement *It is raining* and q the simple statement *John
is at home*, then $p \to q$ represents the compound statement *If it is raining, then
John is at home* or *It is raining only if John is at home*. The truth values of the
compound statements are given in the following table.

Table 3

Component Statements				Compound Statement
Actual Weather Conditions	*It is raining* p	*John is at home* q	John's Actual Location	*If it is raining, then John is at home* $p \rightarrow q$
Raining	T	T	Home	T
Raining	T	F	Not home	F
Not raining	F	T	Home	T
Not raining	F	F	Not home	T

Rows three and four in Table 3 are somewhat bothersome. Why is a compound statement of the form $p \rightarrow q$ considered to be true if p, the antecedent, is false? Another example may clarify that reasoning. Suppose there was an ad in the newspaper that reads, "If you shop in our store on Tuesday, then we will give you a free gift." The antecedent is "you shop in our store on Tuesday." If you actually shopped in the store on Monday, then the antecedent would be false. It would not matter, therefore, if they gave you a free gift or not. Either way you could not call the ad false. Thus, it is considered true.

Biconditional.   Two statements joined by the phrase *if and only if* form a compound statement called a biconditional. The symbol "↔" is used to represent the words *if and only if*. The biconditional *p if and only if q* is denoted symbolically by $p \leftrightarrow q$. It can also be read *p is necessary and sufficient for q*. The truth value of the compound statement $p \leftrightarrow q$, which is determined by the truth value of p and by the truth value of q, has the following property.

Rule 4 (R_4).   The biconditional $p \leftrightarrow q$ is true if both p and q are true or if both p and q are false. That is, the biconditional is true if both component statements have the same truth values. Otherwise, it is false.

Note: The abbreviation of the phrase *if and only if* is *iff*.

Example

Consider the following compound statements.

(a) $1 + 1 = 2$ *if and only if* $2 \times 2 = 4$ $T \leftrightarrow T = T$

(b) $1 + 1 = 2$ *if and only if* $2 \times 2 = 5$ $T \leftrightarrow F = F$

(c) $1 + 1 = 3$ *if and only if* $2 \times 2 = 4$ $F \leftrightarrow T = F$

(d) $1 + 1 = 3$ *if and only if* $2 \times 2 = 5$ $F \leftrightarrow F = T$

The first and fourth compound statements are true since the truth values of the component statements are the same. The second and third compound statements are false.

Example

If p represents the simple statement *It is raining* and q represents the simple statement *John is at home*, then $p \leftrightarrow q$ represents the compound statement *It is raining if and only if John is at home*. The truth values of the compound statement are given in the following table.

Table 4

Component Statements				Compound Statement
Actual Weather Conditions	*It is raining* p	*John is at home* q	John's Actual Location	*It is raining if and only if John is at home* $p \leftrightarrow q$
Raining	T	T	Home	T
Raining	T	F	Not home	F
Not raining	F	T	Home	F
Not raining	F	F	Not home	T

In addition to the four logical connectives there is another operation that will help simplify our investigation of symbolic logic.

Negation.   A statement formed by adding the word *not* to some given statement is called the negation of that statement. The symbol " $\sim$ " is used to represent the word *not*. The negation of the statement p is denoted symbolically by $\sim p$. It is read *not p*. It can also be read *It is false that p is . . .* or *It is not the case that p is . . . .* The truth value of $\sim p$, which is determined by the truth value of p has the following property.

Rule 5 (R$_5$).   The negation of p is true if and only if p is false. That is, *not p* is true if p is false, or *not p* is false if p is true.

Example

Consider the following compound statements.

(a) $1 + 1 \neq 2$ $\sim(T) = F$

(b) $2 \times 2 \neq 5$ $\sim(F) = T$

The first statement is false since it is the negation of a true statement. The second statement is true since it is the negation of a false statement.

Example

If p represents the simple statement *It is raining*, then $\sim p$ represents the statement *It is not raining, It is false that it is raining*, or *It is not the case that it is raining*. The truth values of the negation statements are given in the following table.

Table 5

	Statement	Negation Statement
Actual Weather Conditions	*It is raining* p	*It is not raining* $\sim p$
Raining Not raining	T F	F T

The structure of this table matches the membership table of A′ in Chapter 1.

Symbolizing Statements

In our study of symbolic logic, statements can now be classified as being a conjunction, disjunction, conditional, biconditional, or negation. For example consider the following statements. (In the following examples we will use the underlined letters to symbolize the statements.)

1 *He was at the scene of the crime and was arrested* ($s \wedge a$) is a conjunction.
2 *If he is guilty then he was at the scene of the crime* ($g \rightarrow s$) is a conditional.
3 *He is not guilty or he was arrested* ($\sim g \vee a$) is a disjunction.
4 *It is not the case that he was arrested and is not guilty* $\sim(a \wedge \sim g)$ is a negation.

Even when the compound statement consists of more than two component statements and therefore more than one logical connective, it is possible to classify the compound statement as we did in the preceding examples. However, in symbolizing compound statements which have two or more logical connectives we must be extremely careful. Since a logical connective joins two statements we must make certain that we indicate which statements are to be joined together. This does not present a problem when there are only two component statements and one logical connective. Difficulty can arise when there are two or more connectives. When this occurs we must use parentheses ( ) or some other symbols of grouping, such as brackets [ ] or braces { }, to indicate which component statements are joined by the connectives. To illustrate consider the following examples.

1   *It is cold or it is damp and windy* [$c \lor (d \land w)$] is a disjunction made up of the simple statement c and the compound statement ($d \land w$).
2   *It is cold and damp or it is windy* [$(c \land d) \lor w$] is a disjunction made up of the compound statement ($c \land d$) and the simple statement w.
3   *It is false that if it is windy then it is cold and damp* $\sim [w \to (c \land d)]$ is a negation of the compound statement [$w \to (c \land d)$].
4   *It is either Tom and Bill, or John and Sam* [$(t \land b) \lor (j \land s)$] is a disjunction made up of the compound statements ($t \land b$) and ($j \land s$).

In these four examples we were able to determine how to group the statements by their wording. Placement in the statements of the verbs or subjects, or sometimes the use of punctuation tells us which statements are to be grouped together. Sometimes statements are ambiguous and we are therefore unable to symbolize the statement so that it has a logical meaning. For example consider the following statement.

5   *It is cold or wet and damp* ($c \lor w \land d$).

In this compound statement we are unable to group the component statements and to determine whether it is a conjunction or a disjunction. It is ambiguous and therefore has no logical meaning in our system. We are unable to work with a statement of this type.

To help understand the changing of statements into symbolic form and from symbolic form into words, consider the following illustrative examples. In the following examples p represents the statement *He is a conservative*, q represents the statement *He is a candidate for office*, and r represents the statement *He will be elected.*

1   *He is a conservative or he is not a candidate for office.*
    ($p \lor \sim q$)   a disjunction

2   *It is false that he is not a candidate for office.*
    $\sim(\sim q)$ a negation
3   *He will not be elected if he is not a candidate for office.*
    $(\sim q \to \sim r)$ a conditional
4   *He will be elected if and only if he is a candidate for office.*
    $(r \leftrightarrow q)$ a biconditional
5   *It is false that he is not a candidate for office and he will be elected.*
    $\sim(\sim q \wedge r)$ a negation
6   *If he is a conservative and a candidate for office then he will be elected.*
    $[(p \wedge q) \to r]$ a conditional
7   *He is a candidate for office and, if he is a conservative, he will be elected.*
    $[q \wedge (p \to r)]$ a conjunction
8   *He is a conservative and a candidate for office, or he is a conservative and will not be elected.*
    $[(p \wedge q) \vee (p \wedge \sim r)]$ a disjunction
9   *It is not the case that he will not be elected if he is a conservative and a candidate for office.*
    $\sim[(p \wedge q) \to \sim r]$ a negation
10 *He is a conservative and will not be elected or, it is false that if he is a candidate for office, he will be elected.*
    $[(p \wedge \sim r) \vee \sim(q \to r)]$ a disjunction

Summary

The following table is a summary of the logical connectives and negation. It is essential that you know these definitions before you proceed to the next section.

	Con- junction	Disjunction	Conditional	Biconditional	Negation	
	and	*or*	*if . . . , then . . .*	*if and only if*	*not it is false that . . .*	
p q	$p \wedge q$	$p \vee q$	$p \to q$	$p \leftrightarrow q$	$\sim p$	$\sim q$
T T	T	T	T	T	F	F
T F	F	T	F	F	F	T
F T	F	T	T	F	T	F
F F	F	F	T	T	T	T

Some Translation Hints. It should be noted that if the statement to be translated from English to symbols begins with "It is false that" or "It is not the case that," it is a negation. Begin your translation with a negation sign and translate the rest of the statement inside a pair of brackets. If the statement begins with "Either," it is a disjunction and you should write down two pairs of parentheses separated by a disjunction symbol. Translate everything from the word "Either" up to the word "or" inside the first pair of parentheses, and everything after that, inside the second pair of parentheses.

Exercises 2.1

1. Determine whether the following statements are simple, compound, or neither.

    (a) Dennis is an all-star basketball player.
    (b) Who is Robert going out with this week?
    (c) If Terri strikes out the side, then she will be the winning pitcher.
    (d) Steve was injured playing Rugby, but he still went to the dance.
    (e) It is not the case that Maureen went to sleep and did not finish her book.
    (f) Maybe I will go to the movie.
    (g) Marguerite lost 20 pounds by sticking to her diet.

2. Classify the following compound statements as a conjunction, disjunction, conditional, biconditional, negation, or as ambiguous.

    (a) $p \wedge q$ (b) $\sim p \vee (\sim q \wedge r)$
    (c) $(p \vee r) \wedge t$ (d) $\sim q \wedge r \vee s$
    (e) $(p \vee q) \rightarrow r$ (f) $r \leftrightarrow (p \vee r)$
    (g) $t \rightarrow (a \vee b)$ (h) $\sim [(a \vee b) \rightarrow (p \wedge q)]$
    (i) Chris is sweet and lovable.
    (j) Tony will take Kay fishing if and only if the sun is shining.
    (k) Grandma will make sauerbraten only if Pat makes the potato dumplings.

3. Symbolize the following statements.

    (a) If it rains, we will not go swimming.
    (b) Either you studied hard and passed the test, or you dropped the course.
    (c) The sun came out and we went to the park.
    (d) It is false that he went to school and was sent home by the nurse.
    (e) He will go to jail if and only if he is guilty.
    (f) Hard work is necessary for making money.
    (g) Working hard is sufficient for making money.

As stated earlier in the chapter, the truth value of a compound statement depends on two factors; the truth values of its component simple statements, and the logical connectives used in its formation. Now we will discuss methods of determining the truth value of a compound statement both for the case where the truth values of the component simple statements are known and also for the case when no specific set of truth values of the component simple statements are given.

Evaluating Compound Statements for Specific Truth Values of the Component Statements

In the case where the truth values of the component simple statements are known, the truth value of the compound statement is found by substituting these given truth values (T or F) for the component simple statements and then applying the five rules which govern the logical connectives and negation. The order in which we apply these rules or perform the indicated operations is determined by the symbols of grouping used in the formation of the compound statement.

Consider the following examples.

Example 1

Find the truth value of the statement *It is false that he is not the victor or he quit the race* if *he is the victor* is a true statement and *he quit the race* is a false statement. By symbolizing and using p for *he is the victor* and q for *he quit the race* we have

$$\sim(\sim p \vee q) \qquad \text{with} \qquad p = \text{T and } q = \text{F}$$

$$\sim(\sim p \vee q) \qquad \text{given}$$

$$\sim(\sim \text{T} \vee \text{F}) \qquad \text{substitute T for } p \text{ and F for } q$$

$$\sim(\text{F} \vee \text{F}) \qquad \text{by } R_5 \text{ (negation)}; \ \sim \text{T} = \text{F}$$

$$\sim(\text{F}) \qquad \text{by } R_2 \text{ (disjunction)}; \ \text{F} \vee \text{F} = \text{F}$$

$$\text{T} \qquad \text{by } R_5 \text{ (negation)}; \ \sim \text{F} = \text{T}$$

Therefore, we see that the compound statement is true for the given truth values of the component simple statements.

Example 2

Find the truth value of the statement *It is not the case that if you do not read the newspaper, then you are open-minded and willing to learn* if the statement *you read the newspaper* is false, the statement *you are open-minded* is true, and the statement *you are willing to learn* is false.

Symbolizing using p for *you read the newspaper*, q for *you are open-minded*, and r for *you are willing to learn* we have

$\sim[\sim p \rightarrow (q \wedge r)]$ with $p = F$, $q = T$, and $r = F$

$\sim[\sim p \rightarrow (q \wedge r)]$ given

$\sim[\sim F \rightarrow (T \wedge F)]$ substitute T for q and F for p and r

$\sim[T \rightarrow (T \wedge F)]$ by R_5 (negation); $\sim F = T$

$\sim[T \rightarrow F]$ by R_1 (conjunction); $T \wedge F = F$

$\sim[F]$ by R_3 (conditional); $T \rightarrow F = F$

T by R_5 (negation); $\sim F = T$

Therefore, we see that the compound statement is true for the given truth values of the component simple statements.

The truth values of the above compound statements may also be determined by the following abbreviated procedure.

Example 3

$\sim(\sim \quad p \quad \vee \quad q)$

T   F   T   F   F

4   2   1   3   1   (the numbers indicate the order in which the steps were performed)

Step 1   Substitute T for p and F for q.

Step 2   By R_5 (negation) the negation of p is F.

Step 3   By R_2 (disjunction) the disjunction of the negation of p with q is F.

Step 4   By R_5 (negation) the negation of the statement within the parentheses is T.

Example 4

$$\sim [\sim \quad p \quad \to \quad (q \quad \wedge \quad r)]$$

$$\text{T} \quad \text{T} \quad \text{F} \quad \text{F} \quad \text{T} \quad \text{F} \quad \text{F}$$

$$5 \quad 3 \quad 1 \quad 4 \quad 1 \quad 2 \quad 1$$

Step 1 Substitute T for q and F for p and r.

Step 2 By R_1 (conjunction) the conjunction of q and r is F.

Step 3 By R_5 (negation) the negation of p is T.

Step 4 By R_3 (conditional) the conditional of the negation of p with the conjunction of $(q \wedge r)$ is F.

Step 5 By R_5 (negation) the negation of the statement within the brackets is T.

As a final example consider the statement of question 1 of the introduction (p. 65).

Example 5

Determine the truth or falsity of the statement *It is not the case that if you read the Times, your vote will be prejudiced in the coming election* if in fact you do read the Times and your vote will be prejudiced.

Symbolizing using r for *you read the Times* and p for *your vote will be prejudiced in the coming election* we have

$$\sim \quad [r \quad \to \quad p] \qquad \text{with } r = \text{T and } p = \text{T}$$

$$\text{F} \quad \text{T} \quad \text{T} \quad \text{T}$$

$$3 \quad 1 \quad 2 \quad 1$$

Step 1 Substitute T for r and T for p.

Step 2 By R_3 (conditional); $\text{T} \to \text{T} = \text{T}$

Step 3 By R_5 (negation); $\sim \text{T} = \text{F}$

Therefore, we see that the compound statement is false when the component statements are both true.

In the preceding examples we were able to determine the truth values of the compound statements when given the truth values of all the component simple statements. Sometimes even though we do not know the truth values of all the component simple statements we are still able to determine the truth value of the compound statement. This is true because of the following properties of the logical connectives.

$F \wedge (\text{unknown truth value}) = F$

$T \vee (\text{unknown truth value}) = T$

$F \rightarrow (\text{unknown truth value}) = T$

$(\text{unknown truth value}) \rightarrow T = T$

These properties, which come directly from the rules governing the logical connectives, will be useful in succeeding work.

Example 6

Determine the truth value of the statement *If he was at the scene of the crime and is guilty, then he will be convicted* if we know that *he is not guilty*.

Symbolizing *a* for the statement *he was at the scene of the crime*, *b* for the statement *he is guilty*, and *c* for the statement *he will be convicted* we have

$[(a \wedge b) \rightarrow c]$	with $b = F$
$[(a \wedge b) \rightarrow c]$	given
$[(a \wedge F) \rightarrow c]$	substitute F for b
$[F \rightarrow c]$	by R_1 (conjunction), if one component of a conjunction is false, the conjunction is false
T	by R_3 (conditional), if the antecedent of a conditional is false, the conditional is true

Therefore, the compound statement is true when the statement *he is not guilty* is true.

Example 7

Determine the truth value of the statement *If he is alert, then he is either cautious or not prepared* if the statement *he is prepared* is a false statement.

Symbolizing using *a* for the statement *he is alert*, *c* for the statement *he is cautious*, and *d* for the statement *he is prepared* we have

$[a \rightarrow (c \vee \sim d)]$	with $d = F$
$[a \rightarrow (c \vee \sim d)]$	given
$[a \rightarrow (c \vee \sim F)]$	substitute F for d
$[a \rightarrow (c \vee T)]$	by R_5 (negation); $\sim F$ is T

$[a \rightarrow T]$ by R_2 (disjunction); if one component of a disjunction is true
                    (T), the disjunction is true (T)
T   by R_3 (conditional); if the consequent of a conditional is true
                    (T), the conditional is true (T)

Therefore, the compound statement is true when the statement *he is prepared* is false.

Example 8

Determine the truth value of the statement *If you do not listen to reason, then you are open-minded* if the statement *you listen to reason* is false.

Symbolizing using p for the statement *you do listen to reason* and q for the statement *you are open-minded* we have

$(\sim p \rightarrow q)$ with $p = F$
$(\sim p \rightarrow q)$ given
$(\sim F \rightarrow q)$ substitute F for p
$(T \rightarrow q)$ by R_5 (negation); $\sim F$ is T
?   we cannot determine the truth value of $T \rightarrow q$, since we do not
                    know the truth value of q; $T \rightarrow q$ would be true if q were T and
                    false if q were F—this is by R_3 (conditional)

Therefore, we are unable to determine the truth value of the compound statement when the statement *you listen to reason* is false.

Truth Tables

When the truth values of the component simple statements of some given compound statement are not known, we can investigate the truth values of the compound statement for all of the possible truth values of the component simple statements. This enables us to determine which truth values of the component statements make the compound statement true and which truth values make it false. One of the simplest methods of doing this is by the use of a table called a truth table. We have already used truth tables in the summary of the logical connectives (page 76). The truth table for the logical connective of conjunction shows that $p \wedge q$ is true when both p

p	q	$p \wedge q$
T	T	T
T	F	F
F	T	F
F	F	F

and q are true. It is false for the other truth values of the component simple statements. We will now illustrate the general procedure for constructing truth tables for more complex compound statements.

Constructing Truth Tables

Each simple statement has one of two possible truth values, true or false (T or F). If the compound statement is made up of two component statements, there are four possible sets of truth values for the component simple statements. This can be shown in the following tabular form.

p	q	Compound Statement
T	T	
T	F	
F	T	
F	F	

When the compound statement is made up of three simple component statements there are eight possible sets of truth values of the component simple statements. This can be shown in the following tabular form.

p	q	r	Compound Statement
T	T	T	
T	T	F	
T	F	T	
T	F	F	
F	T	T	
F	T	F	
F	F	T	
F	F	F	

In general, if the compound statement is made up of n component simple statements, there are 2^n possible sets of truth values of the component simple statements. For example, if the compound statement is made up of the four component statements p, q, r, and s, there would be $2^4 = 2 \times 2 \times 2 \times 2 = 16$ possible sets of truth values of the component simple statements.

In the following discussion we will deal with compound statements which are made up of either two or three component statements. The construction of truth tables for compound statements with four or more component statements is time consuming and tedious. All of the procedures developed for compound statements with two or three component statements apply to compound statements made up of four or more component statements.

There are two basic methods that are used in the construction of a truth table. In both methods, we first list all of the possible sets of truth values of the component simple statements. In doing this, it is wise to follow some fixed pattern that will fill in the component truth values quickly. In the table that has two component simple statements there will be four rows (horizontal lines). The first column (vertical line) has two trues and two falses. The second column then has one true, one false, then another one true, one false. This arrangement can be used regardless of what the component letters stand for.

p	q	
T	T	
T	F	
F	T	
T	F	

When the table has eight rows (because there were three components), the first column starts with four trues (always half the number of rows) and four falses. The second column will then have two trues, two falses, followed by another two trues and two falses. Finally, the third column alternates one true, one false, one true, one false, etc. You may recall that we used the system to fill in the membership table in Chapter 1.

p	q	r	
T	T	T	
T	T	F	
T	F	T	
T	F	F	
F	T	T	
F	T	F	
F	F	T	
F	F	F	

We will now illustrate the two basic methods of constructing a truth table. It will not be a good idea to bounce back and forth between the two. To avoid confusion, consider both but then stick with one thereafter.

Method I

Example 1

When is the statement *It is false that he is not a candidate for office or an outspoken critic of the present administration* a true statement?

Symbolizing using p for the statement *he is a candidate for office* and q for the statement *he is an outspoken critic of the present administration* we have $\sim(\sim p \vee q)$. The truth table for this compound statement is constructed as follows.

p	q	$\sim p$	$\sim p \vee q$	$\sim(\sim p \vee q)$
T	T	F	T	F
T	F	F	F	T
F	T	T	T	F
F	F	T	T	F

The first two columns are for the truth values of the component simple statements p and q. There are four rows, one for each of the possible true-false combinations of p and q. The third column contains the truth values of the negation of p. The fourth column contains the truth values of the disjunction of $\sim p$ with q. Finally, the fifth column contains the negation of the truth values of the compound statement $\sim(\sim p \vee q)$.

The truth table shows that the compound statement $\sim(\sim p \vee q)$ is true only when p is true and q is false. It is false for all other truth values of p and q. That is, the statement *It is false that he is not a candidate for office or an outspoken critic of the present administration* is a true statement when *he is a candidate for office* and *he is not a critic of the present administration.*

Example 2

When is the statement *It is not the case that if he is not guilty, then he is a victim and an innocent bystander* a true statement?

Symbolizing using p for the statement *he is guilty*, q for the statement *he is a victim*, and r for the statement *he is an innocent bystander* we have $\sim[\sim p \to (q \wedge r])$. The truth table for this compound statement is constructed as follows.

p	q	r	$\sim p$	$(q \wedge r)$	$\sim p \to (q \wedge r)$	$\sim[\sim p \to (q \wedge r)]$
T	T	T	F	T	T	F
T	T	F	F	F	T	F
T	F	T	F	F	T	F
T	F	F	F	F	T	F
F	T	T	T	T	T	F
F	T	F	T	F	F	T
F	F	T	T	F	F	T
F	F	F	T	F	F	T

The first three columns are for the truth values of the component simple statements p, q, and r. There are eight rows, one for each of the possible true-false combinations of p, q, and r. The fourth column contains the truth values of the negation of p. The fifth column contains the truth values of the conjunction of q and r. The sixth column contains the truth values of the conditional, $\sim p \to (q \wedge r)$. Finally, the seventh column contains the truth values of the negation of $[\sim p \to (q \wedge r)]$.

The truth table shows that the compound statement $\sim[\sim p \to (q \wedge r)]$ is false for all values of p, q, and r except in the last three cases.

In examining the results of the above truth tables we are only concerned with the truth values of the component simple statements and with the truth values of the compound statement under consideration. That is, the truth table of a compound statement consists of the truth values of the component simple statements and the corresponding truth values of the compound statement. The other entries in the truth table are the intermediate steps used in the computation of the truth values of the compound statement. Therefore, the truth tables for the preceding examples are as follows:

p	q	$\sim(\sim p \lor q)$
T	T	F
T	F	T
F	T	F
F	T	F

p	q	r	$\sim[\sim p \to (q \land r)]$
T	T	T	F
T	T	F	F
T	F	T	F
T	F	F	F
F	T	T	F
F	T	F	T
F	F	T	T
F	F	F	T

Method II

The second method is similar to the first but it will enable us to construct truth tables in a more concise manner. We will use this method in our future work.

Example 1

Construct the truth table for the compound statement $\sim(\sim p \lor q)$.

Step 1 Since the compound statement consists of two component simple statements p and q, we construct a table with three columns—one for each of the component simple statements and one for the compound statement under consideration. There will be four rows, one for each of the possible sets of truth values for p and q.

p	q	$\sim(\sim p \vee q)$
T	T	
T	F	
F	T	
F	F	
1	1	

Step 2   Fill in the truth values for p and q in the compound statement.

p	q	$\sim(\sim p \vee q)$	
T	T	T	T
T	F	T	F
F	T	F	T
F	F	F	F
		2	2

Step 3   Fill in the truth values for the negation of p.

p	q	$\sim(\sim$	p	$\vee q)$
T	T	F	T	T
T	F	F	T	F
F	T	T	F	T
F	F	T	F	F
		3		

Step 4   Fill in the truth values for the disjunction of $\sim p$ with q.

p	q	$\sim(\sim$	p	$\vee$	$q)$
T	T	F	T	T	T
T	F	F	T	F	F
F	T	T	F	T	T
F	F	T	F	T	F
				4	

Step 5 Finally, fill in the truth values for the negation of $(\sim p \lor q)$.

p	q	$\sim$	$(\sim$	p	$\lor$	$q)$
T	T	F	F	T	T	T
T	F	T	F	T	F	F
F	T	F	T	F	T	T
F	F	F	T	F	T	F
		5				

This above procedure would be written in the following form with the column numbers indicating the order in which the steps were performed.

p	q	$\sim$	$(\sim$	p	$\lor$	$q)$
T	T	F	F	T	T	T
T	F	T	F	T	F	F
F	T	F	T	F	T	T
F	F	F	T	F	T	F
1	1	5	3	2	4	2

Columns 1 and 5 form the truth table of the compound statement $\sim(\sim p \lor q)$.

Example 2

Construct the truth table for the compound statement $\sim[\sim p \to (q \land r)]$.

p	q	r	$\sim[$	$\sim$	p	$\to$	$(q$	$\land$	$r)]$
T	T	T	F	F	T	T	T	T	T
T	T	F	F	F	T	T	T	F	F
T	F	T	F	F	T	T	F	F	T
T	F	F	F	F	T	T	F	F	F
F	T	T	F	T	F	T	T	T	T
F	T	F	T	T	F	F	T	F	F
F	F	T	T	T	F	F	F	F	T
F	F	F	T	T	F	F	F	F	F
1	1	1	6	3	2	5	2	4	2

Step 1   Since the compound statement consists of three component simple statements *p*, *q*, and *r*, we construct a truth table with four columns, one for
each of the component simple statements and one for the compound
statement under consideration. There are eight rows, one for each of
the possible sets of truth values of *p*, *q*, and *r*.

Step 2   Fill in the truth values for the component simple statements *p*, *q*, and *r*
in the compound statement.

Step 3   Fill in the truth values of the negation of *p*.

Step 4   Fill in the truth values of the conjunction of *q* with *r*.

Step 5   Fill in the truth values of the conditional which has $\sim p$ as the antecedent
and $(q \wedge r)$ as the consequent.

Step 6   Fill in the truth values of the negation of $[\sim p \rightarrow (q \wedge r)]$.

Columns 1 and 6 form the truth table of the compound statement $\sim [\sim p \rightarrow (q \wedge r)]$.

To conclude this section consider an earlier statement, *It is not the case that
if you read the Times, your vote will be prejudiced in the coming election.* When is
this statement a true statement?

Symbolizing using *p* for the statement *you read the Times* and *q* for the statement
your vote will be prejudiced in the coming election we have $\sim (p \rightarrow q)$. The truth
table for this compound statement is constructed as follows.

p	*q*	$\sim$	(*p*	$\rightarrow$	*q*)
T	T	F	T	T	T
T	F	T	T	F	F
F	T	F	F	T	T
F	F	F	F	T	F
1	1	4	2	3	2

The truth table (columns 1 and 4) of the compound statement shows that the
statement is true only when *you read the Times* and *your vote is not prejudiced
in the coming election* are true.

Exercises 2.2

1. Determine the truth values of the compound statements for the indicated truth
   values of the component statements.

   (a) *If Lorraine and Andy come to the party it will be a success. Lorraine and
        Andy came to the party* is a true statement and *the party was a success* is a
        false statement.

(b) *Either Bob was accepted and is happy or he was not accepted and is not happy.*
      He was accepted is a true statement and *he is happy* is a true statement.

(c) *Tom bought a new car but he didn't like the radio. Tom bought a new car* is a
      true statement and *he didn't like the radio* is a false statement.

(d) *It is not the case that Bob loves Margie if and only if Margie loves Bob. Bob*
      loves Margie is a true statement and *Margie loves Bob* is a true statement.

2. If a and d are true statements and b and c are false statements, determine the
   truth values of the following compound statements.

   (a) $a \lor b$ (b) $\sim a \to c$ (c) $(c \land d) \lor a$
   (d) $(a \leftrightarrow b) \land c$ (e) $\sim[(a \lor b) \to \sim c]$
   (f) $\sim a \lor [b \land (c \to \sim d)]$

3. Determine the truth values of the following compound statements. If there is
   insufficient information, state the reason. In each case "*a*" is true.

   (a) $a \to (b \land d)$ (b) $b \lor a$ (c) $(b \land a) \to c$
   (d) $\sim a \land b$ (e) $c \to a$ (f) $b \to (c \lor a)$
   (g) $(\sim a \to b) \lor c$ (h) $a \to b$

4. Construct the truth tables for the following:

   (a) $p \to \sim q$ (b) $\sim p \lor \sim q$
   (c) $\sim(p \land \sim q)$ (d) $(p \land q) \lor \sim r$

5. Determine the truth values of the component statements that make the compound statements false.

   (a) $p \lor \sim q$ (b) $\sim p \to \sim q$
   (c) $\sim(\sim p \land q)$ (d) $\sim(p \leftrightarrow q)$

Earlier in this chapter we classified compound statements as being a
conjunction, disjunction, conditional, biconditional, or negation. Now
we will classify compound statements by their truth values. This will be
done by constructing the truth table for the compound statements and
then examining their truth values.

___

Types of Statements

In general, with regard to their truth values, there are three types of
compound statements; those with all true truth values; those with all false
truth values; and those with some true and some false truth values.
Therefore, we define the following.

Definition 4

Tautology: A tautology is a compound statement which has the truth value true for all of the possible truth values of its component simple statements. That is, it is a statement that is always true.

Consider the following examples.

Example 1

The President vetoes or does not veto the bill.

p	$(p$	$\lor$	$\sim$	$p)$
T	T	T	F	T
F	F	T	T	F
1	2	4	3	2

$p = the\ President\ vetoes\ the\ bill$

Example 2

If he is the President and he is either a Senator or not the President, then he is a Senator.

p	q	$[p$	$\land$	$(q$	$\lor$	$\sim$	$p)]$	$\rightarrow$	q
T	T	T	T	T	T	F	T	T	T
T	F	T	F	F	F	F	T	T	F
F	T	F	F	T	T	T	F	T	T
F	F	F	F	F	T	T	F	T	F
1	1	2	5	2	4	3	2	6	2

$p = he\ is\ the\ President$

$q = he\ is\ a\ Senator$

Examining the truth tables for the compound statements we see that the compound statements are true for all of the truth values of the component statements. Therefore, these compound statements are tautologies.

Definition 5

False Statement: A false statement is a compound statement which has the truth value false for all of the possible truth values of its component simple statements. That is, it is a statement that is always false.

Consider the following examples.

Example 3

The President vetoes and does not veto the bill.

p	p	$\wedge$	$\sim$	p	
T	T	F	F	T	$p =$ *the President vetoes the bill*
F	F	F	T	F	
1	2	4	3	2	

Example 4

He is a Senator or not the President and he is the President and not a Senator.

p	q	[(q	$\vee$	$\sim$	p)	$\wedge$	(p	$\wedge$	$\sim$	q)]	
T	T	T	T	F	T	F	T	F	F	T	$p =$ *he is the President*
T	F	F	F	F	T	F	T	T	T	F	
F	T	T	T	T	F	F	F	F	F	T	$q =$ *he is a Senator*
F	F	F	T	T	F	F	F	F	F	T	
1	1	2	4	3	2	5	2	4	3	2	

Examining the truth tables for the compound statements we see that the compound statements are false for all of the truth values of the component statements. Therefore these compound statements are false statements.

Definition 6

Indeterminate Statements: An indeterminate statement is a compound statement which has the truth value true for some of the possible truth values of its component simple statements and false for the remaining truth values of the component statements. That is, it is a statement that is sometimes true and sometimes false.

Consider the following examples.

Example 5

He is the President and not a Senator.

p	q	p	∧	~	q	
T	T	T	F	F	T	p = he is the President
T	F	T	T	T	F	
F	T	F	F	F	T	q = he is a Senator
F	F	F	F	T	F	
1	1	2	4	3	2	

Example 6

He is not the President and, if he is the President, he is not a Senator.

p	q	[~	p	∧	(p	→	~	q)]	
T	T	F	T	F	T	F	F	T	p = he is the President
T	F	F	T	F	T	T	T	F	
F	T	T	F	T	F	T	F	T	q = he is a Senator
F	F	T	F	T	F	T	T	F	
1	1	3	2	5	2	4	3	2	

Examining the truth tables for the compound statements we see that the compound statements are true for some of the truth values of the component simple statements and false for the other truth values. Therefore, they are indeterminate statements.

Note: The majority of compound statements are indeterminate statements.

Now that we have classified compound statements by their truth values, we can also use their truth values in comparing two compound statements. This will be done by constructing the truth tables for both of the compound statements under investigation and then comparing their respective truth values.

Types of Statement Relations

With regard to their truth values, there are many ways in which two statements can be related. We will restrict our discussion to just four of these relations: when the two statements have identical truth values; when they have opposite truth values; when they have some identical truth values and some opposite truth values; and when they can be joined by the connective of conditional and create a tautology. Therefore, we define the following.

Definition 7

Equivalent Statements: Two statements are equivalent if and only if they have the same truth values. That is, when one statement is true the other is also true; when it is false the other is also false. In other words, *p is equivalent to q* whenever the biconditional $p \leftrightarrow q$ is a tautology.

Note: The symbol $\Leftrightarrow$ may be used to represent *is equivalent to* and distinguish it from $\leftrightarrow$ which stands for the biconditional connective.

Consider the following example.

Example 1

Show that the statement *It is not the case that either he does not listen to reason or he is open-minded* is equivalent to the statement *He listens to reason but is not open-minded*.

Symbolizing and comparing the truth values we see that the two statements have the same truth values and therefore they are equivalent.

p	q	~ (~ p ∨ q)	p ∧ ~ q	
T	T	F F T T T	T F F T	p = he listens to reason
T	F	T F T F F	T T T F	
F	T	F T F T T	F F F T	q = he is open-minded
F	F	F T F T F	F F T F	
1	1	5 3 2 4 2	2 4 3 2	

Definition 8

Contradictory Statements: Two statements are contradictory if and only if they have opposite truth values. That is, when one statement is true the other is false; when it is false the other is true.

Consider the following example.

Example 2

Show that the statement *He does not listen to reason and he is open-minded* is contradictory to the statement *If he is open-minded, then he listens to reason.*

Symbolizing and comparing the truth values we see that the two statements have opposite truth values and therefore they are contradictory.

p	q	~ p ∧ q	q → p	
T	T	F T F T	T T T	p = he listens to reason
T	F	F T F F	F T T	
F	T	T F T T	T F F	q = he is open-minded
F	F	T F F F	F T F	
1	1	3 2 4 2	2 3 2	

Definition 9

Unrelated Statements: Two statements are unrelated if and only if they are neither equivalent nor contradictory. That is, some truth values are the same and some truth values are opposite.

Consider the following example.

Example 3

Show that the statement *He listens to reason and is open-minded* and the statement *He listens to reason or is open-minded* are unrelated.

Symbolizing and comparing the truth values we see that the two statements have some equal truth values and some opposite truth values. Therefore, they are unrelated.

p	q	$p \wedge q$			$p \vee q$			
T	T	T	T	T	T	T	T	$p = $ *he listens to reason*
T	F	T	F	F	T	T	F	
F	T	F	F	T	F	T	T	$q = $ *he is open-minded*
F	F	F	F	F	F	F	F	
1	1	2	3	2	2	3	2	

Note: When we compare the truth values of two statements it is important that they be compared on the same truth table. This will help reduce the possibility of making an error in our comparison.

Definition 10

Implication: If two statements, *p and q*, are joined together by the connective of conditional, and the resulting statement is a tautology, then we may call the *relationship* between *p* and *q* an implication. We may then say that *p implies q*.

Note: The symbol ⇒ may be used to represent an *implication* and distinguish it from → which stands for the conditional connective.

Consider the following example.

Example 4

Show that the statement *It is sunny and I am happy* implies the statement *It is not sunny or I am happy* (i.e., show that $(s \wedge h) \rightarrow (\sim s \vee h)$ is a tautology).

Symbolizing and completing the truth table we see that $(s \wedge h) \rightarrow (\sim s \vee h)$ is a tautology. Therefore, it is an implication and we may write $(s \wedge h) \Rightarrow (\sim s \vee h)$.

s	h				$(s \wedge h) \rightarrow (\sim s \vee h)$					
T	T	T	T	T	T	F	T	T	T	$s =$ It is sunny.
T	F	T	F	F	T	F	T	F	F	
F	T	F	F	T	T	T	F	T	T	$h =$ I am happy.
F	F	F	F	F	T	T	F	T	F	
1	1	2	3	2	5	3	2	4	2	

The next step in our investigation of statement relations is the development of the important procedure of finding a statement equivalent to some given statement.

Discovering Equivalent Statements

To find a statement equivalent to some given statement we first construct the truth table for the given statement in order to determine its truth values. Then we try to construct a statement with the same truth values. We do this by applying the basic definitions of the logical connectives and negation. We will now illustrate this procedure with the following examples.

Example 1

Find an equivalent statement for the statement *It is not the case that the witness told a lie or the judge did not overrule the objection.*

Symbolizing and constructing the truth table we have

p	q	$\sim(p \lor \sim q)$
T	T	F
T	F	F
F	T	T
F	F	F

$p =$ *the witness told a lie*

$q =$ *the judge overruled the objection*

Therefore we are looking for a statement which has the truth values F, F, T, F. We know that the logical connective of conjunction ($\land$) will yield a truth table that contains one true and three false truth values. We examine the row that contains the one true (odd) truth value. This appears in the third row where the component statement p is false and q is true. In order for a conjunction to yield a true truth value, both components must be true. Therefore, if we take $\sim p$, which would be true and conjunct it with q which is true, the third row for the statement $(\sim p \land q)$ would be true. Upon examination of this statement's truth table, we see that it is equivalent to $\sim(p \lor \sim q)$.

p	q	$(\sim p \land q)$
T	T	F
T	F	F
F	T	T
F	F	F

In other words the statement *The witness did not tell a lie and the judge overruled the objection* $(\sim p \land q)$ is equivalent to the given statement *It is not the case that the witness told a lie or the judge did not overrule the objection* $[\sim(p \lor \sim q)]$.

We could also find a conditional equivalent to $\sim(p \lor \sim q)$. We know that a conditional will yield a truth table that contains three true and one false truth values with the false value coming when true implies false. However, our given statement has the values F, F, T, F. In this case we find the contradictory statement to $\sim(p \lor \sim q)$, that is, a statement with the truth values T, T, F, T. Once we have found this contradictory statement, we then negate it to obtain our equivalent statement. To find a statement with truth values T, T, F, T, we examine the third row which has the one false (odd) truth value. Since in this row q is true and p is false, the conditional $q \rightarrow p$ will be false. Upon examination of this statement and its negation, we see that $\sim(q \rightarrow p)$ is equivalent to $\sim(p \vee \sim q)$.

p	q	$q \rightarrow p$	$\sim(q \rightarrow p)$
T	T	T	F
T	F	T	F
F	T	F	T
F	F	T	F

In other words the statement *It is false that if the judge overruled the objection, the witness told a lie* [$\sim(q \rightarrow p)$] is also equivalent to the given statement *It is not the case that the witness told a lie or the judge did not overrule the objection* [$\sim(p \vee \sim q)$].

Another statement equivalent to the given statement would be *It is not the case that if the witness did not tell a lie, then the judge did not overrule the objection* [$\sim(\sim p \rightarrow \sim q)$]. In fact, for each statement there are an unlimited number of equivalent statements. However, there are only a few with the simpler form. The structures will become more complex as we construct more equivalent statements.

Example 2

Find a statement containing a conjunction and disjunction equivalent to the conditional *If he is innocent, he is not in prison.*

Symbolizing using p for *he is innocent* and q for *he is in prison* we have $(p \rightarrow \sim q)$. The truth table and resulting truth values for this statement are

p	q	$(p \rightarrow \sim q)$
T	T	F
T	F	T
F	T	T
F	F	T

We are looking for a statement that has one false and three true truth values (F, T, T, T). Therefore, we first find the disjunction equivalent to the given statement since the logical connective of disjunction ($\vee$) yields a truth table that contains one false and three true truth values. We examine the row that contains the one false (odd) truth value. This appears in the first row where both of the component statements p and q are true. In order for a disjunction to yield a false truth value, both components must be false. Therefore, if we take $\sim p$, which is false, and disjunct it with $\sim q$, which is also false, the first row for the compound statement $(\sim p \vee \sim q)$ would be false. Upon examination of this statement's truth table, we see that it is equivalent to $(p \rightarrow \sim q)$.

p	q	$(\sim p \vee \sim q)$
T	T	F
T	F	T
F	T	T
F	F	T

Therefore, *He is not innocent or not in prison* is a disjunction equivalent to the given statement.

We now find a conjunction equivalent to the given statement $(p \rightarrow \sim q)$. We know that a conjunction will yield a truth table that contains three false and one true truth values with the true value coming when both components of the conjunction are true. However, our given statement has the values F, T, T, T. Here we find a conjunction contradictory to the given statement, that is, a statement with the truth values, T, F, F, F. Once we find this contradictory statement, we negate it to obtain our equivalent statement. To find a conjunction with truth values T, F, F, F, we examine the first row which has the one true (odd) truth value. Since in this row both p and q are true, the conjunction $(p \wedge q)$ is also true. Upon examination of this statement and its negation, we see that $\sim(p \wedge q)$ is equivalent to the given statement $(p \rightarrow \sim q)$.

p	q	$(p \wedge q)$	$\sim(p \wedge q)$
T	T	T	F
T	F	F	T
F	T	F	T
F	F	F	T

Actually, we were unable to find a conjunction equivalent to the given statement. We found a negation of a conjunction *It is not the case that he is innocent and in prison* equivalent to the given statement.

Exercises 2.3

1. By the use of truth tables, determine whether each of the following compound statements is a tautology, false statement, or indeterminate statement.

   (a) $(p \lor q) \lor (p \land q)$ (b) $(p \land q) \to q$
   (c) $\sim[(p \lor q) \land \sim p]$ (d) $[p \land (p \to q)] \to q$
   (e) $[(p \lor q) \land \sim p] \to \sim q$
   (f) If Ed yells again the referee will call a technical foul, but if the referee calls a technical foul, Ed will yell again.

2. Using truth tables determine whether the following pairs of statements are equivalent, contradictory, or unrelated.

   (a) $\sim p \lor q$ and $\sim(p \land \sim q)$ (b) $(p \to q)$ and $\sim(\sim p \lor q)$
   (c) $(\sim q \to p)$ and $(\sim p \to q)$ (d) $p \land q$ and $p \lor q$
   (e) $(p \land q) \to r$ and $\sim(p \land q) \lor r$ (f) $(p \lor q) \land r$ and $p \lor (q \land r)$
   (g) $p \to q$ and $\sim p \lor q$ (h) $\sim p \lor q$ and $\sim(p \land \sim q)$

3. Using the truth tables from problem 2, determine if the first statement *implies* the second statement (a) to (h).

Logically Equivalent Statements

In the section on discovering equivalent statements we saw that it is possible to find equivalent statements for a given statement. It was possible to take a statement containing either a conjunction, disjunction, or conditional and express it in forms containing the other two connectives.

In this section we will list nine general equivalence rules that will allow us to find equivalent statements for a given statement without going through the time consuming procedure of the previous section. In the discussion of the equivalence relations, when there is more than one part to the equivalence relation, we will prove only the first by truth tables. The proof of the remaining relations will be left as exercises.

The nine equivalence relations are as follows:

Double Negation Equivalence (E$_1$).   The negation of the negation of a statement is equivalent to the statement.

$$\sim(\sim p) \text{ is equivalent to } p$$

p	$\sim(\sim p)$
T	T
F	F

Examples

$\sim[\sim(p \lor r)]$ is equivalent to $(p \lor r)$
$\sim[\sim(p \land \sim q)]$ is equivalent to $(p \land \sim q)$

Idempotent Equivalences (E$_2$).   The conjunction or disjunction of a statement with itself is equivalent to itself.

$$(p \lor p) \text{ is equivalent to } p$$

$$(p \land p) \text{ is equivalent to } p$$

p	$p \lor p$
T	T
F	F

Examples

$(\sim r \land \sim r)$ is equivalent to $\sim r$
$[(p \land \sim q) \lor (p \land \sim q)]$ is equivalent to $(p \land \sim q)$

Commutative Equivalences (E$_3$).   The order of a conjunction or disjunction does not affect the truth value of the compound statement.

$(p \lor q)$ is equivalent to $(q \lor p)$

$(p \land q)$ is equivalent to $(q \land p)$

p	q	$(p \lor q)$	$(q \lor p)$
T	T	T	T
T	F	T	T
F	T	T	T
F	F	F	F

Examples

$\sim(\sim p \lor q)$ is equivalent to $\sim(q \lor \sim p)$
$[p \lor (\sim q \land r)]$ is equivalent to $[(\sim q \land r) \lor p]$

Associative Equivalences (E$_4$).   The truth value of the disjunction or conjunction of three statements whose order is preserved does not depend on the manner in which the statements are grouped.

$[p \lor (q \lor r)]$ is equivalent to $[(p \lor q) \lor r]$

$[p \land (q \land r)]$ is equivalent to $[(p \land q) \land r]$

p	q	r	$[p \lor (q \lor r)]$	$[(p \lor q) \lor r]$
T	T	T	T	T
T	T	F	T	T
T	F	T	T	T
T	F	F	T	T
F	T	T	T	T
F	T	F	T	T
F	F	T	T	T
F	F	F	F	F

Examples

$[\sim p \lor (\sim q \lor r)]$ is equivalent to $[(\sim p \lor \sim q) \lor r]$
$[p \lor q] \lor (r \land \sim p)$ is equivalent to $p \lor [q \lor (r \land \sim p)]$

Distributive Equivalences (E$_5$).   The conjunction of a statement with the disjunction of two other statements is equivalent to the disjunction of the conjunctions of the first with the second and the first with the third. This is also true for disjunction distributed over conjunction.

$[p \wedge (q \vee r)]$ is equivalent to $[(p \wedge q) \vee (p \wedge r)]$

$[p \vee (q \wedge r)]$ is equivalent to $[(p \vee q) \wedge (p \vee r)]$

p	q	r	$[p \wedge (q \vee r)]$	$[(p \wedge q) \vee (p \wedge r)]$
T	T	T	T	T
T	T	F	T	T
T	F	T	T	T
T	F	F	F	F
F	T	T	F	F
F	T	F	F	F
F	F	T	F	F
F	F	F	F	F

Examples

$[\sim p \vee (q \wedge \sim r)]$ is equivalent to $[(\sim p \vee q) \wedge (\sim p \vee \sim r)]$
$(p \wedge q) \vee (\sim p \wedge r)$ is equivalent to $[(p \wedge q) \vee \sim p] \wedge [(p \wedge q) \vee r]$

DeMorgan's Equivalence (E$_6$).   The disjunction (conjunction) of two statements is equivalent to a conjunction (disjunction) of the two statements if and only if the signs of the two statements and of their compound statement are changed.

$\sim(p \vee q)$ is equivalent to $(\sim p \wedge \sim q)$

$\sim(p \wedge q)$ is equivalent to $(\sim p \vee \sim q)$

p	q	$\sim(p \vee q)$	$(\sim p \wedge \sim q)$
T	T	F	F
T	F	F	F
F	T	F	F
F	F	T	T

Examples

$\sim (p \lor \sim q)$ is equivalent to $(\sim p \land q)$
$\sim [\sim (p \land q) \lor (p \to q)]$ is equivalent to $[(p \land q) \land \sim (p \to q)]$

Conditional Equivalence (E_7).   A conditional is equivalent to a disjunction if and only if the disjunction is the negation of the antecedent, with the consequent of the conditional.

$$(p \to q) \text{ is equivalent to } (\sim p \lor q)$$

p	q	$(p \to q)$	$(\sim p \lor q)$
T	T	T	T
T	F	F	F
F	T	T	T
F	F	T	T

Examples

$(p \to \sim q)$ is equivalent to $(\sim p \lor \sim q)$
$\sim [\sim p \to (q \land r)]$ is equivalent to $\sim [p \lor (q \land r)]$

Contrapositive Equivalence (E_8).   A conditional is equivalent to another conditional if and only if the negation of the antecedent of the first is the consequent of the second and the negation of the consequent of the first is the antecedent of the second.

$$(p \to q) \text{ is equivalent to } (\sim q \to \sim p)$$

p	q	$(p \to q)$	$(\sim q \to \sim p)$
T	T	T	T
T	F	F	F
F	T	T	T
F	F	T	T

Examples

$(\sim p \to q)$ is equivalent to $(\sim q \to p)$
$[\sim p \to (q \land r)]$ is equivalent to $[\sim (q \land r) \to p]$

Biconditional Equivalence (E_9).   The conjunction of two conditionals, which has the antecedent of the first conditional as the consequent of the second and the antecedent of the second conditional as the consequent of the first, is equivalent to a biconditional statement formed by the antecedent and consequent of the first conditional.

$$(p \leftrightarrow q) \text{ is equivalent to } [(p \rightarrow q) \wedge (q \rightarrow p)]$$

p	q	$(p \leftrightarrow q)$	$[(p \rightarrow q) \wedge (q \rightarrow p)]$
T	T	T	T
T	F	F	F
F	T	F	F
F	F	T	T

Examples

$[(p \rightarrow \sim q) \wedge (\sim q \rightarrow p)]$ is equivalent to $(p \leftrightarrow \sim q)$
$[(\sim p \rightarrow q) \wedge (q \rightarrow \sim p)]$ is equivalent to $[\sim p \leftrightarrow q]$

Of the nine equivalence relations listed above, DeMorgan's and Conditional equivalences are the two that will be used most frequently in succeeding work since they enable us to express a statement containing either a conjunction, disjunction, or conditional in forms containing the other two. For example, the statement *It is not the case that if you listen to a political speech, your vote will be prejudiced in the coming election* $[\sim (l \rightarrow p)]$ can now be written *It is not the case that you do not listen to a political speech or your vote will be prejudiced in the coming election* $[\sim (\sim l \vee p)]$ or more simply *You listen to a political speech and your vote will not be prejudiced in the coming election* $[l \wedge \sim p]$.
   The above is true since:

$$\sim (l \rightarrow p) \quad \overset{\text{by}}{\leftrightarrow} \quad \sim (\sim l \vee p) \quad \overset{\text{by}}{\leftrightarrow} \quad (l \wedge \sim p)$$
$$\text{Conditional} \qquad \text{DeMorgan's}$$
$$\text{equivalence} \qquad \text{equivalence}$$

To conclude this discussion of equivalence relations, consider the following examples.

Example 1

By applying DeMorgan's equivalence (E$_6$) write a statement equivalent to *It is not the case that he is a <u>c</u>onservative or not a <u>R</u>epublican.*

Symbolizing using the letters indicated we have $\sim(c \lor \sim r)$. By (E$_6$)

$$\sim(c \lor \sim r) \leftrightarrow (\sim c \land r)$$

Therefore, the statement *He is not a conservative and he is a republican* is equivalent to the given statement.

Example 2

By applying the Distributive equivalence (E$_5$) write a statement equivalent to *He is not a <u>R</u>epublican and he is either a <u>l</u>iberal or not a <u>c</u>onservative.*

Symbolizing using the letters indicated we have $\sim r \land (l \lor \sim c)$. By (E$_5$)

$$[\sim r \land (l \lor \sim c)] \leftrightarrow [(\sim r \land l) \lor (\sim r \land \sim c)]$$

Therefore, the statement *Either he is not a Republican and a liberal or he is not a Republican and not a conservative* is equivalent to the given statement.

Example 3

By applying the Double Negation equivalence (E$_1$) write a statement equivalent to *It is not the case that he is not a <u>R</u>epublican.*

Symbolizing using the letter indicated we have $[\sim(\sim r)]$. By (E$_1$)

$$[\sim(\sim r)] \leftrightarrow (r)$$

Therefore, the statement *He is a Republican* is equivalent to the given statement.

Example 4

By applying the Biconditional equivalence (E$_9$) write a statement equivalent to *If he is a <u>l</u>iberal, he is not a <u>R</u>epublican and if he is not a <u>R</u>epublican, he is a liberal.*

Symbolizing using the letters indicated we have $[(l \to \sim r) \land (\sim r \to l)]$. By (E$_9$)

$$[(l \to \sim r) \land (\sim r \to l)] \leftrightarrow [l \leftrightarrow \sim r]$$

Therefore, the statement *He is a liberal if and only if he is not a Republican* is equivalent to the given statement.

Example 5

By applying the Conditional equivalence (E_7) write a statement equivalent to *He is a Republican or not a conservative.*

Symbolizing using the letters indicated we have $(r \lor \sim c)$. By (E_7)

$$(r \lor \sim c) \leftrightarrow (\sim r \to \sim c)$$

Therefore, the statement *If he is not a Republican, he is not a conservative* is equivalent to the given statement.

Converse, Inverse, and Contrapositive

Thus far we have discussed statements that are logically equivalent. We will now investigate the relationships of several forms of a conditional, some of which are equivalent and some which are not. With regard to the conditional $p \to q$ (*If p, then q*) we define the following related conditionals.

Converse. The converse of the conditional $p \to q$ (*If p, then q*) is the conditional $q \to p$ (*If q, then p*) that is, the converse of a conditional is formed by interchanging the antecedent and consequent of the conditional.

Inverse. The inverse of the conditional $p \to q$ (*If p, then q*) is the conditional $\sim p \to \sim q$ (*If not p, then not q*) that is, the inverse of a conditional is formed by negating both the antecedent and consequent of the conditional.

Contrapositive. The contrapositive of the conditional $p \to q$ (*If p, then q*) is the conditional $\sim q \to \sim p$ (*If not q, then not p*) that is, the contrapositive of a conditional is formed by both negating and interchanging the antecedent and consequent of the conditional.

To illustrate these relationships, consider the following:

(a)   Conditional: *If you listen to reason, you are open-minded.*

(b)   Converse: *If you are open-minded, you listen to reason.*

(c)   Inverse: *If you do not listen to reason, you are not open-minded.*

(d)   Contrapositive: *If you are not open-minded, you do not listen to reason.*

As a further illustration consider the following:

(a)   Conditional: *If he is a Russian leader, he is a member of the Communist party.*

(b)   Converse: *If he is a member of the Communist party, he is a Russian leader.*

(c)   Inverse: *If he is not a Russian leader, he is not a member of the Communist party.*

(d)   Contrapositive: *If he is not a member of the Communist party, he is not a Russian leader.*

The relationships between a conditional, its converse, its inverse, and its contrapositive are shown in the following truth tables:

		Conditional	Converse	Inverse	Contrapositive
p	q	$p \rightarrow q$	$q \rightarrow p$	$\sim p \rightarrow \sim q$	$\sim q \rightarrow \sim p$
T	T	T	T	T	T
T	F	F	T	T	F
F	T	T	F	F	T
F	F	T	T	T	T

Examining the above illustrative examples and truth tables we see that:

1   A conditional and its converse are unrelated statements.

2   A conditional and its inverse are unrelated statements.

3   A conditional and its contrapositive are equivalent statements.

4   The converse and inverse of a conditional are equivalent statements. This is true since they are the contrapositives of one another $(q \rightarrow p) \leftrightarrow (\sim p \rightarrow \sim q)$.

Therefore, when a conditional is true its converse and inverse are not necessarily true. They are sometimes true and sometimes they are false.

Exercises 2.4

Using the indicated equivalence relations, write a logically equivalent statement for each of the following.

1. $\sim[\sim(p \wedge q)]$ (Double Negation)   2. $p \vee p$ (Idempotent)

3. q (Idempotent)   4. $(p \to q) \vee r$ (Commutative)

5. $(p \vee q) \vee r$ (Associative)   6. $\sim p \to q$ (Contrapositive)

7. $a \leftrightarrow b$ (Biconditional)   8. $\sim p \vee q$ (Conditional equivalence)

9. $p \to \sim q$ (Conditional equivalence) 10. $p \vee (\sim q \wedge r)$ (Distributive)

11. $r \wedge (\sim p \vee q)$ (Distributive)   12. $\sim p \wedge q$ (DeMorgan's)

13. It is not the case that the sun is shining and the fishing is good. (DeMorgan's)

14. If Joan has lobster for dinner or there is a basketball game, Bill will come home early. (Contrapositive)

15. Don and Gladys go to the beach if and only if the sun is shining. (Biconditional)

16–18 For each of the following conditionals find the indicated conditional.

16. $\sim p \to q$ (Inverse)   17. $(p \wedge q) \to r$ (Converse)

18. $p \to (\sim q \vee r)$ (Contrapositive)

A Rationale for the Definition
of the Truth Values of the
Logical Connectives of Conditional

To conclude this section we will use the notion of equivalent statements to provide a rationale for our definition of the truth values of the logical connective of conditional.

There are two approaches to providing a rationale for this definition. They are as follows.

Rationale I

We can intuitively accept the definition of $T \to T = T$ and $T \to F = F$.

However, providing a rationale for the truth values that we assign to $F \to T$ and $F \to F$ presents a problem.

If we did not assign truth values to these two cases the truth table for conditional would be as follows:

p	q	*if . . . , then . . .* $p \to q$
T	T	T
T	F	F
F	T	?
F	F	?

There are four possible combinations of T-F values that can be assigned to the ? in the table. Listing these four possibilities and the values for $T \to T$ and $T \to F$ which already have been agreed upon, we have:

p	q	$p \to q$ Possibility I	$p \to q$ Possibility II	$p \to q$ Possibility III	$p \to q$ Possibility IV
T	T	T	T	T	T
T	F	F	F	F	F
F	T	T	T	F	F
F	F	T	F	T	F

If we accept the definitions that $T \to T = T$ and $T \to F = F$ then the above table lists the only four possible definitions for $p \to q$. Therefore, if we want a definition of $p \to q$ that assigns truth values to $F \to T$ and $F \to F$, it must be one of these four possibilities.

Examining the four possibilities we see that II has the same truth values as statement q, III has the same truth values as the statement $(p \leftrightarrow q)$, and IV has the same truth values as the statement $(p \wedge q)$. Since $(p \rightarrow q)$ has a different meaning than p, $(p \leftrightarrow q)$, and $(p \wedge q)$ we would not want it to have a truth value definition equivalent to any of these. If we accept this statement, then the only possible definition for $p \rightarrow q$ is possibility I, which defines both $F \rightarrow T$ and $F \rightarrow F$ as T.

Rationale II

If we are given the statement *If it is raining, then John is at home* $(p \rightarrow q)$ we know that in our system of symbolic logic there are only two possible weather conditions with regard to rain. *Either it is not raining or it is raining* $(\sim p \vee p)$. Examining the two weather possibilities we see that if it is not raining, we are unable to draw any conclusions about John's location. However, if it is raining the conditional tells us that John is at home. Therefore, when we say *if it is raining, then John is at home* $(p \rightarrow q)$, it is the same as saying *it is not raining or John is at home* $(\sim p \vee q)$. In other words the two statements are equivalent and therefore their truth tables should be identical. Constructing the truth tables for $(\sim p \vee q)$ we have

p	q	$(\sim p \vee q)$
T	T	T
T	F	F
F	T	T
F	F	T

and therefore the truth table and the corresponding truth values for $p \rightarrow q$ are

p	q	$p \rightarrow q$
T	T	T
T	F	F
F	T	T
F	F	T

Exercises

1. Determine whether the following statements are simple, compound, or neither.

   (a) *He is not guilty.*
   (b) *He is a liberal political leader.*
   (c) *Will he be successful, if he studies?*
   (d) *If $a = b$ and $b = c$, then $a = c$.*
   (e) *Please help those in need.*
   (f) *What is truth?*
   (g) *Write a letter to the president of the company and complain.*
   (h) *He was found guilty but not sent to prison or he is innocent.*
   (i) *He is a victim of social injustice.*
   (j) *It is true or false.*

2. Classify the following compound statements as a conjunction, disjunction, conditional, biconditional, negation, or as ambiguous.

   (a) $\sim(p \wedge \sim q)$ (b) $p \vee (q \wedge \sim r)$
   (c) $p \wedge (q \leftrightarrow r)$ (d) $p \wedge q \to \sim r$
   (e) $\sim(p \wedge \sim q) \vee (r \vee s)$ (f) $(p \vee \sim q) \leftrightarrow (r \wedge s)$
   (g) $[p \to (q \wedge r)] \wedge s$ (h) $\sim[p \to (q \wedge r)]$
   (i) $\sim[p \vee (q \wedge r)] \to \sim s$ (j) $(p \leftrightarrow q) \to \sim r$

3. If p stands for *Joe is smart*, q for *Joe studies*, and r for *Joe is passing math*, then write the statements symbolized by:

   (a) $(p \wedge q) \to r$
   (b) $p \vee (\sim q \to \sim r)$
   (c) $\sim[(p \wedge \sim q) \vee r]$
   (d) $(q \wedge r) \leftrightarrow p$
   (e) $(p \wedge q) \vee (p \wedge r)$

4. Symbolize the following statements, using the letters indicated and then classify them as a conjunction, disjunction, conditional, biconditional, or negation.

   (a) *If prices are lower, then we save more money.* (p, s)
   (b) *Prices are lower if taxes are higher.* (p, t)
   (c) *Either taxes are higher and prices are lower or, there is a recession and prices are not lower.* (t, p, r)
   (d) *Consumer goods are scarce or expensive but there is full employment.* (s, e, f)
   (e) *It is not the case that if you do not listen to reason, you are open-minded.* (r, m)
   (f) *If Jim works hard, he earns money; but if he does not work hard, he has time to relax.* (w, e, r)

(g) *Either I invest my money or, if I do not invest it, it does not earn interest for me.* (*i, e*)

(h) *It is not the case that Jim is interested in school politics and if his grades are high he will run for office.* (*i, g, r*)

(i) *He will major in either mathematics and physics or mathematics and chemistry.* (*m, p, c*)

(j) *If he is a Russian leader, he is a Communist, but he is not a Communist.* (*r, c*)

5. The definition of the truth values of the logical connective of conjunction could be represented in the following tabular form

$\wedge$	T	F
T	T	F
F	F	F

Construct similar tables for the logical connectives of disjunction, conditional, and biconditional.

6. Exclusive disjunction denoted by $p \veebar q$ means *p or q but not both.* Construct a truth table for this logical connective.

7. For what truth values of p and q do $p \wedge q$ and $p \rightarrow q$ have the same truth values?

8. If a and b are true statements and c and d are false statements, determine the truth values of the following compound statements.

(a) $(\sim a \vee c)$

(b) $[c \vee (a \rightarrow d)]$

(c) $[\sim a \vee \sim (c \wedge d)]$

(d) $\sim [b \leftrightarrow (\sim a \wedge d)]$

(e) $(a \leftrightarrow b) \vee (b \rightarrow \sim a)$

(f) $(a \rightarrow b) \wedge (\sim c \rightarrow d)$

(g) $\sim \{\sim [\sim b \wedge (a \wedge \sim d)]\}$

(h) $\sim a \vee [b \wedge (c \rightarrow \sim d)]$

(i) $(\sim a \vee c) \rightarrow (b \rightarrow d)$

(j) $\sim [\sim a \vee (b \wedge d)] \vee (c \rightarrow \sim d)$

(k) $\sim b \wedge [(\sim c \vee d) \leftrightarrow (a \rightarrow \sim b)]$

(l) $\sim [a \rightarrow (b \rightarrow c)] \rightarrow d$

9. Determine the truth values of the statements of problem 4 (a–e) for the indicated truth values of the component statements.

(a) *Prices are lower* is a true statement and *we save more money* is a false statement.

(b) *Taxes are higher* is a false statement and *prices are lower* is a true statement.

(c) *Taxes are higher* is a true statement, *prices are lower* is a false statement, and *there is a recession* is a true statement.

(d) *Consumer goods are scarce* is a true statement, *consumer goods are expensive* is a true statement, and *there is full employment* is a false statement.

(e) *You listen to reason* and *you are open-minded* are true statements.

10. Determine the truth values of the following compound statements. In the cases where there is insufficient information, state the reason.

(a) $\sim a \rightarrow (b \lor \sim c)$, given that c is false.

(b) $(a \leftrightarrow b) \lor (b \rightarrow \sim a)$, given that b is false.

(c) $a \land [\sim b \rightarrow (c \land d)]$, given that a is false.

(d) $a \rightarrow (b \land d)$, given that a is true.

(e) $(\sim a \lor c) \rightarrow (b \rightarrow d)$, given that b is false.

(f) $(a \rightarrow c) \land d$, given that c is true.

(g) $a \rightarrow (c \lor d)$, given that c is true.

(h) $(a \land b) \leftrightarrow c$, given that b is true.

(i) $[(a \rightarrow b) \land a] \rightarrow \sim a$, given that a is false.

(j) $a \rightarrow (\sim b \land c)$, given that b is false and c is true.

(k) *He will not be elected, if he is not a candidate for office* if we know that he is a candidate for office.

(l) *It is false that he is not a candidate for office and he will be elected* if we know that he will not be elected.

(m) *If he is a conservative and a candidate for office then he will be elected* if we know that he will be elected.

(n) *He is a conservative or a candidate for office, and he is a conservative or will not be elected* if we know that he is a candidate for office.

(o) *He is a conservative and will not be elected or, it is false that if he is a candidate for office, he will be elected* if we know that he is a candidate for office but he will not be elected.

11. Construct the truth tables for the following compound statements.

(a) $(\sim p \lor \sim q)$ (b) $\sim (\sim p \rightarrow q)$

(c) $[p \land \sim (q \rightarrow \sim r)]$ (d) $[\sim p \rightarrow (q \rightarrow \sim r)]$

(e) $[(p \rightarrow \sim q) \land (r \lor \sim p)]$ (f) $[(p \rightarrow q) \rightarrow (q \rightarrow p)]$

(g) $[\sim (\sim p \lor q) \leftrightarrow (p \rightarrow q)]$ (h) $[(p \land \sim q) \lor (\sim p \land q)]$

(i) $\{(p \land q) \rightarrow [\sim p \land (q \lor r)]\}$ (j) $\sim p \land [(p \rightarrow \sim q) \land q]$

12. Determine the truth values of the component statements that make the following compound statements true.

(a) *He is guilty, and he is a suspect or has avoided the police.*

(b) *If he is a Justice of the Supreme Court, it is not the case that he is an elected official and not a lawyer.*

(c) *Either there is social injustice and an inadequate welfare system, or it is false that if there is no social injustice, then we are not concerned.*

(d) *Bill's answer is incorrect, or if Bill's answer is correct, then John's and Sam's answers are incorrect.*

(e) *Either if it is A, then it is B, or it is not the case that if it is not B, then it is A or not C.*

13. By use of truth tables, determine whether each of the following compound statements is a tautology, false statement, or indeterminate statement.

   (a) $\sim(p \lor \sim q)$
   (b) $p \to (q \to p)$
   (c) $(\sim p \lor p) \land (q \land \sim q)$
   (d) $p \lor \sim(q \land r)$
   (e) $p \land [(q \to r) \land \sim p]$
   (f) *He is not guilty or, if he is in prison, he is guilty.*
   (g) *Either his theory is correct, or if he made an error, he will re-examine his premises.*
   (h) *It is not the case that the club either won or lost the pennant, or they played in the World Series.*

14. Using truth tables determine whether the following pairs of statements are equivalent, contradictory, or unrelated.

   (a) $(p \lor q)$ and $\sim(\sim p \to q)$
   (b) $\sim(p \land \sim q)$ and $(\sim p \to \sim q)$
   (c) $(\sim p \to \sim q)$ and $(q \to p)$
   (d) $r \land (p \to q)$ and $[(\sim p \land r) \lor (q \land r)]$
   (e) $(\sim p \leftrightarrow \sim q)$ and $[(\sim p \land q) \lor \sim(\sim q \to \sim p)]$
   (f) *Congress did not pass the bill or the President vetoed it* and the statement *If the President did not veto the bill, then Congress did not pass it.*
   (g) *If his reasoning and premises are correct, we must accept the conclusion* and the statement *His reasoning and premises are correct, if we accept the conclusion.*
   (h) *It is not the case that either he is in jail or if he is guilty, his trial has not yet been conducted* and the statement *If he is not in jail, then either he is not guilty or his trial has not yet been conducted.*

15. Find statements containing symbols of conjunction, disjunction, and conditional for each of the following sets of truth values.

Conjunction						
Disjunction						
Conditional						
p	q	(a)	(b)	(c)	(d)	(e)
T	T	F	F	T	F	T
T	F	T	F	T	T	T
F	T	T	F	T	F	F
F	F	T	T	F	F	T

Examine the statements to determine whether or not general rules exist that enable us to convert a statement containing a conjunction to a statement containing a disjunction and to convert a statement containing a disjunction to a statement containing a conditional. If such rules are discovered, state them.

16. For each of the following compound statements find the simplest logically equivalent statement.

(a) $[p \lor (\sim q \land \sim p)]$
(b) $p \to [q \land (p \lor \sim q)]$
(c) $\sim [(p \to q) \land (\sim q \to \sim p)]$
(d) $\sim [(p \land \sim q) \lor (p \land \sim r)]$
(e) *Either he is honest and sincere or he is not honest and not sincere.*
(f) *He is either tall or not strong and he is either heavy or not strong.*

17. Using truth tables prove the following equivalence relations.

(a) $(p \land p) \leftrightarrow p$
(b) $(p \land q) \leftrightarrow (q \land p)$
(c) $[p \land (q \land r)] \leftrightarrow [(p \land q) \land r]$
(d) $[p \lor (q \land r)] \leftrightarrow [(p \lor q) \land (p \lor r)]$
(e) $\sim (p \land q) \leftrightarrow (\sim p \lor \sim q)$

18. Using the indicated equivalence relation, write a logically equivalent statement for each of the following:

(a) Double Negation equivalence
$\sim [\sim (p \lor q)]$
$\sim [\sim (\sim p \land \sim q)]$
$(p \land q)$

(b) Identity equivalence
$(\sim p \lor \sim p)$
$(p \to q) \lor (p \to q)$
$\sim [\sim (p \to q) \land \sim (p \to q)]$

(c) Commutative equivalence
$(\sim d \land n) \lor s$
$(s \land l) \land \sim (d \to s)$
$(n \to s) \land (n \lor l)$

(d) DeMorgan's equivalence
$\sim (\sim p \lor \sim q)$
$[(a \lor b) \land \sim (c \to d)]$
$(c \to d) \land \sim a$

(e) Associative equivalence
$c \lor (\sim a \lor \sim b)$
$a \land [(c \to d) \land (d \lor c)]$
$[(a \lor b) \land (a \to c)] \land (a \lor d)$

(f) Distributive equivalence
$\sim s \land (\sim a \lor d)$
$a \lor [\sim (b \land c) \land (c \to d)]$
$(b \to c) \land [(c \to a) \lor (a \lor d)]$

(g) Contrapositive equivalence
$\sim p \to \sim (q \land r)$
$(p \lor q) \to \sim r$
$(p \to q) \to \sim (r \to s)$

(h) Conditional equivalence
$\sim p \to \sim (q \land r)$
$p \lor (q \to r)$
$\sim (\sim p \lor q)$

(i) Biconditional equivalence
$[(\sim p \to \sim q) \land (\sim q \to \sim p)]$
$[(p \land q) \to \sim r] \land [\sim r \to (p \land q)]$
$(p \leftrightarrow \sim q)$

19. Find a logically equivalent statement for each of the following compound statements using the indicated equivalence relation.

(a) *If we do not save more money, then prices are not lower.* (Contrapositive)

(b) *It is not the case that you do not listen to reason or you are open-minded.*
     (DeMorgan's)
(c) *Jim works hard and does a good job or Jim works hard and does not have*
     time to relax. (Distributive)
(d) *Prices are lower if taxes are not higher.* (Conditional)
(e) *If he works hard, he earns a good wage and if he earns a good wage, he*
     works hard. (Biconditional)
(f) *It is not the case that if you listen to reason, you are not open-minded.*
     (Conditional, DeMorgan's)

20. Find a contradictory statement, other than just the negation of the given
     statement, for each of the following.

(a) *If he is a Russian leader, he is a Communist.*
(b) *He does not listen to reason or he is not open-minded.*
(c) *It is not the case that if he works hard, he will not succeed.*

21. For each of the following conditionals find the indicated conditional.

(a) $p \rightarrow \sim q$ (contrapositive)
(b) $\sim q \rightarrow \sim p$ (inverse)
(c) $\sim p \rightarrow (q \wedge \sim r)$ (converse)
(d) $(p \wedge q) \rightarrow \sim r$ (inverse of the contrapositive)
(e) $\sim (p \rightarrow \sim q)$ (converse of the inverse)
(f) *If he is a senator, he is a politician.* (converse)
(g) *If he is elected to office, he is not a Supreme Court judge or not the*
     Secretary of State. (inverse)
(h) *If he receives a pension, then he is retired.* (contrapositive)
(i) *If his reasoning and premises are correct, the conclusion is correct.*
     (inverse, converse, contrapositive)

Review Test

1. The statement *He is a humanitarian if and only if he is not selfish* stated
   symbolically is

   (a) $h \rightarrow \sim s$ (b) $\sim h \rightarrow s$ (c) $h \leftrightarrow \sim s$
   (d) $h \wedge \sim s$ (e) none of these

2. The statement *It is false that he will pass if he studies* stated symbolically is

   (a) $\sim (p \rightarrow a)$ (b) $\sim p \rightarrow s$ (c) $\sim s \rightarrow p$
   (d) $\sim (s \rightarrow p)$ (e) none of these

3. The statement *Either he is addicted to drugs or if he is not addicted to drugs, he is a self-reliant individual* stated symbolically is

(a) $a \lor \sim a \to s$
(b) $a \lor (\sim a \to s)$
(c) $(a \lor \sim a) \to s$
(d) $[a \lor \sim (a \to s)]$
(e) none of these

Complete problems 4–6 with respect to the following information: $p = symbolic logic is the science of reasoning$; $q = symbolic logic is an exact science$; and $r = science is systemized knowledge$.

4. The word form of the statement $r \to p$ is:

(a) *If science is systemized knowledge, then symbolic logic is an exact science of reasoning.*
(b) *If symbolic logic is the science of reasoning, then science is systemized knowledge.*
(c) *Science is systemized knowledge, if symbolic logic is the science of reasoning.*
(d) *Science is systemized knowledge if and only if symbolic logic is the science of reasoning.*
(e) none of these

5. The word form of the statement $(p \to q) \lor \sim r$ is:

(a) *Either, if symbolic logic is the science of reasoning, then it is an exact science, or science is not systemized knowledge.*
(b) *If symbolic logic is the science of reasoning, then symbolic logic is either an exact science or science is not systemized knowledge.*
(c) *Either, if symbolic logic is an exact science, then it is the science of reasoning, or science is not systemized knowledge.*
(d) *If symbolic logic is an exact science, then either it is the science of reasoning or science is not systemized knowledge.*
(e) none of these

6. The word form of the statement $\sim [\sim r \to (p \land q)]$ is:

(a) *It is not the case that if symbolic logic is an exact science of reasoning, then science is not systemized knowledge.*
(b) *If science is not systemized knowledge, then symbolic logic is an exact science of reasoning.*
(c) *It is false that symbolic logic is an exact science of reasoning, if science is not systemized knowledge.*
(d) *Symbolic logic is an exact science of reasoning, if science is not systemized knowledge.*
(e) none of these

7. The statement $\sim p \land (q \to \sim r)$ is

(a) a conjunction
(b) a disjunction
(c) a conditional
(d) a negation
(e) ambiguous

8. The statement $\sim [p \to (\sim q \land r)]$ is

    (a)   a conjunction   (b)   a disjunction   (c)   a conditional
    (d)   a negation   (e)   ambiguous

9. The statement *He is a Republican or a Democrat and a liberal* is

    (a)   a conjunction   (b)   a disjunction   (c)   a conditional
    (d)   a negation   (e)   ambiguous

10. The statement *He is a conservative and a Republican if he is not a Democrat* is

    (a)   a conjunction   (b)   a disjunction   (c)   a conditional
    (d)   a negation   (e)   ambiguous

For problems 11–15, p and q are true statements and r is a false statement.

11. The statement $(\sim p \lor r)$ is

    (a)   true   (b)   false   (c)   a statement whose truth value cannot be
                                               determined due to insufficient information

12. The statement $\sim (\sim p \to q)$ is

    (a)   true   (b)   false   (c)   a statement whose truth value cannot be
                                               determined due to insufficient information

13. The statement $[p \to (q \land \sim r)]$ is

    (a)   true   (b)   false   (c)   a statement whose truth value cannot be
                                               determined due to insufficient information

14. The statement $[(p \land r) \lor \sim s]$ is

    (a)   true   (b)   false   (c)   a statement whose truth value cannot be
                                               determined due to insufficient information

15. The statement $(q \to r) \leftrightarrow [p \land \sim (q \lor r)]$ is

    (a)   true   (b)   false   (c)   a statement whose truth value cannot be
                                               determined due to insufficient information

16. If p is a true statement, then the statement $\sim p \to (q \land r)$ is

    (a)   true   (b)   false   (c)   a statement whose truth value cannot be
                                               determined due to insufficient information

17. If q is a true statement, then the statement $\sim p \to (q \land r)$ is

    (a)   true   (b)   false   (c)   a statement whose truth value cannot be
                                               determined due to insufficient information

18. If r is a false statement, then the statement $[(p \to r) \land q] \lor \sim r$ is

   (a) true   (b) false   (c) a statement whose truth value cannot be determined due to insufficient information

19. If p is a true statement, then the statement $(\sim p \land q) \to (p \lor r)$ is

   (a) true   (b) false   (c) a statement whose truth value cannot be determined due to insufficient information

20. The statement $\sim(\sim p \lor q)$ is a

   (a) tautology   (b) false statement   (c) indeterminate statement

21. The statement $[\sim p \lor (p \lor q)]$ is a

   (a) tautology   (b) false statement   (c) indeterminate statement

22. The statement $[\sim p \to (q \land r)]$ is a

   (a) tautology   (b) false statement   (c) indeterminate statement

23. The statement $\sim [q \to (p \to q)]$ is a

   (a) tautology   (b) false statement   (c) indeterminate statement

24. The statements $(p \to \sim q)$ and $\sim(q \to \sim p)$ are

   (a) logically equivalent   (b) contradictory   (c) unrelated

25. The statements $(p \lor \sim q)$ and $(\sim p \lor q)$ are

   (a) logically equivalent   (b) contradictory   (c) unrelated

26. The statements $p \lor (q \land r)$ and $(\sim p \to q) \land (\sim p \to r)$ are

   (a) logically equivalent   (b) contradictory   (c) unrelated

27. The statement $\sim a \to (b \lor c)$ is logically equivalent to the statement

   (a) $\sim a \lor (b \lor c)$ (b) $a \lor \sim(b \lor c)$
   (c) $a \lor (\sim b \land \sim c)$ (d) $a \land (b \lor c)$
   (e) none of these

28. The statement $\sim(\sim p \lor q)$ is contradictory to the statement

   (a) $(p \land \sim q)$ (b) $\sim(p \to q)$
   (c) $\sim(p \lor \sim q)$ (d) $(p \to q)$
   (e) none of these

29. The truth table below is for the statement

p	q	?
T	T	T
T	F	T
F	T	F
F	F	T

(a) $p \lor q$
(b) $p \land q$
(c) $p \to q$
(d) $q \to p$
(e) none of these

30. The truth table below is for the statement

p	q	?
T	T	F
T	F	T
F	T	F
F	F	F

(a) $p \lor {\sim} q$
(b) $p \to q$
(c) ${\sim} p \to q$
(d) ${\sim} p \land q$
(e) none of these

31. The truth table below is for the statement

p	q	?
T	T	F
T	F	T
F	T	T
F	F	F

(a) $p \to q$
(b) $q \to p$
(c) $p \leftrightarrow q$
(d) ${\sim} p \leftrightarrow {\sim} q$
(e) none of these

32. A simpler statement logically equivalent to $p \lor ({\sim} p \land {\sim} r)$ is

(a) $p \lor {\sim} r$ (b) $p \land {\sim} r$ (c) $p \lor {\sim} p$
(d) ${\sim} r$ (e) r

33. A simpler statement logically equivalent to $\{q \lor [(r \land {\sim} r) \lor p]\}$ is

(a) $r \lor {\sim} r$ (b) $q \lor p$ (c) $q \lor r$
(d) $q \land p$ (e) q

34. If the statement $(a \land b)$ is false, then the statement $(a \lor {\sim} a)$ is

(a) true   (b) false   (c) a statement whose truth value cannot be
                              determined due to insufficient information

35. If the statement $(a \lor {\sim} b)$ is true, then the statement $({\sim} a \to {\sim} b)$ is

(a) true   (b) false   (c) a statement whose truth value cannot be
                              determined due to insufficient information

36. If the statement $[(a \rightarrow b) \rightarrow c]$ is false, then the statement $[a \rightarrow (b \rightarrow c)]$ is

    (a) true   (b) false   (c) a statement whose truth value cannot be
                                            determined due to insufficient information

37. By DeMorgan's equivalence (E_6), a logically equivalent statement to the
    statement $\sim(\sim p \vee \sim q)$ is

    (a) $(p \vee q)$ (b) $\sim(\sim p \wedge q)$ (c) $(p \wedge q)$
    (d) $(p \rightarrow q)$ (e) none of these

38. By the Associative equivalence (E_4), a logically equivalent statement to the
    statement $[a \vee (\sim b \vee c)]$ is

    (a) $[a \wedge (\sim b \vee c)]$ (b) $[a \rightarrow (\sim b \rightarrow c)]$
    (c) $[(a \vee \sim b) \vee c]$ (d) $[a \vee (\sim b \wedge c)]$
    (e) none of these

39. By the Conditional equivalence (E_7), a logically equivalent statement to the
    statement $[(p \wedge q) \vee \sim r]$ is

    (a) $[(p \wedge q) \rightarrow r]$ (b) $[r \rightarrow (p \wedge q)]$
    (c) $[\sim(p \wedge q) \vee r]$ (d) $[\sim(p \wedge q) \rightarrow \sim r]$
    (e) none of these

40. The statement $[(\sim p \rightarrow q) \wedge (q \rightarrow \sim p)]$ is logically equivalent to the statement
    $(\sim p \leftrightarrow q)$ by the

    (a) Contrapositive equivalence   (b) Biconditional equivalence
    (c) DeMorgan's equivalence   (d) Distributive equivalence
    (e) Conditional equivalence

41. The statement $(\sim p \rightarrow \sim q)$ is logically equivalent to the statement $(p \vee \sim q)$
    by the

    (a) Contrapositive equivalence   (b) Biconditional equivalence
    (c) DeMorgan's equivalence   (d) Distributive equivalence
    (e) Conditional equivalence

42. The statement $[p \wedge \sim(q \vee r)]$ is logically equivalent to the statement
    $\sim[\sim p \vee (q \vee r)]$ by the

    (a) Contrapositive equivalence   (b) Biconditional equivalence
    (c) DeMorgan's equivalence   (d) Distributive equivalence
    (e) Conditional equivalence

43. The inverse of the conditional $r \rightarrow q$ is

    (a) $q \rightarrow r$ (b) $\sim q \rightarrow \sim r$ (c) $\sim q \rightarrow r$
    (d) $\sim r \rightarrow \sim q$ (e) none of these

44. The converse of the conditional $r \rightarrow q$ is

(a) $q \rightarrow r$ (b) $\sim q \rightarrow \sim r$ (c) $\sim q \rightarrow r$
(d) $\sim r \rightarrow \sim q$ (e) none of these

45. The converse of the statement *If $x = y$, then $y = x$* is

(a) *If $x \neq y$, then $y \neq x$*
(b) *If $y \neq x$, then $x \neq y$*
(c) *If $y = x$, then $x = y$*
(d) *$x = y$ if and only if $y = x$*
(e) none of these

46. A conditional and its converse are

(a) logically equivalent   (b) contradictory   (c) unrelated

47. A conditional and its inverse are

(a) logically equivalent   (b) contradictory   (c) unrelated

48. The converse of a conditional and the inverse of the same conditional are

(a) logically equivalent   (b) contradictory   (c) unrelated

3

ARGUMENTS AND PROOFS

Earlier we were concerned with discovering the truth values of individual statements and with determining whether or not two statements are related. In this section we will examine sequences or groups of statements called arguments.

In question 4 of the introduction (to Chapter 2) we asked the question: if we agree with the statement *If Mr. Dixon is truly concerned about the welfare of his country, he votes in every election* and we know for a fact that *Mr. Dixon votes in every election* must we then agree with the statement *Mr. Dixon is truly concerned about the welfare of his country*? These three statements when taken together form a structure called an argument. They are generally arranged in the following form:

1 *If Mr. Dixon is truly concerned about the welfare of his country, he votes in every election.*

2 *Mr. Dixon votes in every election.*

3 *Therefore, Mr. Dixon is truly concerned about the welfare of his country.*

The question that we will try to answer in this section is whether or not this and similar arguments are logical.

What do we mean by *logical*? Does it mean that the conclusion that has been reached is true, or does it mean that correct reasoning has been used in reaching the conclusion?

To answer these questions we will develop a method of studying arguments to see if they are logical or not, but first we must define what we mean by an argument.

127

Definition 1

Argument: An argument is a series of statements called premises with a final statement called the conclusion. The conclusion is said to follow from the premises.

In our example the first two statements are the premises and the last statement is the conclusion. In examining this and all following arguments, the term *logical* will not be used. Instead, we will determine whether the argument is valid or invalid. We use the word *valid* instead of *logical* in order to avoid any preconceived definitions of the word logical.

Definition 2

Valid argument: A valid argument is one in which true premises always lead to a true conclusion.

Note: The word *always* is extremely important. It means that in examining the argument we cannot say "valid" simply because we find one example where all the premises are true and the conclusion is true. We must make sure that the conclusion is true each and every time that all the premises are true (i.e., for all possible combinations of truth values of the components of the premises that make the premises true). In some arguments, true premises will lead to a conclusion that is sometimes true and sometimes false. In this case, the argument is invalid.

Strangely enough, when there is no circumstance where all the premises are true simultaneously, we say that the premises are contradictory (or inconsistent) and the argument is *valid*.

In speaking about the validity of an argument, we are not concerned with the actual truth or falsity of the premises and conclusion. We are only concerned with the *structure* of the argument itself. In most cases we are not able to determine the actual truth or falsity of the premises.

In the above example, since we do not know Mr. Dixon we do not know if the statement *Mr. Dixon votes in every election* is true or false. What we are concerned with is, that if the two given premises are true, is the conclusion also true. If so, then the argument is valid.

Consider the following arguments. (The statements concerning their validity will be verified in the next section.)

Example 1

If men have landed on the moon, then the space age has begun.
The space age has begun.
Therefore, men have landed on the moon.

This argument is invalid because there is a set of circumstances under which the premises are all true and the conclusion is false. When we state these circumstances, we are giving a counter-example. In this case, the counter-example is the set of circumstances when the statement *Men have landed on the moon* is false, and the statement *The space age has begun* is true.

Example 2

Men have landed on the moon or the space age has begun.
Men have not landed on the moon.
Therefore, the space age has begun.

The argument is valid. The structure is such that true premises would always lead to a true conclusion. The fact that the second premise is actually false does not matter.

Example 3

If men have landed on the moon, then the space age has begun.
Men have not landed on the moon.
Therefore, the space age has not begun.

The argument is invalid. The counter-example is: The statement *Men have landed on the moon* is false and *The space age has begun* is true (thus, the first premise is true), *Men have not landed on the moon* is true (the second premise is true), *The space age has not begun* is false (the conclusion is false). True premises led to a false conclusion. Once again, we have to be careful not to let the actual condition of the premises influence our judgment of the structure of the argument.

Example 4

If men have landed on the moon, then the space age has begun.
The space age has not begun.
Therefore, men have not landed on the moon.

The argument is valid.

In studying the above arguments to determine whether they are valid or invalid, it is best that we do not examine them in their original form since we can be deceived by the wording of the statements. To avoid this danger we shall study the symbolic representation of the arguments. The four examples shown above have the following symbolic form.

$$
\begin{array}{llll}
(1)\ \ p \to q & (2)\ \ p \lor q & (3)\ \ p \to q & (4)\ \ p \to q \\
\quad\ \ q & \quad\ \ \sim p & \quad\ \ \sim p & \quad\ \ \sim q \\
\hline
\quad \therefore\ p & \quad \therefore\ q & \quad \therefore\ \sim q & \quad \therefore\ \sim p
\end{array}
$$

By studying the argument's symbolic form we avoid being misled by the practical truth or falsity of the statements and concern ourselves only with the structure of the argument itself.

Test for Validity
By Assuming True Premises

Since by definition, a valid argument is one in which true premises always lead to a true conclusion, we will test an argument's validity by determining what the truth value of the conclusion would be if we assume the premises to be true.

To illustrate, we will examine Examples 1 and 2.

Example 1

$$
\begin{array}{c}
p \to q \\
q \\
\hline
\therefore\ p
\end{array}
$$

We set the two premises $p \to q$ and q equal to true. Since q is true, the first premise becomes ($p \to$ true). We now try to determine the truth value of p. If p were false, we would have (false $\to$ true) and if p were true, we would have (true $\to$ true). This shows that the first premise will be true no matter what truth value we substitute for p. Therefore, the argument is invalid since the conclusion (p) can be either true or false when the premises are both true.

Example 2

$$
\begin{array}{c}
p \lor q \\
\sim p \\
\hline
\therefore\ q
\end{array}
$$

We set the two premises $p \lor q$ and $\sim p$ equal to true. Since the second premise $\sim p$ is true, p is false. The first premise becomes (false $\lor q$). Since the first premise is true, and we know that in order for a disjunction to be true, one of its components must be true, we can conclude that q must be true. Therefore, this argument is valid since true premises always lead to a true conclusion. This method can become quite difficult when the premises are more complicated and their number increased. However, this difficulty can be overcome by using truth tables to test the validity of an argument.

Test of Validity by Truth Tables

An alternate definition for a valid argument is to say that an argument is valid if the conjunction of the premises implies the conclusion (i.e., $(p_1 \land p_2 \land p_3 \land \cdots \land p_n) \to c$ is an implication). With this definition in mind we can construct a truth table to test the following argument.

Example 1

$p \to q$
q
$\overline{}$
$\therefore p$

p	q	$[(p$	$\to$	$q)$	$\land$	$q]$	$\to$	p
T	T	T	T	T	T	T	T	T
T	F	T	F	F	F	F	T	T
F	T	F	T	T	T	T	F	F
F	F	F	T	F	F	F	T	F
		1	2	1	3	1	4	1

The argument is invalid because the conjunction of the premises does not imply the conclusion, (Column 4 representing the truth values of the implication are not all T's.)

Upon further consideration, we can simplify our work by constructing a truth table that contains columns for the individual premises and the conclusion. Here we only have to concern ourselves with those rows in which all the premises are true. If the conclusion is false in any such row, the argument is invalid. Otherwise, it is valid.

Let us consider the same argument in this way:

$$p \to q$$
$$\frac{q}{\therefore p}$$

	Premises			Conclusion
p	q	$p \to q$	q	p
T	T	T	T	T
T	F	F	F	T
F	T	T	T	F
F	F	T	F	F

In the first row of the truth table, both the premises and the conclusion are true but in the third row where both premises are true, we have a false conclusion. This contradicts our definition of validity that true premises always lead to a true conclusion. Therefore, the argument is invalid.

If we symbolize the argument that introduced this section, we have

$$c \to v \qquad c = \textit{Mr. Dixon is truly concerned about}$$
$$v \qquad\qquad \textit{the welfare of his country}$$
$$\therefore c \qquad v = \textit{Mr. Dixon votes in every election}$$

This argument has exactly the same structure as the previous example, with c replacing p and v replacing q. Since the first argument is invalid and the structures of the two arguments are identical, the second argument is also invalid. This is verified by the following truth table:

	Premises			Conclusion
c	v	$c \to v$	v	c
T	T	T	T	T
T	F	F	F	T
F	T	T	T	F
F	F	T	F	F

This argument is invalid since in the third row, true premises have a false conclusion.

Example 2

$$p \vee q$$
$$\sim p$$
$$\therefore q$$

	Premises			Conclusion
p	q	$p \vee q$	$\sim p$	q
T	T	T	F	T
T	F	T	F	F
F	T	T	T	T
F	F	F	T	F

In this argument there is only one row, the third, in which both premises are true. Since the conclusion is also true, the argument is valid.

Example 3

For a third and final example of this procedure, consider the following argument.

He is guilty or the judge declared a mistrial. If the judge declared a mistrial, then he will have a new trial. He will not have a new trial. Therefore he is guilty.

Symbolically we have

$$g \vee m \qquad g = \text{\textit{he is guilty}}$$
$$m \to n \qquad m = \text{\textit{the judge declared a mistrial}}$$
$$\sim n \qquad\ \ n = \text{\textit{he will have a new trial}}$$
$$\therefore g$$

	Premises					Conclusion
g	m	n	$g \vee m$	$m \to n$	$\sim n$	g
T	T	T	T	T	F	T
T	T	F	T	F	T	T
T	F	T	T	T	F	T
T	F	F	T	T	T	T
F	T	T	T	T	F	F
F	T	F	T	F	T	F
F	F	T	F	T	F	F
F	F	F	F	T	T	F

This argument is valid since in the fourth row, which is the only case where all three premises are true, the conclusion is also true.

Note: If the truth table of the argument contains no row in which all of the premises are true, this indicates that the premises contradict one another. That is, the argument contains at least two premises that cannot be simultaneously true. In this case we define the argument to be valid.

For example consider the following argument. *If he is guilty he will go free. It is not the case that he is not guilty or he will go free. Therefore he is guilty.* Symbolizing we have

$$p \rightarrow q \qquad\qquad p = he\ is\ guilty$$
$$\frac{\sim(\sim p \vee q)}{\therefore p} \qquad\qquad q = he\ will\ go\ free$$

		Premises		Conclusion
p	q	$p \rightarrow q$	$\sim(\sim p \vee q)$	p
T	T	T	F	T
T	F	F	T	T
F	T	T	F	F
F	F	T	F	F

There are no rows in which both premises are true. Therefore, the argument is valid due to contradictory premises. This conclusion is also confirmed by our alternate definition of a valid argument. If two of the premises cannot be true simultaneously, then one of them must be false. If a conjunction of premises has one premise false, then the entire conjunction is false. If the conjunction represents the antecedent of a conditional, then the conditional must be true. Since it will be true under all circumstances it is a tautology and the argument is valid.

An even more obvious example of an argument with contradictory premises is:

$$p$$
$$\frac{\sim p}{\therefore q}$$

As strange as it may seem this argument is valid even though the conclusion "q" stands for any statement whatsoever. Perhaps we would like to call such arguments "trivial," "nonsensical," or even "ridiculous"; but, since it satisfies our definition, we must also call it "valid." In that respect it is similar to our system of justice in the United States. A person is presumed innocent until proven guilty. Here we presume an argument is valid unless it is shown to be invalid. The only way to do that is to show that all the premises are true and the conclusion is false.

Test of Validity by TRUTH-MAN

The TRUTH-MAN thrives on the truth. He starts his hunt on the first row of a truth table at the truth value of the first premise (truth values of component statements are ignored). He swallows the letter if it is a "T" and moves to the right to the truth value of the next premise. If he encounters an "F" under any of the premises, he retreats and goes on to the beginning of the next row. (A man of truth should never be forced to swallow a false premise.)

So long as he meets "T's" under the premises, he swallows them and keeps moving to the right. If he passes through all the premises in this fashion, he attacks the value under the conclusion. If it is a "T" he swallows it and goes on to the next row, but if it is an "F", he is swallowed and the game is over. (The argument is invalid—a man of truth cannot survive after reaching a false conclusion.)

Should the TRUTH-MAN go through the entire table without being eaten by a false conclusion, even though he may never get through to the conclusion (i.e., he was always forced to retreat from a false premise), then the argument is valid. (Note: He must try every row. It is not sufficient just to reach one True conclusion.)

Example 1

$$p \rightarrow q$$
$$\sim p$$
$$\therefore \sim q$$

p	q	$p \longrightarrow q$	$\sim p$	$\sim q$
T	T	T	F	F
T	F	F	F	T
F	T	T — T — F	T	T
F	F	T	T	T

TRUTH-MAN "eaten" Argument Invalid

Example 2

$$p \lor q$$
$$\sim q$$
$$\therefore p$$

p	q	$p \lor q$	$\sim q$	$\sim p$
T	T	T	F	T
T	F	T — T — T	T	
F	T	T	F	F
F	F	F	T	F

TRUTH-MAN made it—Argument Valid

Example 3

$p \to q$
$p \land \sim q$
$\therefore p \lor q$

p	q	$p \longrightarrow q$	$p \land \sim q$	$p \lor q$
T	T	T	F	T
T	F	F	T	T
F	T	T	F	T
F	F	T	F	F

Argument Valid

Example 4

$p \lor \sim q$
$\sim p \to \sim r$
$\sim r \land \sim q$
$\therefore p$

p	q	r	$p \lor \sim q$	$\sim p \longrightarrow \sim r$	$\sim r \land \sim q$	p
T	T	T	T	T	F	T
T	T	F	T	T	F	T
T	F	T	T	T	F	T
T	F	F	T	T	T	T
F	T	T	F	F	F	F
F	T	F	F	T	F	F
F	F	T	T	F	F	F
F	F	F	T	T	T	(F)

Sorry, TRUTH-MAN—Argument Invalid

Exercises 3.1

1. Using truth tables (any method) test the validity of the following arguments.

(a) $p \to \sim q$
$\sim q$
$\therefore \sim p$

(b) $p \to \sim q$
$\sim q \to r$
$\therefore p \to r$

(c) $(p \land q) \lor r$
$\sim r$
$\therefore p \land q$

(d) $\sim p \lor r$
$\sim p \lor \sim r$
$\therefore \sim p$

(e) $q \to \sim r$
$\sim q$
$\therefore \sim r$

(f) $\sim r \to p$
$\sim r$
$\therefore p$

2. Test the validity of the arguments in Example 1 by assuming that the premises are true.

The preceding method of testing the validity of an argument is direct and relatively easy to work with when the number of component simple statements that make up the premises and conclusion is small. However, if there are five or six component simple statements in the structure of the argument, a truth table of 32 or 64 rows, respectively, would be required to make the test. To handle arguments of this type we will use another and more concise method.

Test of Validity by Flow Charting

An alternate method to truth tables for testing the validity of an argument is flow charting. A flow chart is a chart or illustration of the steps or reasoning used to test the validity of an argument. In this method we directly apply the definition of a valid argument by determining whether or not we obtain a true conclusion when the premises are true. That is, we set the premises equal to true and then apply the definitions of the logical connectives and negation to determine the truth value of the conclusion.

There are two basic methods of proof: the direct method and the indirect method. These two methods will now be discussed separately.

The Direct Method of Proof

In using the direct method of proof we set the premises equal to true and then using these true premises try to determine the truth value of the conclusion. If we can show the conclusion to be true when the premises are true, the argument is valid. If true premises lead to a conclusion that is false or if the conclusion's truth value cannot be determined, then the argument is invalid. Finally, if by setting the premises true we arrive at a contradictory statement, such as $T = F$, the premises are contradictory and therefore the argument is valid. Consider the following examples.

Example 1

*If men have landed on the moon, then the space age has begun. The space age has not begun. Therefore, men have **not** landed on the moon.*

Symbolizing we have

$$p \to q \qquad p = \textit{men have landed on the moon}$$
$$\underline{\sim q} \qquad q = \textit{the space age has begun}$$
$$\sim p$$

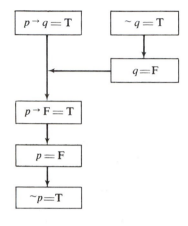

1. Set the premises equal to true.

2. Since $\sim q = T$, we have by negation $q = F$.

3. Substituting $q = F$ in the statement $p \to q = T$, we have $p \to F = T$.

4. By the definition of conditional, $p \to F$ can only equal T if $p = F$.

5. Since $p = F$, we have $\sim p = T$. This shows the argument to be valid since true premises led to a true conclusion.

Example 2

If he is guilty, he is in prison. He is not guilty. Therefore, he is not in prison.

Symbolizing we have

$$p \to q \qquad p = \textit{he is guilty}$$
$$\underline{\sim p} \qquad q = \textit{he is in prison}$$
$$\sim q$$

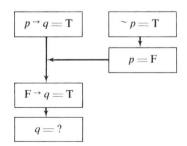

1. Set the premises equal to true.

2. Since $\sim p = T$, we have by negation, $p = F$.

3. Substituting $p = F$ in the statement $p \to q = T$, we have $F \to q = T$.

4. We cannot conclude whether q is true or false since by the definition of conditional $F \to T = T$ and $F \to F = T$. If q can be either true or false, then the conclusion $\sim q$ can also be either true or false and therefore the argument is invalid.

Example 3

The President vetoed the bill or it was not passed by Congress. If the President vetoed the bill, Congress will override the veto. Congress will not override the veto. Therefore the bill was not passed by Congress.

Symbolizing we have

$$p \lor \sim q \qquad p = \textit{the President vetoed the bill}$$
$$p \to r \qquad q = \textit{the bill was passed by Congress}$$
$$\underline{\sim r} \qquad r = \textit{Congress will override the veto}$$
$$\therefore \sim q$$

1. Set the premises equal to true.

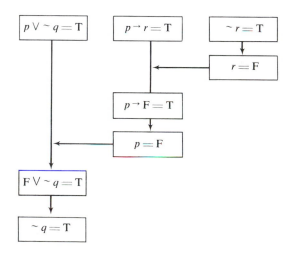

2. Since $\sim r = T$, we have by negation $r = F$.

3. Substituting $r = F$ in the statement, $p \to r = T$, we have $p \to F = T$.

4. By the definition of conditional $p \to F$ only equals T if $p = F$.

5. Substituting $p = F$ in the statement $p \lor \sim q = T$, we have, $F \lor \sim q = T$.

6. By the definition of disjunction $F \lor \sim q$ can only equal T if $\sim q = T$. This shows the argument to be valid since true premises led to a true conclusion.

Example 4

He is a Republican or a Democrat. He is a Republican or not a Democrat. Therefore, he is a Republican.

Symbolizing we have

$$p \lor q \qquad p = he\ is\ a\ Republican$$
$$\underline{p \lor \sim q} \qquad q = he\ is\ a\ Democrat$$
$$p$$

1. Set the premises equal to true.

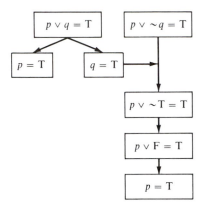

2. By the definition of disjunction if the statement $p \lor q = T$ then either $p = T$ or $q = T$. If $p = T$ the argument is valid since true premises led to a true conclusion.

3. If $q = T$ substituting in the statement $p \lor \sim q = T$, we have $p \lor \sim T = T$.

4. By negation we have $p \lor F = T$.

5. By the definition of disjunction $p \lor F$ can only equal T, if $p = T$. This shows the argument to be valid since true premises led to a true conclusion.

Example 5

If he is correct, then I am wrong. It is not the case that he is not correct or I am wrong. Therefore, he is correct.

Symbolizing we have

$$p \rightarrow q \qquad p = he\ is\ correct$$
$$\underline{\sim (\sim p \lor q)} \qquad q = I\ am\ wrong$$
$$p$$

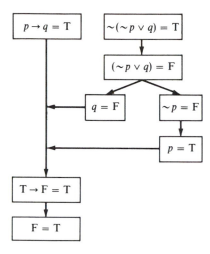

1. Set the premises equal to true.

2. Since its negation is T, $(\sim p \vee q) = $ F.

3. By the definition of disjunction if the statement $(\sim p \vee q)$ = F, then $\sim p = $ F and $q = $ F.

4. Since $\sim p = $ F, we have by negation $p = $ T.

5. Substituting $p = $ T and $q = $ F in that statement $p \rightarrow q = $ T, we have $T \rightarrow F = $ T.

6. By the definition of conditional $T \rightarrow F = $ F. By setting the premises equal to true we obtained a contradiction. Therefore the argument is valid due to contradictory premises.

A slight variation of the above method is needed when the conclusion of the argument is a conditional. When this occurs, the form of proof that is used is called a conditional proof. In doing a proof of this type, the antecedent of the conclusion is taken as a premise. It is called a conditional premise. We try to prove the consequent of the conditional to be true. That is, we assume the antecedent to be true and then try to prove the consequent true. (If the antecedent were false, then the conclusion would be true anyway.) Consider the following example.

Example 6

Prices are higher or there is a recession. If there is a recession, there is high un-employment. Therefore, if prices are not higher, there is high unemployment.

Symbolizing we have

$$p \lor q \qquad p = prices \ are \ higher$$
$$\underline{q \rightarrow r} \qquad q = there \ is \ a \ recession$$
$$\sim p \rightarrow r \qquad r = there \ is \ high \ unemployment$$

Since the conclusion is a conditional, we will assume the antecedent $\sim p$ to be true and try to show the conclusion r to be true.

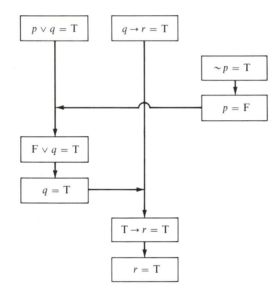

1. Set the premises equal to true.

2. Set the antecedent of the conclusion (conditional premise) equal to true.

3. Since $\sim p = T$, we have by negation $p = F$.

4. Substituting $p = F$ in the statement $p \lor q = T$, we have $F \lor q = T$.

5. By the definition of disjunction the statement $F \lor q$ can only equal T if $q = T$.

6. Substituting $q = T$ in the statement $q \rightarrow r = T$, we have $T \rightarrow r = T$.

7. By the definition of conditional $T \rightarrow r$ can only equal T if $r = T$.

Therefore, the conclusion $\sim p \rightarrow r$ is true, since $\sim p = T$ leads to $r = T$ and $T \rightarrow T = T$. We do not have to consider the possibility of the conditional premise $\sim p$ being false since if $\sim p = F$ the conclusion $\sim p \rightarrow r$ would be true. That is, $F \rightarrow r = T$.

The Indirect Method of Proof

Often, the establishing of the validity of an argument by the direct method is difficult. An alternate method exists which does not require the precise chain of reasoning that is required when the direct method is employed. This second method is called the *indirect method* of proof.

In the indirect method we list all the possible solutions of the problem and then proceed to eliminate those possibilities that are inconsistent or that contradict already established laws. The one remaining possibility is the solution. By this process we do not directly arrive at our result, but rather obtain it indirectly.

For example, suppose a robbery has occurred and we are certain that either Mr. Jones, Mr. Smith, or Mr. White has committed the crime but from the available evidence we are unable to directly prove which one committed the robbery. However, the evidence enables us to prove that Mr. White and Mr. Smith could not have committed the crime. Therefore, we have now indirectly arrived at the conclusion that Mr. Jones is the robber.

This technique is ideally suited for testing the validity of arguments, since there are only two possible solutions to each problem. That is, the conclusion of the argument is either true or false.

Since our objective in proving the validity of an argument is to show the conclusion true when the premises are true, we will try to eliminate the other possible solution—the conclusion is false. We do this by assuming the conclusion false and then using it to obtain a contradiction to one of the premises. If a contradiction is obtained it eliminates the conclusion is false as a possible solution and therefore leaves the other alternative (the conclusion is true) as our solution.

Note: In using the indirect method in the following problems we assume that the premises are not contradictory. Contradictory premises will always lead to a contradiction.

Example 1

The premises are true or the conclusion is true. If the premises are not true the conclusion is not true. Therefore, the premises are true.

Symbolizing we have

$$p \lor q \qquad p = \text{the premises are true}$$
$$\underline{\sim p \rightarrow \sim q} \qquad q = \text{the conclusion is true}$$
$$p$$

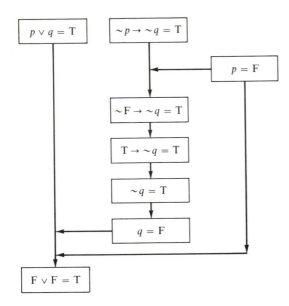

1. Set the premises equal to true.

2. Set the conclusion equal to false.

3. Substituting $p = $ F in the statement $\sim p \to \sim q = $ T, we have $\sim$F $\to \sim q = $ T.

4. By negation we have T $\to \sim q = $ T.

5. By the definition of conditional T $\to \sim q$ only equals T if $\sim q = $ T.

6. By negation we have $q = $ F.

7. Substituting $p = $ F and $q = $ F in the statement $p \lor q = $ T, we have F $\lor$ F $= $ T.

Using the indirect premise $p = $ F, we have arrived at the contradiction F $\lor$ F $= $ T. Therefore, $p = $ F is eliminated, leaving the other possibility $p = $ T as the solution. The argument is valid.

Example 2

It is either A or B. If it is B, then it is C. It is not C. Therefore, it is A.

Symbolizing we have

$$
\begin{array}{ll}
a \lor b & a = \textit{it is A} \\
b \to c & b = \textit{it is B} \\
\underline{\sim c} & c = \textit{it is C} \\
a
\end{array}
$$

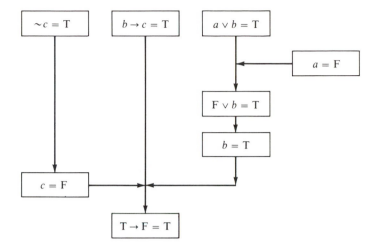

1. Set the premises equal to true.

2. Set the conclusion equal to false.

3. Substituting $a = F$ in the statement $a \lor b = T$ we have $F \lor b = T$.

4. By definition of disjunction $F \lor b$ is T only if $b = T$.

5. By negation we have $c = F$.

6. Substituting $b = T$ and $c = F$ in statement $b \to c = T$ we have $T \to F = T$.

Using the indirect premise $a = F$, we have arrived at the contradiction $T \to F = T$. Therefore, $a = F$ is eliminated leaving the other possibility $a = T$ as the solution. The argument is valid.

Example 3

The drug is effective or there are severe side effects. If there are severe side effects, the drug will be recalled. Therefore, if the drug is not effective, it will be recalled.

Symbolizing we have

$$p \lor q \qquad p = \text{the drug is effective}$$
$$\underline{q \to r} \qquad q = \text{there are severe side effects}$$
$$\sim p \to r \qquad r = \text{the drug will be recalled}$$

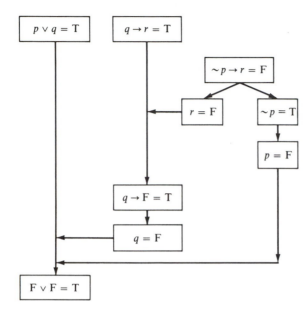

1. Set the premises equal to true.

2. Set the conclusion equal to false.

3. By the definition of conditional $\sim p \to r$ is F only if $\sim p = T$ and $r = F$.

4. By negation we have $p = F$.

5. Substituting $r = F$ in the statement $q \to r = T$, we have $q \to F = T$.

6. By the definition of conditional $q \to F$ is T only if $q = F$.

7. Substituting $p = F$ and $q = F$ in the statement $p \lor q = T$, we have $F \lor F = T$.

Using the indirect premise $\sim p \to r = F$, we have arrived at the contradiction $F \lor F = T$. Therefore, $\sim p \to r = F$ is eliminated leaving the other possibility $\sim p \to r = T$ as the solution. The argument is valid.

The indirect method is an efficient method of testing the validity of an argument since there are only two possible solutions. However, when a problem has a large number of solutions the indirect method becomes impractical. In fact, it is often impossible to list all of the possible solutions to a problem and therefore, impossible to use the indirect method of proof.

Exercises 3.2

Using flow charts, test the validity of the following arguments. Do each problem by the direct method and the indirect method.

1. $p \lor q$ 2. $p \to q$ 3. $\sim p \to q$
   $\underline{\sim q}$ $\underline{\sim q}$ $\underline{q \to \sim r}$
   $\therefore p$ $\therefore \sim p$ $\therefore \sim p \to \sim r$

4. $\sim r \to \sim p$ 5. $p \to q$
   $\underline{\sim(\sim p \land q)}$ $\underline{\sim p}$
   $\sim r$ $\therefore \sim q$
   $\therefore \sim q$

So far in this chapter we have tested the validity of arguments by assuming that the premises are true, by truth tables, and by flow charts. Now we will develop our fourth and final method of proving the validity or invalidity of an argument. We will examine arguments by dividing them into a series of smaller arguments whose validity will have already been established. We will also use the logical equivalence relations established in Chapter 2. This procedure, which is called a formal proof of the validity of an argument, enables us to prove the validity of an argument in the most concise form.

Definition

Formal Proof of Validity: A formal proof of the validity of an argument is a proof of the truth of the conclusion, derived from true premises by subdividing the argument into a series of already proven simpler valid arguments and or logically equivalent statements.

Basic Valid Argument Forms

Before we can prove the validity of arguments by the formal method, we must first establish the eight basic valid argument forms that we will use. The basic equivalence relations have already been developed in Chapter 2.

In listing the eight basic arguments their validity will be established by truth tables. The traditional name for each argument will be given but the argument will be primarily referred to in later work by its number. The eight basic valid argument forms are as follows.

Modus Ponens (A$_1$).   Given that a conditional is true and the antecedent is true, we can conclude that the consequent is true. That is:

antecedent → consequent $p \to q$
antecedent   or p
∴ consequent   ∴ q

p	q	$p \to q$	p	q
T	T	T	T	T
T	F	F	T	F
F	T	T	F	T
F	F	T	F	F

Examples

(a) $\sim p \to \sim q$ (b) $(p \wedge q) \to \sim r$
     $\sim p$ $(p \wedge q)$
     ∴ $\sim q$ ∴ $\sim r$

(c) *If he is a Communist, he is an atheist.*
     He is a Communist.
     Therefore, he is an atheist.

Modus Tollens (A$_2$).   Given that a conditional is true and the negation (opposite) of the consequent is true, we can then conclude that the negation (opposite) of the antecedent is true. That is:

antecedent → consequent $p \to q$
not consequent   or $\sim q$
∴ not antecedent   ∴ $\sim p$

p	q	$p \to q$	$\sim q$	$\sim p$
T	T	T	F	F
T	F	F	T	F
F	T	T	F	T
F	F	T	T	T

Examples

(a) $\sim p \to \sim q$ (b) $(p \wedge q) \to r$
     q $\sim r$
     ∴ p ∴ $\sim (p \wedge q)$

(c) *If he is a Russian leader, he is a Communist.*
     He is not a Communist.
     _____
     Therefore, he is not a Russian leader.

Disjunctive Syllogism (A_3).   Given that a disjunction is true and the negation (opposite) of one of the disjuncts is true, we can then conclude that the other disjunct is true. That is:

disjunct 1 $\vee$ disjunct 2 $p \vee q$
not disjunct 1 $\sim p$
_____   _____
$\therefore$ disjunct 2 $\therefore q$

p	q	$p \vee q$	p	q
T	T	T	F	T
T	F	T	F	F
F	T	T	T	T
F	F	F	T	F

Examples

(a) $\sim p \vee \sim q$
     q
     _____
     $\therefore \sim p$

(b) $(p \to r) \vee (q \wedge s)$
     $\sim (q \wedge s)$
     _____
     $\therefore (p \to r)$

(c) *He is a Russian leader or a Communist.*
     He is not a Russian leader.
     _____
     Therefore, he is a Communist.

Hypothetical Syllogism (A_4).   Given that two conditionals are true with the consequent of the first being the antecedent of the second, we can then conclude that the conditional, whose antecedent is the antecedent of the first conditional and whose consequent is the consequent of the second, is true. That is:

antecedent 1 $\to$ consequent 1 $p \to q$
consequent 1 $\to$ consequent 2   or $q \to r$
_____   _____
$\therefore$ antecedent 1 $\to$ consequent 2 $\therefore p \to r$

p	q	r	$p \to q$	$q \to r$	$p \to r$
T	T	T	T	T	T
T	T	F	T	F	F
T	F	T	F	T	T
T	F	F	F	T	F
F	T	T	T	T	T
F	T	F	T	F	T
F	F	T	T	T	T
F	F	F	T	T	T

Examples

(a) $\sim r \to q$
     $q \to \sim p$
     $\overline{\therefore \ \sim r \to \sim p}$

(b) $(p \wedge \sim q) \to (r \wedge s)$
     $(r \wedge s) \to (q \wedge \sim p)$
     $\overline{\therefore \ (p \wedge \sim q) \to (q \wedge \sim p)}$

(c) *If he is a Russian leader, then he is a Communist.*
     If he is a Communist, then he is an atheist.
     $\overline{\textit{Therefore, if he is a Russian leader, he is an atheist.}}$

Conjunctive Addition (A_5).   Given two statements to be true, we can then conclude that the conjunction of the two statements is true. That is:

conjunct 1
conjunct 2
$\overline{\therefore \ \text{conjunct 1} \wedge \text{conjunct 2}}$ or

p
q
$\overline{\therefore \ p \wedge q}$

p	q	$p \wedge q$
T	T	T
T	F	F
F	T	F
F	F	F

Examples

(a) $\sim p$
     $\sim q$
     $\overline{\therefore \ \sim p \wedge \sim q}$

(b) $(p \to q)$
     $(q \vee r)$
     $\overline{\therefore \ (p \to q) \wedge (q \vee r)}$

(c) *He is a Russian leader.*
     He is a Communist.
     _____
     Therefore, he is a Communist and a Russian leader.

Conjunctive Simplification (A_6).   Given a conjunction that is true, we can then conclude that both conjuncts are true. That is:

conjunct 1 ∧ conjunct 2 $p \wedge q$
_____   or   _____
∴ conjunct 1   ∴ p
or   or
∴ conjunct 2   ∴ q

p	q	$p \wedge q$	p or q
T	T	T	T T
T	F	F	T F
F	T	F	F T
F	F	F	F F

Examples

(a) $p \wedge \sim q$ (b) $(p \vee q) \wedge \sim r$
     _____   _____
     ∴ p ∴ $(p \vee q)$
     or   or
     ∴ $\sim q$ ∴ $\sim r$

(c) *He is a Russian leader and not an atheist.*
     _____
     Therefore, he is a Russian leader.
     or
     Therefore, he is not an atheist.

Disjunctive Addition (A_7).   Given a statement to be true, we can then conclude that the statement disjuncted with any other statement is true. That is:

disjunct 1 p
_____   or   _____
∴ disjunct 1 ∨ disjunct?   ∴ $p \vee q$

p	q	p	$p \vee q$
T	T	T	T
T	F	T	T
F	T	F	T
F	F	F	F

Examples

(a) $\sim p$

   $\therefore \sim p \vee q$

(b) $(p \wedge q)$

   $\therefore (p \wedge q) \vee (r \rightarrow s)$

(c)   *He is a Russian leader.*

   Therefore, he is a Russian leader or a martian.

Disjunctive Simplification (A_8).   Given a statement disjuncted with a second statement to be true and also the statement disjuncted with the negation of the second statement to be true, we can then conclude that the statement must be true. That is:

disjunct 1 ∨ disjunct 2 $p \vee q$

disjunct 1 ∨ not disjunct 2   or $p \vee \sim q$

$\therefore$ disjunct 1 $\therefore p$

p	q	$p \vee q$	$p \vee \sim q$	p
T	T	T	T	T
T	F	T	T	T
F	T	T	F	F
F	F	F	T	F

Examples

(a) $(\sim p \vee q)$

   $(\sim p \vee \sim q)$

   $\therefore \sim p$

(b) $p \vee (q \rightarrow r)$

   $p \vee \sim (q \rightarrow r)$

   $\therefore p$

(c)   *He is a Russian leader or he is a Communist.*

   He is a Russian leader or he is not a Communist.

   Therefore, he is a Russian leader.

Now that we have established the validity of these eight basic argument forms we will use them along with the nine equivalence relations in our work with formal proofs.

The following tables are a summary of the logically equivalent statements and valid argument forms.

Summary of the Equivalence Relations

(E$_1$) *Double Negation Equivalence:*   (E$_5$) *Distributive Equivalences:*

$$\sim(\sim p)\leftrightarrow p$$

$$[p \vee (q \wedge r)]\leftrightarrow[(p \vee q) \wedge (p \vee r)]$$

(E$_2$) *Idempotent Equivalences:*

$$[(p \wedge (q \vee r)]\leftrightarrow[(p \wedge q) \vee (p \wedge r)]$$

$$p\leftrightarrow(p \vee p)$$

(E$_6$) *DeMorgan's Equivalences:*

$$p\leftrightarrow(p \wedge p)$$

$$\sim(p \wedge q)\leftrightarrow(\sim p \vee \sim q)$$

(E$_3$) *Commutative Equivalences:*

$$\sim(p \vee q)\leftrightarrow(\sim p \wedge \sim q)$$

$$(p \vee q)\leftrightarrow(q \vee p)$$

(E$_7$) *Conditional Equivalence:*

$$(p \wedge q)\leftrightarrow(q \wedge p)$$

$$(p \rightarrow q)\leftrightarrow(\sim p \vee q)$$

(E$_4$) *Associative Equivalences:*   (E$_8$) *Contrapositive Eqivalence:*

$$[p \vee (q \vee r)]\leftrightarrow[(p \vee q) \vee r]$$

$$(p \rightarrow q)\leftrightarrow(\sim q \rightarrow \sim p)$$

$$[p \wedge (q \wedge r)]\leftrightarrow[(p \wedge q) \wedge r]$$

(E$_9$) *Biconditional Equivalence:*

$$(p \leftrightarrow q)\leftrightarrow[(p \rightarrow q) \wedge (q \rightarrow p)]$$

Summary of the Basic Valid Argument Forms

(A$_1$) *Modus Ponens:*   (A$_5$) *Conjunctive Addition:*

$$p \rightarrow q, p \therefore q$$

$$p, q \therefore p \wedge q$$

(A$_2$) *Modus Tollens:*   (A$_6$) *Conjunctive Simplification:*

$$p \rightarrow q, \sim q \therefore \sim p \qquad\qquad p \wedge q \therefore p$$

(A$_3$) *Disjunctive Syllogism:*   (A$_7$) *Disjunctive Addition:*

$$p \vee q, \sim p \therefore q \qquad\qquad p \therefore p \vee q$$

(A$_4$) *Hypothetical Syllogism:*   (A$_8$) *Disjunctive Simplification:*

$$p \rightarrow q, q \rightarrow r \therefore p \rightarrow r \qquad\qquad (p \vee r), (p \vee \sim r) \therefore p$$

Exercises 3.3

If valid, name the valid argument form for each of the following arguments. Otherwise write invalid.

1. $p \rightarrow \sim q$
   $\sim q$
   $\therefore p$

2. $p \vee q$
   $\sim q$
   $\therefore p$

3. $\sim p \rightarrow q$
   $\sim p$
   $\therefore \sim q$

4. $p \rightarrow (r \wedge s)$
   $\sim (r \wedge s)$
   $\therefore \sim p$

5. $p \wedge (r \wedge s)$
   $\therefore r \wedge s$

6. $p \rightarrow r$
   $r \rightarrow s$
   $\therefore \sim s$

7. $m \rightarrow r$
   m
   $\therefore r$

8. $p \rightarrow \sim q$
   $\sim q \rightarrow r$
   $\therefore p \rightarrow r$

9. $(p \wedge q)$
   $\therefore (p \wedge q) \vee r$

10. $(r \vee t) \rightarrow (p \wedge m)$
    $\sim (p \wedge m)$
    $\therefore \sim (r \vee t)$

11. $(r \vee s)$
    $\therefore (r \vee s) \wedge m$

12. p
    $r \vee m$
    $\therefore p \wedge (r \vee m)$

13. $(p \wedge q) \vee \sim r$
    $\sim (p \wedge q) \vee \sim r$
    $\therefore \sim r$

Formal Proofs.   In doing formal proofs there are two methods; the direct and the indirect. The concepts are similar to those used for flow charting technique.

The Direct Method of Formal Proof

In using the direct method of formal proof we begin by setting or assuming the premises true. Then by applying the rules for the basic valid argument forms and logically equivalent statements we attempt to prove the conclusion true. That is, we try to show that true premises lead to a true conclusion.

Consider the following example whose validity we want to prove.

Example 1

If he contracts smallpox, then he was not vaccinated. Either he has malaria or he has contracted smallpox and is in danger of dying. He does not have malaria. Therefore, he was not vaccinated.

Symbolizing we have

1.	$s \to {\sim} v$	$s =$ *he has contracted smallpox*
2.	$m \lor (s \land d)$	$v =$ *he was vaccinated*
3.	${\sim} m$	$m =$ *he has malaria*
	$\overline{{\sim} v}$	$d =$ *he is in danger of dying*

We now try to show that the conclusion v will be true if the premises are true. Take premises 2 and 3 and apply argument form A_3 (disjunctive syllogism)

$$
\begin{array}{ll}
2. & m \lor (s \land d) \\
3. & {\sim} m \\
\hline
4. & (s \land d)
\end{array}
$$

Since premises 2 and 3 are assumed true, $(s \land d)$ the conclusion of the valid argument is also true .

We now apply argument form A_6 (conjunctive simplification) to premise 4.

$$
\begin{array}{ll}
4. & (s \land d) \\
\hline
5. & s
\end{array}
$$

The conclusion of the valid argument s is true since the premise $(s \land d)$ is true. Applying argument form A_1 (modus ponens) to premises 1 and 5 which are true we have

$$
\begin{array}{ll}
1. & s \to {\sim} v \\
5. & s \\
\hline
6. & {\sim} v
\end{array}
$$

We have now proven that the conclusion $\sim v$ is true, when the three premises are true.

The procedure just described to explain the steps and reasons involved is quite lengthy. In our future work the proofs will be constructed in the following concise form.

$$
\begin{array}{llll}
1. & s \rightarrow \sim v & & \text{premise} \\
2. & m \lor (s \land d) & & \text{premise} \\
3. & \sim m & & \text{premise} \\
\hline
4. & (s \land d) & 2, 3 & \text{A}_3 \\
5. & s & 4 & \text{A}_6 \\
6. & \sim v & 1, 5 & \text{A}_1 \\
\end{array}
$$

The question now arises; how do we know what argument forms or equivalence relations to use, and what step to start with? Before we begin to answer this question *it is essential that you know the basic argument forms and equivalence relations.* Otherwise the following work will be unnecessarily difficult.

In developing the formal proof for the above argument, the following reasoning was employed. We want to prove $\sim v$ true given that the three premises are true. The only premise in which $\sim v$ appears is the first. It is the consequent of a conditional. The A_1 (modus ponens) argument form tells us that the consequent of a conditional will be true if the antecedent is true. Therefore, in order to have $\sim v$ true we must first show that s is true; s appears in the second premise. We know that if we have the conjunct $(s \land d)$ true, s would be true by use of argument form A_6 (conjunctive simplification). Since $(s \land d)$ is a conjunct, it would be true if the negation of the other disjunct were true. This is argument form A_3 (disjunctive syllogism). Premise 3 tells us that $\sim m$ the negation of the other disjunct is true. What we have done is work in reverse. In reasoning out our proof we work backwards through the argument and then reverse the process in formalizing the argument.

Example 2

If he is wounded, he was sent home. It is false that he was sent home and not discharged. He was not discharged but he is on leave. Therefore, he is not wounded.

Symbolizing we have

$$
\begin{array}{lll}
1. & w \rightarrow h & \qquad w = \textit{he is wounded} \\
2. & \sim(h \land \sim d) & \qquad h = \textit{he was sent home} \\
3. & \sim d \land l & \qquad d = \textit{he was discharged} \\
\hline
& \sim w & \qquad l = \textit{he is on leave} \\
\end{array}
$$

We want to prove that $\sim w$ is true given that the three premises are true. The antecedent of premise 1 is w. The A_2 (modus tollens) argument tells us that we will have the negation of the antecedent ($\sim w$) true if we have the negation of the consequent ($\sim h$) true. Now h appears in premise 2. To deal with a premise which is the negation of a compound statement, it is best to replace it by its positive logical equivalent. In this case we will replace premise 2, $\sim(h \wedge \sim d)$, by its DeMorgan's equivalent ($\sim h \vee d$). The disjunct $\sim h$ can be shown to be true by argument form A_3 (disjunctive syllogism) if we have the negation of the other disjunct ($\sim d$) true. If we apply the argument of conjunctive simplification to premise 3 ($\sim d \wedge l$) we have $\sim d$ true. Reversing this process and writing it in the formal form we have:

1.	$w \rightarrow h$	premise	
2.	$\sim(h \wedge \sim d)$	premise	
3.	$\sim d \wedge l$	premise	
4.	$\sim d$	3	A_6
5.	$\sim h \vee d$	2	E_6
6.	$\sim h$	4, 5	A_3
7.	$\sim w$	1, 6	A_2

As in the flow charting method, a slight variation of the above method is needed when the conclusion of the argument is a conditional. In doing a proof of this type, the antecedent of the conclusion is taken as a premise. Using this conditional premise we try to prove the consequent of the conclusion true. That is, we assume the antecedent to be true and then try to prove the consequent true. Consider the following example.

Example 3

Either it is solid but porous or it is liquid. If it is liquid, it is not gaseous. Therefore, if it is not solid it is not gaseous.

Symbolizing we have

1.	$(s \wedge p) \vee l$	$s = $ *it is solid*
2.	$l \rightarrow \sim g$	$p = $ *it is porous*
	$\sim s \rightarrow \sim g$	$l = $ *it is liquid*
		$g = $ *it is gaseous*

Since the conclusion is a conditional, we assume the antecedent $\sim s$ to be true. We now try to prove the consequent $\sim g$ true. Argument form A_1 (modus ponens) states that the consequent ($\sim g$) of premise 2 ($l \rightarrow \sim g$) will be true if the antecedent (l) is true. We see that l appears as one of the disjuncts of premise 1, $[(s \wedge p) \vee l]$. Argument form A_3 (disjunctive syllogism) states that the disjunct l will be true if the negation of the other disjunct $\sim(s \wedge p)$ is true. By E_6 (De-Morgan's equivalence) $\sim(s \wedge p)$ is equivalent to ($\sim s \vee \sim p$). Therefore, $\sim(s \wedge p)$

will be true if $(\sim s \lor \sim p)$ is true. As our conditional premise we took $\sim s$ to be true. Therefore, by argument form A_7 (disjunctive addition) $\sim s \lor \sim p$ is true. Reversing this reasoning process and writing it in the formal form we have:

1.	$(s \land p) \lor l$	premise
2.	$l \to \sim g$	premise
3.	$\sim s$	conditional premise
4.	$\sim s \lor \sim p$	3 A_7
5.	$\sim (s \land p)$	4 E_6
6.	l	1, 5 A_3
7.	$\sim g$	2, 6 A_1
8.	$\sim s \to \sim g$	3, 7 conditional form of the conclusion

This proof shows that if $\sim s$ is true, then $\sim g$ is true and by definition $(T \to T)$ is true. If $\sim s$ is false, the statement $\sim s \to \sim g$ is still true since $(F \to T)$ and $(F \to F)$ are true. We write conditional form of the conclusion as our reason for this step.

Often, the establishing of the validity of an argument by the direct method is difficult. This is especially true for arguments which do not have an obvious starting point. When this occurs, we can use the indirect method of proof which does not require the precise chain of reasoning that is required in using the direct method.

The Indirect Method of Formal Proof

As previously stated the indirect method of proof is ideally suited for proving the validity of an argument, since there are only two possible solutions to each problem. That is, the conclusion of the argument is either true or false. In our discussion we will say that either the conclusion is true or the negation (opposite) of the conclusion is true.

Since our objective in proving the validity of an argument is to show the conclusion true when the premises are true, we will try to eliminate the other possible solution, namely the negation of the conclusion. We do this by assuming the negation (opposite) of the conclusion as a true premise and then use it to obtain a contradiction in the formal proof of the argument. That is, we assume the opposite of the conclusion to be true and then try to find some contradiction. If a contradiction is obtained it would eliminate the negation of the conclusion as a possible solution and leave the other alternative, the conclusion is true, as our solution.

Consider the following examples.

Example 1

If he drinks or uses drugs he is dependent. He uses drugs or is dependent. Therefore, he is dependent.

Symbolizing we have

1.	$(d \lor t) \to a$	$d = $ *he drinks*
2.	$(d \lor a)$	$t = $ *he uses drugs*
	a	$a = $ *he is dependent*

To prove this argument by the indirect method, we take the opposite of the conclusion ($\sim a$) as a third premise and then apply appropriate argument forms and logically equivalent statements in hopes of obtaining a contradiction.

1.	$(d \lor t) \to a$	premise
2.	$(d \lor a)$	premise
3.	$\sim a$	premise for the indirect proof
4.	d	2, 3 A_3
5.	$d \lor t$	4 A_7
6.	a	1, 5 A_1

We see that steps 3 ($\sim a$) and 6 (a) contradict. Since premises 1 and 2 are true and we used valid arguments in our proof, the only error would be in our assumption that ($\sim a$) is true. This contradiction shows that ($\sim a$) is false and therefore the other alternative (a) must be true.

There is no definite plan of attack in setting up the indirect proof as there is when we do it directly. We apply various argument forms and equivalent statements to the premises until we obtain a contradiction.

Another contradiction could have been obtained for the above problem. To illustrate:

1.	$(d \lor t) \to a$	premise
2.	$(d \lor a)$	premise
3.	$\sim a$	premise for the indirect proof
4.	d	2, 3 A_3
5.	$\sim(d \lor t)$	1, 3 A_2
6.	$\sim d \land \sim t$	5 E_6
7.	$\sim d$	6 A_6

In this case steps 4 (d) and 7 ($\sim d$) contradict, indicating that the assumed premise $\sim a$ is false. Therefore a is true.

As another illustration of the indirect method, we shall consider an argument which has a conditional conclusion. Consider Example 3 on p. 157.

Example 2

$$(s \wedge p) \vee l$$
$$l \to \sim g$$
$$\overline{\sim s \to \sim g}$$

$s = $ *it is solid*
$p = $ *it is porous*
$l = $ *it is liquid*
$g = $ *it is gaseous*

1.	$(s \wedge p) \vee l$	premise
2.	$l \to \sim g$	premise
3.	$\sim(\sim s \to \sim g)$	premise for the indirect proof
4.	$\sim(s \vee \sim g)$	3 E_8
5.	$(\sim s \wedge g)$	4 E_6
6.	$\sim s$	5 A_6
7.	g	5 A_6
8.	$\sim l$	2, 7 A_2
9.	$(s \wedge p)$	1, 8 A_3
10.	s	9 A_6

Step 6 ($\sim s$) and 10 (s) contradict, indicating that the assumed premise $\sim(\sim s \to \sim g)$ is false. Therefore, $(\sim s \to \sim g)$ is true.

In the case where the indirect method is applied to an argument that has a conditional conclusion, the denial of the conditional is the assumed premise. In working a proof of this type, the next step is to apply the Conditional equivalence to this assumption and then apply DeMorgan's equivalence. We then proceed until we obtain the desired contradiction.

Counter-Example
and Invalid Arguments

If no contradiction arises when we assume the negation of the conclusion as a premise, we cannot assume that the argument is invalid. When this occurs, we must prove the argument to be invalid by means of a counter-example. That is, we must find a case where the premises of the argument are true and the conclusion is false. For example, if we attempt to prove the validity of the argument:

$$p \vee q$$
$$q \to r$$
$$\frac{\sim r}{\sim p}$$

by the indirect method we would be unable to obtain a contradiction. This would lead us to suspect that the argument is invalid. The invalidity of this argument is proven by the following counter-example.

Component Statements			Premises			Conclusion
p	q	r	$p \lor q$	$q \to r$	$\sim r$	$\sim p$
T	F	F	T	T	T	F

This counter-example shows that when p is true and q and r are false, the argument has true premises and a false conclusion. Therefore, it is invalid.

Finding a counter-example is not always a simple task. In the above example we want the conclusion ($\sim p$) to be false; therefore, the component statement (p) must be true. The premise ($\sim r$) must be true; therefore, the component statement (r) is false. Finally, we found the truth value of the remaining component statement (q) by a trial and error examination of the premises. That is, since the premise ($q \to r$) must be true and (r) is false, we see that (q) must also be false. Since if (q) were true, the premise ($q \to r$) would be (T $\to$ F), which is false. Of course, we could also find a counter-example by preparing a complete truth table and noting the truth values of the component statements that make all the premises true and the conclusions false.

Proving Arguments with Biconditional Conclusions

The one form of the conclusion that we have not discussed is the bi-conditional form. The direct formal proof of an argument with a bicon-ditional conclusion is similar to that of an argument with a conditional conclusion, with the exception that there are two distinct parts to the proof. If we are given an *if and only if* statement to prove, we first assume the *if* statement true and then prove the *only if* statement true. Then we assume the *only if* statement true and prove the *if* statement true. Consider the following example.

Example 1

If he is on trial, he was arrested. If he is a falsely accused victim, he is on trial. If he was arrested, he is a murderer. He is a falsely accused victim and not a murderer. Therefore, he was arrested if and only if he is a falsely accused victim.

Symbolizing we have

$$t \rightarrow a \qquad t = \textit{he is on trial}$$
$$v \rightarrow t \qquad a = \textit{he was arrested}$$
$$a \rightarrow m \qquad v = \textit{he is a falsely accused victim}$$
$$\underline{v \vee \sim m} \qquad m = \textit{he is a murderer}$$
$$a \leftrightarrow v$$

Therefore,

1.	$t \rightarrow a$	premise	
2.	$v \rightarrow t$	premise	
3.	$a \rightarrow m$	premise	
4.	$v \vee \sim m$	premise	
5a.	a	conditional premise	
6a.	m	3, 5a	A_1
7a.	v	4, 6a	A_3
8a.	$a \rightarrow v$	5a, 7a	conditional form of conclusion
5b.	v	conditional premise	
6b.	t	2, 5b	A_1
7b.	a	1, 6b	A_1
8b.	$v \rightarrow a$	5b, 7b	conditional form of conclusion
9.	$(a \rightarrow v) \wedge (v \rightarrow a)$	8a, 8b	A_5
10.	$(a \leftrightarrow v)$	9	E_9

It is usually very difficult to prove the validity of an argument with a biconditional conclusion by the indirect method.

Illustrative Examples of Formal Proofs

We will conclude this section with the following illustrative examples.

Example 2

He will be promoted, if he is successful. If he is promoted then he works hard or found a new job. He is successful and he does not work hard. Therefore, he found a new job.

Direct Method:

1.	$s \rightarrow p$	premise	
2.	$p \rightarrow (w \lor j)$	premise	

Prove: (j)

3.	$s \land \sim w$	premise	
4.	s	3	A_6
5.	p	1, 4	A_1
6.	$w \lor j$	2, 5	A_1
7.	$\sim w$	3	A_6
8.	j	6, 7	A_3

Indirect Method:

1.	$s \rightarrow p$	premise	

Prove: (j)

2.	$p \rightarrow (w \lor j)$	premise	
3.	$s \land \sim w$	premise	
4.	$\sim j$	premise for the indirect proof	
5.	s	3	A_6
6.	$\sim w$	3	A_6
7.	$s \rightarrow (w \lor j)$	1, 2	A_4
8.	$w \lor j$	5, 7	A_1
9.	w	4, 8	A_3

Steps 9 (w) and 6 $(\sim w)$ contradict. The assumed premise $(\sim j)$ is false and therefore j is true.

Example 3

It is false that he will leave school and enter the army. If he does not enter the army, then he will get a job. Therefore, if he leaves school, then he will get a job.

Direct Method:

1.	$\sim (s \land e)$	premise	

Prove: $(s \rightarrow g)$

2.	$\sim e \rightarrow g$	premise	
3.	s	conditional premise	
4.	$\sim s \lor \sim e$	1	E_6
5.	$\sim e$	3, 4	A_6
6.	g	2, 5	A_1
7.	$s \rightarrow g$	3, 6	conditional form of the conclusion

Indirect Method:

    1. $\sim(s \wedge e)$ premise
    2. $\sim e \rightarrow g$ premise
    3. $\sim(s \rightarrow g)$ premise for the indirect proof
    4. $\sim(\sim s \vee g)$ 3 E$_8$
    5. $s \wedge \sim g$ 4 E$_6$
    6. s 5 A$_6$
    7. $\sim g$ 5 A$_6$
    8. e 2, 7 A$_2$
    9. $\sim s \vee \sim e$ 1 E$_6$
  10. $\sim s$ 8, 9 A$_3$

Steps 6 (s) and 10 ($\sim s$) contradict. The assumed premise $\sim(s \rightarrow g)$ is false, therefore, $(s \rightarrow g)$ is true.

Example 4

If he is arrested, he will be locked up. He is not locked up or, he robbed the bank and did not see the judge. If he robbed the bank, he would be arrested. It is false that he did not see the judge and is not arrested or, he robbed the bank. Therefore, he saw the judge if and only if he was not arrested.

Biconditional Conclusion:

    1. $a \rightarrow u$ premise

                                      Prove: $(j \leftrightarrow \sim a)$
    2. $\sim u \vee (b \wedge \sim j)$ premise
    3. $b \rightarrow a$ premise
    4. $\sim(\sim j \wedge \sim a) \vee b$ premise

  5a. j conditional premise
  6a. $j \vee \sim b$ 5a A$_7$
  7a. $\sim b \vee j$ 6a E$_3$
  8a. $\sim(b \wedge \sim j)$ 7a E$_6$
  9a. $\sim u$ 2, 8a A$_3$
 10a. $\sim a$ 1, 9a A$_2$
 11a. $j \rightarrow \sim a$ 5a, 10a conditional form of the conclusion

  5b. $\sim a$ conditional premise
  6b. $\sim b$ 3, 5b A$_2$
  7b. $\sim(\sim j \wedge \sim a)$ 4, 6b A$_3$
  8b. $j \vee a$ 7b E$_6$
  9b. j 5b, 8b A$_3$
 10b. $\sim a \rightarrow j$ 5b, 9b conditional form of the conclusion

  11. $(j \rightarrow \sim a) \wedge (\sim a \rightarrow j)$ 11a, 10b A$_5$
  12. $(j \leftrightarrow \sim a)$ 11 E$_9$

Exercises 3.4

1. If the conclusion to an argument is $(p \wedge q) \to r$, what would you use (a) as a conditional premise? (b) as the premise for the indirect proof?

2. In an indirect proof we are hoping to get a _____.

3–6 Furnish the steps for the following formal proofs.

3.
(1)	$r \vee s$	premise
(2)	$r \to q$	premise
(3)	$\sim s$	premise
(4)	r	(Step 4)
(5)	q	(Step 5)
(6)	$r \wedge q$	(Step 6)

4.
(1)	$a \vee \sim p$	premise
(2)	$(\sim p \wedge \sim r) \to s$	premise
(3)	$\sim(a \vee r)$	premise
(4)	$\sim a \wedge \sim r$	(Step 4)
(5)	$\sim a$	(Step 5)
(6)	$\sim p$	(Step 6)
(7)	$\sim r$	(Step 7)
(8)	$\sim p \wedge \sim r$	(Step 8)
(9)	s	(Step 9)

5.
(1)	$p \to \sim q$	premise
(2)	$s \to (q \vee \sim r)$	premise
(3)	r	premise
(4)	p	conditional premise
(5)	$\sim q$	(Step 5)
(6)	$\sim q \wedge r$	(Step 6)
(7)	$\sim(q \vee \sim r)$	(Step 7)
(8)	$\sim s$	(Step 8)
(9)	$p \to \sim s$	conditional proof

6.
(1)	$s \to (r \wedge p)$	premise
(2)	$s \vee \sim q$	premise
(3)	$\sim q \to p$	premise
(4)	$\sim p$	premise for the indirect method
(5)	q	(Step 5)
(6)	s	(Step 6)
(7)	$r \wedge p$	(Step 7)
(8)	p	(Step 8)
(9)	$\therefore p$	(Step 9)

7. Give a formal proof for the following argument.

(1)	$\sim(p \vee \sim q)$	premise
(2)	$\sim p \to r$	premise
(3)	$t \to \sim r$	premise

$$\therefore t \wedge q$$

8–9 Furnish a counter-example for the following invalid arguments.

8. $p \to q$
   $\sim p$
   $\therefore \sim q$

9. $\sim p \vee q$
   $p \to r$
   $\sim r \wedge s$
   $\therefore q \wedge s$

Chapter Exercises

1. Test the validity of the following arguments using (a) truth tables and (b) by assuming true premises.

(a) $q \rightarrow \sim p$
     $\sim(\sim p)$
     $\therefore q$

(b) $p \rightarrow \sim q$
     $p \vee q$
     $\therefore p \leftrightarrow \sim q$

(c) $\sim r \rightarrow \sim p$
     $\sim(\sim p \wedge q)$
     $\sim r$
     $\therefore \sim q$

(d) $a \rightarrow (b \wedge c)$
     $b \rightarrow \sim a$
     $\therefore \sim a$

(e) $\sim(p \wedge q)$
     $q \rightarrow r$
     $p \wedge (p \rightarrow q)$
     $\therefore q \wedge r$

(f) $p \rightarrow \sim q$
     $q \rightarrow \sim r$
     $p \wedge \sim r$
     $\therefore \sim q \wedge \sim r$

(g) *If science is systemized knowledge, then logic is the science of reasoning. Logic is an exact science or it is not the science of reasoning. Science is systemized knowledge. Therefore, logic is an exact science.*

(h) *If it is not four-sided, then it is not a rectangle. It has more than four sides if it is not a rectangle. It does not have more than four sides. Therefore, it is four-sided.*

(i) *If $x = y$, then $2x = 2y$. It is false that $2x = 2y$ and $y \neq z$. Therefore, if $y \neq z$, then $x \neq y$.*

(j) *The tumor is either malignant or benign. If the tumor is malignant, then it can be cured if it is detected in its early stages. The tumor was not benign and it is detected in its early stages. Therefore, it can be cured.*

(k) *If the temperature is above $32°$ and there are dark clouds, then it will rain. There are dark clouds and it will not rain. Therefore, the temperature is not above $32°$.*

2. Using flow charts test the validity of the following arguments.

(a) $q \rightarrow \sim p$
     $\sim(\sim p)$
     $\therefore \sim q$

(b) $(p \rightarrow \sim q)$
     $(p \wedge q)$
     $\therefore q$

(c) $\sim p \vee q$
     $q \rightarrow \sim r$
     r
     $\therefore \sim p$

(d) $a \rightarrow b$
     $\sim(\sim c \wedge b)$
     $\therefore \sim c \rightarrow \sim a$

(e) $p \lor \sim r$
      $p \to q$
      $q \to r$
      $\underline{\sim s}$
      $\therefore \sim r$

(f) $(p \to q) \lor \sim r$
      $\sim r \to s$
      $\underline{\sim s \lor q}$
      $\therefore q$

(g)   *If a subject is not interesting, then it is boring. The subject is not useful if it is boring. The subject is not interesting. Therefore, it is not useful.*

(h)   *If $1 + x = 1$, then $x = 0$. Either $x \neq 0$ or $2x = 0$. Therefore, if $2x \neq 0$, then $1 + x \neq 1$.*

(i)   *It was murder or suicide. There was no weapon found at the scene of the crime and if it were murder, there would be a motive. If there were a motive, then there would be a weapon at the scene of the crime. Therefore, it was suicide.*

(j)   *If the bill came out of committee, then Congress voted on it. If Congress voted on the bill and passed it then the President vetoed the bill. It is false that the President vetoed the bill and it did not come out of committee. Therefore, Congress did not pass the bill.*

(k)   *He did not cross the picket line or he quit his job. If he quit his job, he did not strike. He did cross the picket line or he did not strike. Therefore, he did not strike.*

3.   Test the validity of the following arguments by either truth tables or flow charts.

(a)   *If he is poor and cannot find a job, then he needs assistance. He cannot find a job and either he is economically independent or does not need assistance. Therefore, if he is not economically independent, he is poor.*

(b)   *If he is open-minded he listens to reason and is willing to learn. He is open-minded but does not listen to reason. Therefore, he is not willing to learn.*

(c)   *If his reasoning is logical, then his conclusion is true, if his premises are true. His reasoning is logical but his conclusion is false. Therefore, his premises are false.*

(d)   *The satellite will orbit the earth if and only if it reaches orbital velocity. Either the satellite did not reach orbital velocity or it reached escape velocity and is on its way to the moon. The satellite is not on its way to the moon. Therefore, the satellite will not orbit the earth.*

(e)   *If he is an atheist, he does not believe in God. Either he believes in life after death or he believes in God or the immortality of the soul. He is an atheist and he believes either in the immortality of the soul or in no life after death. Therefore, he believes in the immortality of the soul.*

4. Using a flow chart show that the following argument has contradictory premises. (Use *both* the direct and indirect methods.)

$$p \rightarrow q$$
$$\sim (\sim p \vee q)$$
$$\therefore p$$

5. Using a flow chart show that the premises of the following argument do not contain sufficient information from which to derive the conclusion. (Use *both* the direct and indirect methods.)

$$\sim p \rightarrow q$$
$$\sim (q \vee \sim r)$$
$$\therefore p \wedge s$$

6. Using truth tables show that the following argument is invalid.

If today is Monday, then tomorrow is Tuesday.
Tomorrow is Tuesday.
Therefore, today is Monday.

Why is this seemingly logical argument invalid?

7. Name the valid argument form of each of the following arguments.

(a) $\sim (c \wedge d) \rightarrow e$
     $\sim (c \wedge d)$
     $\therefore e$

(b) $f \rightarrow (b \wedge d)$
     $\sim (b \wedge d)$
     $\therefore \sim f$

(c) $(p \wedge \sim q) \vee (q \wedge \sim r)$
     $\sim (p \wedge \sim q)$
     $\therefore q \wedge \sim r$

(d) $d \wedge (a \vee \sim b)$
     $\therefore a \vee \sim b$

(e) $\sim r \rightarrow \sim s$
     s
     $\therefore r$

(f) $\sim (a \wedge b)$
     $(c \rightarrow a)$
     $\therefore \sim (a \wedge b) \wedge (c \rightarrow a)$

(g) $b \rightarrow c$
     $\therefore (b \rightarrow c) \vee \sim d$

(h) $a \rightarrow \sim b$
     $\sim b \rightarrow c$
     $\therefore a \rightarrow c$

(i) $(a \wedge b) \vee \sim c$
     $(a \wedge b) \vee c$
     $\therefore a \wedge b$

(j) $a \rightarrow (b \rightarrow c)$
     a
     $\therefore b \rightarrow c$

(k) $(a \rightarrow c) \vee (d \vee e)$
     $\sim (d \vee e)$
     $\therefore a \rightarrow c$

(l) $r \rightarrow \sim (p \vee q)$
     $(p \vee q)$
     $\therefore \sim r$

(m) $a \land \sim c$
      $\therefore \sim c$

(n) $(\sim a \to \sim b) \lor c$
      $\sim(\sim a \to \sim b)$
      $\therefore c$

(o) $\sim s \to (t \land r)$
      $\sim(t \land r)$
      $\therefore s$

8. Supply the missing statement in each of the following valid argument forms or logical equivalences.

(a) $(r \land p) \to \sim q$
      q
      $\therefore$?

   (1) $r \land \sim p$ (2) p
   (3) $\sim r \land \sim p$ (4) $\sim(r \land p)$

(b) $a \to (b \to c)$
      ?
      $\therefore \sim a$

   (1) $b \to c$ (2) $\sim b \to \sim c$
   (3) a (4) $\sim(b \to c)$

(c) $(a \land \sim b) \lor (b \land \sim c)$
      ?
      $\therefore a \land \sim b$

   (1) $b \lor \sim c$ (2) $b \land \sim c$
   (3) $b \to c$ (4) $\sim(b \land \sim c)$

(d) $(a \lor \sim b) \land d$
      ?

   (1) $(a \lor \sim b)$ (2) $d \lor a$
   (3) $(d \lor \sim b)$ (4) $\sim(d \lor \sim b)$

(e) $\sim(\sim a \to \sim b) \lor c$
      $\sim c$
      $\therefore$?

   (1) $\sim(a \land b)$ (2) $\sim(\sim a \to \sim b)$
   (3) $\sim a \to \sim b$ (4) $a \lor b$

(f) $\sim p \lor q$
      $\therefore$?

   (1) $\sim p$ (2) q
   (3) $\sim(\sim p \land q)$ (4) $\sim(p \land \sim q)$

(g) $d \land (a \lor \sim b)$
      $\therefore$?

   (1) $d \land a$ (2) $d \land \sim b$
   (3) $a \land \sim b$ (4) $(d \land a) \lor (d \land \sim b)$

(h) $\sim(\sim f \lor \sim a)$
      $\therefore$?

   (1) $f \lor a$ (2) $f \land a$
   (3) $f \land \sim a$ (4) $\sim f \land a$

(i) $a \to (b \land c)$
      ?
      $\therefore a \to \sim d$

   (1) $a \to \sim b$ (2) $(b \land c) \to \sim d$
   (3) $a \to c$ (4) $a \land \sim d$

(j) $(a \land b)$
      $\therefore$?

   (1) $a \lor b$ (2) $\sim a \land \sim b$
   (3) a (4) $\sim(a \to b) \lor \sim d$

9. For each of the following formal proofs give the numbers of the previous steps used and the valid arguments forms and/or logical equivalences used in each step of the proof.

(a) (1) $p \to q$ premise
   (2) $\sim r \to \sim q$ premise
   (3) $\sim(\sim p \lor \sim s)$ premise
   (4) $p \land s$
   (5) p
   (6) q
   (7) r
   (8) s
   (9) $\therefore r \land s$

(b) (1) $l \to \sim e$ premise
   (2) $\sim(\sim e \land \sim s)$ premise
   (3) l premise
   (4) $e \lor s$
   (5) $\sim e$
   (6) $\therefore s$

(c) (1) $a \to (b \to c)$ premise
   (2) $(c \land d) \to e$ premise
   (3) $f \to (b \land d)$ premise
   (4) $\sim(\sim f \lor \sim a)$ premise
   (5) $f \land a$
   (6) f
   (7) $b \land d$
   (8) a
   (9) $b \to c$
  (10) b
  (11) c
  (12) d
  (13) $c \land d$
  (14) $\therefore e$

(d) (1) $(p \land \sim q) \lor (q \land \sim r)$ premise
   (2) $p \to s$ premise
   (3) $\sim s \lor t$ premise
   (4) $\sim t$ premise
   (5) $\sim s$
   (6) $\sim p$
   (7) $\sim p \lor q$
   (8) $\sim(p \land \sim q)$
   (9) $q \land \sim r$
  (10) $\therefore q$

(e) (1) $\sim(\sim a \to \sim b) \lor c$ premise
   (2) $(d \land a) \lor (d \land \sim b)$ premise
   (3) $c \to (d \to b)$ premise
   (4) $d \land (a \lor \sim b)$
   (5) $a \lor \sim b$
   (6) $(\sim a \to \sim b)$
   (7) c
   (8) $(d \to b)$
   (9) d
  (10) b
  (11) $\therefore a$

(f) (1) $\sim(a \to b) \land \sim c$ premise
   (2) $(a \to c) \lor (\sim d \lor e)$ premise
   (3) $\sim e$ premise
   (4) $\sim(a \to b)$
   (5) $\sim(\sim a \lor b)$
   (6) $a \land \sim b$
   (7) a
   (8) $\sim c$
   (9) $a \land \sim c$
  (10) $\sim(\sim a \lor c)$
  (11) $\sim(a \to c)$
  (12) $\sim d \lor e$
  (13) $\therefore \sim d$

(g) (1) $\sim b \to \sim a$ premise
   (2) $b \to (c \lor d)$ premise
   (3) $a \land \sim c$ conditional premise
   (4) a
   (5) b
   (6) $c \lor d$
   (7) $\sim c$
   (8) d
   (9) $\therefore (a \land \sim c) \to d$

(h) (1) $r \to \sim(p \lor q)$ premise
   (2) $s \to p$ premise
   (3) $\sim(s \to \sim r)$ premise for the indirect method
   (4) $\sim(\sim s \lor \sim r)$
   (5) $s \land r$
   (6) s
   (7) p
   (8) r
   (9) $\sim(p \lor q)$

$$(10) \quad \sim p \land \sim q$$
$$(11) \quad \sim p$$
$$(12) \quad \therefore \ s \to \sim r$$

(i)
(1) $\sim (b \land c) \to \sim a$ premise
(2) $a \to (\sim b \land d)$ premise
(3) a premise for the indirect proof
(4) $(\sim b \land d)$
(5) $\sim b$
(6) $b \land c$
(7) b
(8) $\therefore \ \sim a$

(j)
(1) $a \to (b \to c)$ premise
(2) $(a \land d) \lor (a \land e)$ premise
(3) $\sim (\sim b \lor d)$ conditional premise
(4) $a \land (d \lor e)$
(5) a
(6) $b \to c$
(7) $b \land \sim d$
(8) b
(9) c
(10) $d \lor e$
(11) $\sim d$
(12) e
(13) $c \land e$
(14) $\therefore \ \sim (\sim b \lor d) \to (c \land e)$

10. Give a formal proof for each of the following arguments. Give the numbers of the previous steps used and the reason for each step.

(a)
$\sim r \to \sim p$
$\sim (\sim p \land q)$
$\sim r$
$\overline{\therefore \ \sim q}$

(b)
$a \to (b \land c)$
$b \to \sim a$
$\overline{\therefore \ \sim a}$

(c)
$\sim (p \land \sim q)$
$q \to \sim s$
$\overline{\therefore \ p \to \sim s}$

(d)
$\sim p \lor \sim q$
$(r \to p) \lor t$
$\sim (\sim r \lor t)$
$\overline{\therefore \ \sim q}$

(e)
$p \to \sim q$
$p \lor q$
$\overline{\therefore \ p \leftrightarrow \sim q}$

(f)
$a \lor (b \lor \sim c)$
$d \to \sim a$
$\sim (b \land \sim e)$
$\overline{\therefore \ (d \land \sim e) \to \sim c}$

(g) *If he has blind obedience for authority, he does not think for himself. He has confidence in his own ability or he has blind obedience for authority. He does not have confidence in his own ability. Therefore, he does not think for himself.*

(h) *It should not be believed if it is not true. If it should not be believed, it should be rejected. It should not be rejected. Therefore, it is true.*

(i)   *If we do not accept the hypothesis, then we reject the theory. It is not the case that we accept the hypothesis and state a new hypothesis. We do not reject the theory. Therefore, we do not state a new hypothesis.*

(j)   *If formal education is the key to success, then going to school is important. If going to school is important, then studying and learning a large number of facts is important. Learning a large number of facts is not important. Therefore, formal education is not the key to success.*

(k)   *He received an impartial trial or he did not receive justice. The jury was not selected properly and if he received an impartial trial, his rights were protected. If his rights were protected, the jury was selected properly. Therefore he did not receive justice.*

(l)   *If he is a reformer, he wants change and an improved political system. It is not the case that he wants an improved political system or is not a reformer. Therefore, either he wants change or he is just a political opportunist.*

(m) *If he expresses concern and a desire to help his fellow man, he is either sincere or a hypocrite. He expresses concern and is not a hypocrite. Therefore, if he expresses a desire to help his fellow man, he is sincere.*

(n)   *If the disadvantaged are being helped to improve their situation, then it is false that our present welfare system needs reform and more trained workers. Our present welfare system needs more trained workers and either our present welfare system needs reform or more funds are needed. More funds are not needed. Therefore, the disadvantaged are not being helped to improve their situation.*

(o)   *If organized labor has helped the working man, then collective bargaining and arbitration are intelligent ways to reach agreement. It is not the case that strikes are not good for society and collective bargaining is always honestly conducted. Therefore, if organized labor has helped the working man and strikes are not good for society, then arbitration is an intelligent way to reach agreement.*

(p)   *Privacy is the right of each individual or, the government has the right to wiretap and maintain surveillance of its citizens. If the government can wiretap, then it has assumed the role of a watchdog and overseer of its citizenry. Therefore, if the government has not assumed the role of a watchdog of its citizenry, privacy is the right of each individual.*

(q)   *If he is speeding and is a drunken driver, he will be arrested. He is speeding and either he is stopped by the police or not arrested. He is a drunken driver. Therefore, he is stopped by the police.*

(r)   *If it is an abc, then it is either an efg or an hij. Either it is not an efg or it is a pqr. If it is an xyz, then it is not an hij. It is an xyz and not a pqr. Therefore, it is not an abc.*

(s)   *If the rights of society are placed above the rights of the individual, then the society does not have liberty for all. If the society has liberty for all, then it is not a totalitarian society. Either the rights of society are placed above the rights of the individual or the society is totalitarian. Therefore, the society does not have liberty for all.*

(t)   *He treats his fellow man with compassion, if he believes in human rights or the dignity of man. If he does not believe in human rights, he treats his fellow man with compassion. Therefore, he treats his fellow man with compassion.*

(u)   *If his fingerprints were found at the scene of the crime, he is the murderer or a witness of the crime. If he is the murderer, he had a motive or is insane. It is not the case that if he did not have a motive, then his fingerprints were not found at the scene of the crime. He is a friend of the victim or not a witness to the crime and he owned a gun similar to the murder weapon. Therefore, if he did not own a gun similar to the murder weapon or he is not a friend of the victim, then he is insane.*

11.   Find a counter-example for the following *invalid* arguments.

(a) $\sim p \rightarrow \sim q$
      $\sim p \lor r$
      $\underline{\sim r \land s}$
      q

(b) $a \lor (b \land \sim c)$
      $\underline{a \rightarrow d}$
      $\sim d \rightarrow \sim b$

(c)   *He is a liberal or not a Democrat. If he is a conservative, he votes Republican. He votes Republican or is not a liberal. He is a conservative. Therefore, he is a Democrat.*

Review Test

1.   The following truth table shows the argument to be

p	q	Premise	Premise	Conclusion
T	T	F	T	T
T	F	F	T	T
F	T	F	T	T
F	F	T	F	F

(a)   valid
(b)   invalid
(c)   none of these

2. The following truth table shows the argument to be

p	q	Premise	Premise	Conclusion
T	T	T	T	F
T	F	F	T	T
F	T	T	F	F
F	F	T	F	T

(a) valid
(b) invalid
(c) none of these

3. The following truth table shows the argument to be

p	q	Premise	Premise	Conclusion
T	T	T	F	F
T	F	F	T	F
F	T	F	T	T
F	F	T	T	T

(a) valid
(b) invalid
(c) none of these

4. The following truth table shows the argument to be

p	q	r	Premise	Premise	Premise	Conclusion
T	T	T	T	T	T	T
T	T	F	T	T	T	T
T	F	T	F	F	T	F
T	F	F	F	T	T	F
F	T	T	F	T	F	T
F	T	F	T	T	T	F
F	F	T	F	F	F	T
F	F	F	T	T	T	T

(a) valid
(b) invalid
(c) none of these

5. Which set of truth values makes the following argument valid?

p	q	Premise	Premise	Conclusion	(a)	(b)	(c)	(d)
T	T	T	F	?	T	T	T	T
T	F	T	T	?	T	F	T	F
F	T	T	T	?	F	T	T	F
F	F	F	F	?	F	F	F	T

6. Which set of truth values makes the following argument valid?

p	q	Premise	Premise	Conclusion	(a)	(b)	(c)	(d)
T	T	?	F	F	T	F	T	F
T	F	?	T	T	F	T	F	F
F	T	?	F	T	F	T	T	T
F	F	?	T	F	T	F	T	T

7. The flow chart shows the validity of which of the following arguments:

(a) $\sim a \lor \sim b$ (b) $\sim a \lor \sim b$ (c) $\sim a \lor \sim b$ (d)   none of
     $\sim a \to \sim b$ $\sim a \to \sim b$ $\sim a \to \sim b$ these
     $\therefore b$ $\therefore \sim b \to \sim a$ $\therefore \sim b$

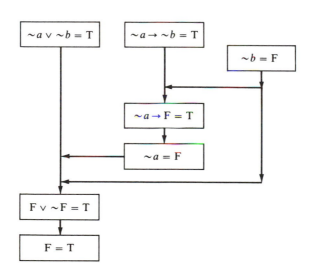

8. The flow chart for testing the validity of the argument

$(\sim p \to \sim q)$
$(q \lor r) \land \sim r$ is
$\therefore p$

(a)

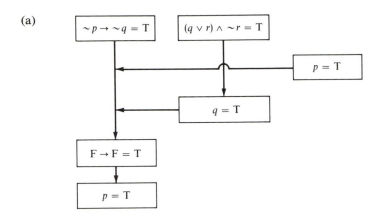

(b)

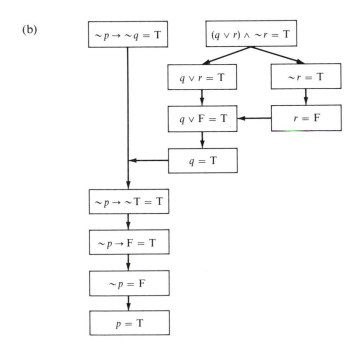

(c)

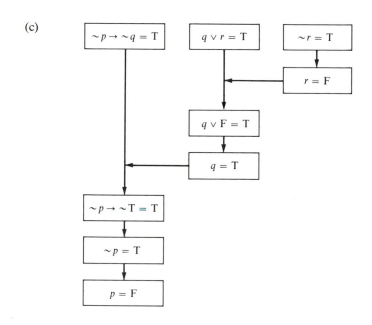

(d)   none of these

9.   The flow chart for testing the validity of the argument

$$p \lor \sim q$$
$$q \land r \qquad \text{is}$$
$$\overline{\therefore p}$$

(a)   (b)

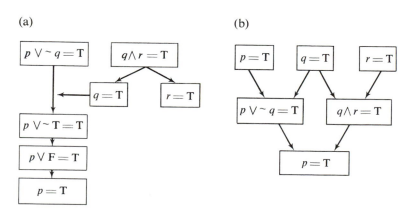

(c)

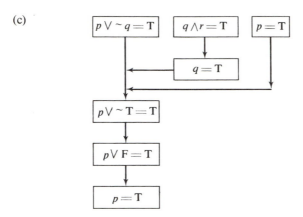

(d)   none of these

10.   The flow chart shows the validity of which of the following arguments?

(a) $\sim a \rightarrow \sim b$
      $b \lor \sim c$
      c
      _____
      $\therefore \sim a$

(b) $\sim a \rightarrow \sim b$
      $b \lor \sim c$
      c
      _____
      $\therefore a$

(c) $\sim a \rightarrow \sim b$
      $b \lor \sim c$
      c
      _____
      $\therefore a \land \sim c$

(d)   none of these

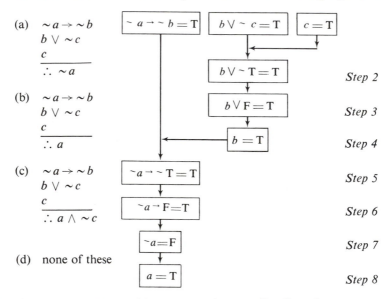

Complete problems 11–14 with respect to the preceding flow chart.

11.   The reason for step 3 is the definition of

      (a)   conjunction   (b)   disjunction   (c)   conditional

      (d)   negation   (e)   none of these

12.   The reason for step 4 is the definition of

      (a)   conjunction   (b)   disjunction   (c)   conditional

      (d)   negation   (e)   none of these

13. The reason for step 5 is the definition of

(a) conjunction   (b) disjunction   (c) conditional

(d) negation   (e) none of these

14. The reason for step 6 is the definition of

(a) conjunction   (b) disjunction   (c) conditional

(d) negation   (e) none of these

15. The following argument is an example of

$p \rightarrow \sim q$ (a) A_1 (b) A_2 (c) A_2
p (d) A_5 (e) none of these
$\overline{\therefore \ \sim q}$

16. The following argument is an example of

$\sim q \lor \sim r$ (a) A_1 (b) A_2 (c) A_3
r (d) A_5 (e) none of these
$\overline{\therefore \ \sim q}$

17. The following argument is an example of

$\sim q \land \sim r$ (a) A_1 (b) A_2 (c) A_3
$\overline{\therefore \ \sim q}$ (d) A_4 (e) none of these

18. The following argument is an example of

$\sim a \rightarrow \sim b$ (a) A_1 (b) A_2 (c) A_3
b (d) A_5 (e) none of these
$\overline{\therefore \ a}$

19. Which of the following conclusions makes the argument valid?

$(r \land p) \rightarrow \sim q$ (a) $(r \land \sim p)$ (b) p
q (c) $(\sim r \lor \sim p)$ (d) $(r \land p)$
$\overline{\therefore \text{ conclusion}}$ (e) none of these

20. Which of the following conclusions makes the argument valid?

$p \rightarrow r$ (a) q (b) $\sim r$
$q \rightarrow \sim r$ (c) $q \land s$ (d) $\sim(q \lor s)$
p (e) none of these
$\overline{\therefore \text{ conclusion}}$

21. Which of the following statements makes the argument valid?

$\sim p \rightarrow \sim q$ (a) $\sim q \lor p$ (b) r
$\sim p \lor r$ (c) $q \rightarrow p$ (d) $\sim r \land \sim p$
premise   (e) none of these
$\overline{\therefore\ \sim q}$

22. Which of the following statements makes the argument valid?

$(p \land q) \rightarrow r$ (a) $(\sim r \lor \sim s)$ (b) $q \rightarrow r$
$(\sim r \lor s) \land q$ (c) $\sim s$ (d) s
premise   (e) none of these
$\overline{\therefore\ \sim p}$

Complete problems 23–27 with respect to the following formal proof.

(1) $p \lor (q \land \sim r)$ premise
(2) $q \rightarrow r$ premise
(3) $(p \lor \sim r) \rightarrow (\sim p \lor s)$ premise
(4) $\sim q \lor r$ step 4
(5) $\sim (q \land \sim r)$ step 5
(6) p step 6
(7) $p \lor \sim r$ step 7
(8) $\sim p \lor s$ step 8
(9) $\therefore\ s$ step 9

23. The reason for step 4 is

(a) A_3 (b) A_6 (c) E_3 (d) E_6 (e) none of these

24. The reason for step 5 is

(a) A_6 (b) E_3 (c) E_6 (d) E_8 (e) none of these

25. The reason for step 6 is

(a) A_3 (b) A_6 (c) E_6 (d) E_7 (e) none of these

26. The reason for step 7 is

(a) A_3 (b) A_6 (c) E_6 (d) E_8 (e) none of these

27. The reason for step 8 is

(a) A_2 (b) A_4 (c) E_6 (d) E_8 (e) none of these

Complete problems 28–32 with respect to the following formal proof.

(1) $\sim (b \land c) \rightarrow \sim a$ premise
(2) $a \rightarrow (\sim b \land d)$ premise
(3) a premise for the indirect proof

(4)	$\sim b \wedge d$	step 4
(5)	$\sim b$	step 5
(6)	$b \wedge c$	step 6
(7)	b	step 7
(8)	$\therefore \sim a$	step 8

28. The reason for step 4 is

    (a) A_1 (b) A_2 (c) A_3 (d) A_4 (e) none of these

29. The reason for step 5 is

    (a) E_6 (b) E_8 (c) A_5 (d) A_6 (e) none of these

30. The reason for step 6 is

    (a) A_1 (b) A_2 (c) A_6 (d) A_8 (e) none of these

31. The reason for step 7 is

    (a) E_6 (b) E_8 (c) A_5 (d) A_6 (e) none of these

32. The reason for step 8 is

    (a) A_1 (b) A_3 (c) E_6 (d) E_8 (c) none of these

33. Is the following argument valid? *In order to be a scholar, it is necessary for you to know Latin. You do not know Latin. Therefore, you are not a scholar.*

    (a) yes   (b) no   (c) the validity cannot be determined from the above information

34. Is the following argument valid? *If today is March 1st, then yesterday was February 28th. Yesterday was February 28th. Therefore, today is March 1st.*

    (a) yes   (b) no   (c) the validity cannot be determined from the above information

35. Is the following argument valid? *Either he is free or his rights are limited. If his rights are limited, he is free. Therefore, he is free.*

    (a) yes   (b) no   (c) the validity cannot be determined from the above information.

36. Is the following argument valid? *If he is a Russian leader, he is a member of the Communist party. He is a member of the Communist party, if he is an atheist. Therefore, if he is a Russian leader, he is an atheist.*

    (a) yes   (b) no   (c) the validity cannot be determined from the above information

4

LOGIC AND SETS

We have seen that a close relationship exists between the laws of set theory and the rules governing logic. In this chapter we shall show how these two areas of mathematics complement one another. That is, set theory will be used to test the validity of arguments, and symbolic logic will be used to prove the laws of set theory. As a final application, symbolic logic will be used in the theory of electrical switching circuits. This last idea is extremely important since electrical switching circuits are an essential component of high-speed electronic computers.

Truth Sets and Arguments

We shall now show how set theory, especially Venn diagrams, can be used in testing the validity of an argument. Consider the simple statement "*He is tall.*" This statement, which we shall denote by p, is either true or false depending on the "he" to whom we are referring. For some people it is true and for others it is false. This simple statement p can be joined to other statements to form compound statements. Let U be the universal set whose elements are all the possible compound statements containing the simple statement p. This universal set can be subdivided into two disjoint subsets: the set P, whose elements are the compound statements in which the simple statement p is true, and the set P', whose elements are the compound statements in which the simple statement p is false. We call P the truth set of statement p since, for all the elements of P, statement p is true. We call P' the truth set of $\sim p$ since it contains all the compound statements in which p is false. The following Venn diagram illustrates this concept:

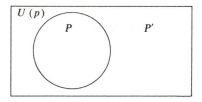

We see that $U(p) = P \cup P'$.

If we carry this idea one step further and consider compound statements which contain two simple statements p and q, we use the following standard Venn diagram:

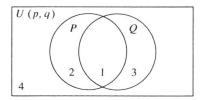

Here U is the universal set of all compound statements containing the simple statements p and q. The set P is the truth set of p, and Q is the truth set of q. However, in the above figure we see that there are four disjoint subsets, each of which has distinct truth values for p and q. In area 1 $(P \cap Q)$, since we are within both truth sets P and Q, the simple statements p and q are both true. In area 2 $(P \cap Q')$, since we are in the truth set P but outside the truth set Q, the simple statement p is true and q is false. Using the same reasoning, in area 3 $(P' \cap Q)$ p is false and q is true, and in area 4 $(P' \cap Q')$ both p and q are false. That is,

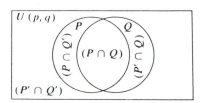

This relationship can be shown by the following table:

Table 1

Statement		Area	Truth Set
p	q		
T	T	1	$P \cap Q$
T	F	2	$P \cap Q'$
F	T	3	$P' \cap Q$
F	F	4	$P' \cap Q'$

Table 1 or the Venn diagram enables us to determine the truth values of the statements p and q for any of the four disjoint subsets. For example, area 3, which is the set $P' \cap Q$, is the set of compound statements for which p is false and q is true.

For compound statements consisting of three simple statements, p, q, and r, we use the Venn diagram shown below and the following table.

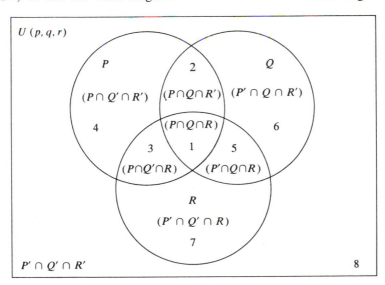

Since it is difficult to construct Venn diagrams for four or more sets, we shall limit our discussion to compound statements made up of three or less component statements.

Table 2

Statement			Area	Truth Set
p	q	r		
T	T	T	1	$P \cap Q \cap R$
T	T	F	2	$P \cap Q \cap R'$
T	F	T	3	$P \cap Q' \cap R$
T	F	F	4	$P \cap Q' \cap R'$
F	T	T	5	$P' \cap Q \cap R$
F	T	F	6	$P' \cap Q \cap R'$
F	F	T	7	$P' \cap Q' \cap R$
F	F	F	8	$P' \cap Q' \cap R'$

Definition 1

Truth Set: The truth set of a compound statement made up of component statements, $p, q, r, \ldots$ is a subset of the universal set $U(p, q, r, \ldots)$, all of whose elements make the compound statement true. Elements from any other set of U make the compound statement false.

To illustrate this definition we shall examine the truth sets for negation and the four basic logical connectives. We first construct the truth tables for the connectives and also include the truth sets for the four disjoint subsets.

Table 3

p	q	Area	Truth Sets	$\sim p$	$p \lor q$	$p \land q$	$p \to q$	$p \leftrightarrow q$
T	T	1	$P \cap Q$	F	T	T	T	T
T	F	2	$P \cap Q'$	F	T	F	F	F
F	T	3	$P' \cap Q$	T	T	F	T	F
F	F	4	$P' \cap Q'$	T	F	F	T	T

It is now a simple matter to construct a Venn diagram and shade in the truth set. For example, suppose we want to find the truth set of $p \lor q$. We first see that it is true for any of the elements in the sets $P \cap Q, P \cap Q'$, and $P' \cap Q$. We then shade in these areas (1, 2, and 3) to obtain the truth set of $p \lor q$. The truth sets for the other connectives are also found by shading in those areas for which the compound statements are true. Therefore, we have the following:

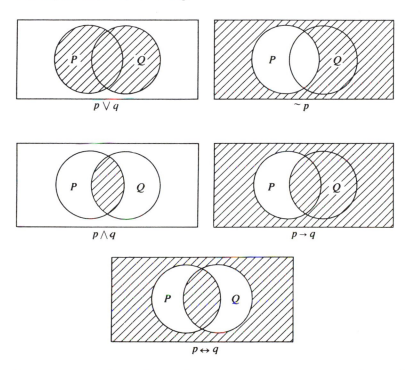

Examining the above Venn diagrams we see that the truth sets for negation and the four logical connectives can be represented in terms of sets P and Q as follows:

Table 4

Statement	Truth Set
$\sim p$	P'
$p \lor q$	$P \cup Q$
$p \land q$	$P \cap Q$
$p \rightarrow q$	$(P \cap Q')'$ or $(P' \cup Q)$
$p \leftrightarrow q$	$(P \cap Q) \cup (P \cup Q)'$

Note: Since $(p \rightarrow q) \leftrightarrow (\sim p \lor q)$, the simplest form of the truth set for $p \rightarrow q$ is $P' \cup Q$.

Using these relationships we can now write the truth set for any given compound statement by converting negation ($\sim$) to complement ($'$), disjunction ($\lor$) to union ($\cup$), conjunction ($\land$) to intersection ($\cap$), and conditional ($\rightarrow$) to the union of the complement of the antecedent with the consequent.

Example 1

Find the truth sets and construct Venn diagrams for the following compound statements:

$$\text{(a)} \quad p \land (p \rightarrow q) \qquad \text{(b)} \quad p \lor (q \land \sim r)$$

(a)   To find the truth set for $p \land (p \rightarrow q)$, we convert it to set notation by use of Table 4. Therefore, we have $P \cap (P' \cup Q)$. We can now construct the Venn diagram for this truth set, or we can first simplify the form of the truth set by applying the laws of Boolean algebra.

(1)	$P \cap (P' \cup Q)$	Given truth set
(2)	$(P \cap P') \cup (P \cap Q)$	1, Distributive
(3)	$\phi \cup (P \cap Q)$	2, Inverse
(4)	$(P \cap Q)$	3, Identity

We see that $P \cap (P' \cup Q) = P \cap Q$. This leads us to conclude that since $p \land (p \rightarrow q)$ and $p \land q$ have the same truth sets, they must be equivalent.

The following is the Venn diagram for the truth set $P \cap (P' \cup Q)$:

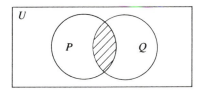

(b)   To find the truth set for $p \lor (q \land \sim r)$, we convert it to set notation by use of Table 4. Therefore, we have $P \cup (Q \cap R')$ or $P \cup (Q - R)$. This truth set is in simplest form and cannot be reduced by applying the laws of Boolean algebra. The following is the Venn diagram for the truth set $P \cup (Q \cap R')$:

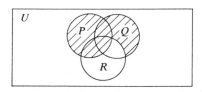

Since we can write truth sets for compound statements, it is now possible to use truth sets to test the validity of an argument.

An argument, which is a series of statements called premises and a final statement called the conclusion, is said to be valid if, whenever the premises are true, the conclusion is also true. That is, true premises must always yield a true conclusion for an argument to be valid. To apply truth sets to this definition, we find the truth set whose elements make the premises all true. Once we have found this truth set of the premises, we then determine whether all of its elements also make the conclusion true. If so, the argument is valid. (If any of the elements of the truth set of the premises yield a false conclusion, the argument is invalid.) In other words, we determine whether or not the truth set for which the premises are all true is a subset of the truth set of the conclusion. When it is, we have true premises always yielding a true conclusion (a valid argument). Therefore, we have the following:

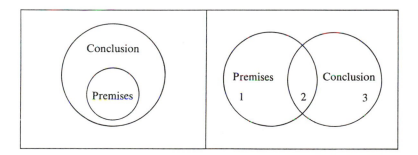

Valid argument: (truth set of the premises) ⊆ (truth set of the conclusion).

Invalid argument: (truth set of the premises) ⊄ (truth set of the conclusion).

In the second case, we see that the argument is invalid since there are elements for which the premises are true and the conclusion is false (area 1).

The truth set of the premises (the set for which all the premises are true) is the intersection of the truth sets of the individual premises. Since conjunction and intersection are related operations, it is also true that the truth set of the premises is the truth set of the conjunction of the premises. For example, the truth set of the premises $p \rightarrow q$ and $\sim q$ can be written as

$$(p \to q) \land \sim q,$$

which is $(P' \cup Q) \cap Q'$

Definition 2

Valid Argument: An argument is valid if the truth set of the conjunction of the premises is a subset of the truth set of the conclusion.

Example 2

Test the validity of the following arguments:

$$\begin{array}{ll}
\text{(a)} \quad p \to q & \qquad \text{(b)} \quad \sim p \to q \\
\qquad \sim q & \qquad \qquad \quad r \lor \sim q \\
\cline{1-1}
\qquad \therefore p & \qquad \qquad \quad \sim r \\
& \qquad \qquad \quad \therefore p
\end{array}$$

(a)   We construct the Venn diagrams for the truth set of the conjunction of the premises and the truth set of the conclusion.

Premises	*Conclusion*
$(p \to q) \land \sim q$	p
$(P' \cup Q) \cap Q'$	P
by Boolean algebra	
$(P \cup Q)'$	

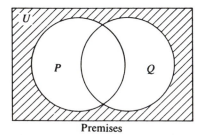

Premises

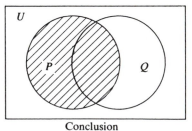

Conclusion

Examining the diagrams we see that the truth set of the premises is not a subset of the truth set of the conclusion; that is, $[(P \cup Q) \cap Q'] \not\subseteq P$. Therefore, the argument is invalid.

(b) We construct the Venn diagrams for the truth set of the conjunction of the premises and the truth set of the conclusion.

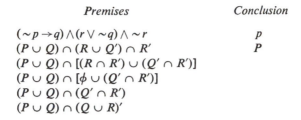

$$(\sim p \rightarrow q) \wedge (r \vee \sim q) \wedge \sim r$$
$$(P \cup Q) \cap (R \cup Q') \cap R'$$
$$(P \cup Q) \cap [(R \cap R') \cup (Q' \cap R')]$$
$$(P \cup Q) \cap [\phi \cup (Q' \cap R')]$$
$$(P \cup Q) \cap (Q' \cap R')$$
$$(P \cup Q) \cap (Q \cup R)'$$

Conclusion

$$p$$
$$P$$

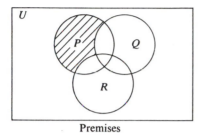

Premises

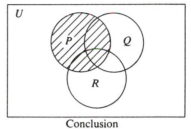

Conclusion

Examining the diagrams we see that the truth set of the premises is a subset of the truth set of the conclusion; that is, $[(P \cup Q) \cap (R \cup Q') \cap R'] \subseteq P$. Therefore, the argument is valid.

Exercises 4.1

1. Shade the truth set for the following compound statements:

(a) $\sim q \rightarrow p$

(b) $\sim p \vee (p \rightarrow q)$

(c) $\sim p \wedge (q \vee \sim r)$

(d) $\sim p \rightarrow (q \wedge \sim r)$

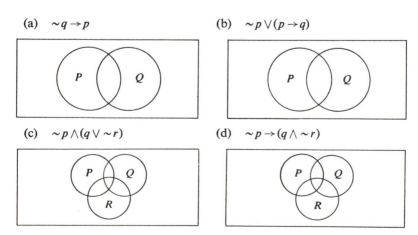

2. By examining the Venn diagram, determine whether the following arguments are valid or invalid:

(a) $p \vee \sim q$
$\sim q$
$\overline{}$
$\therefore p$

Premises

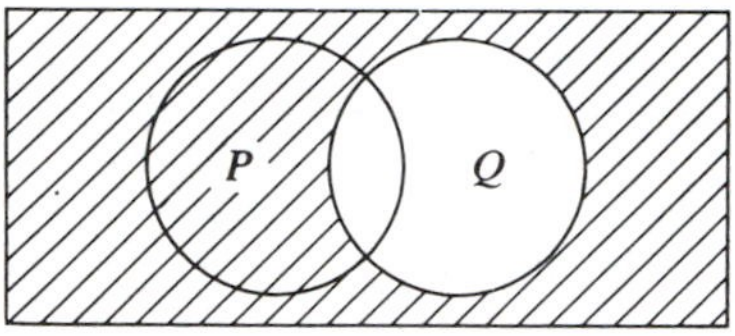

$(P \cup Q') \cap Q'$

Conclusion

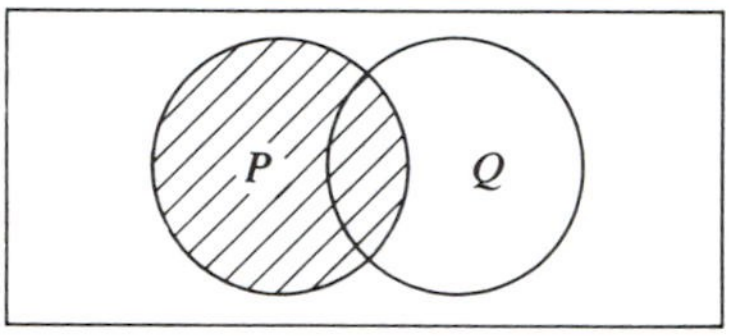

P

(b) $\sim p \to q$
$\sim p \vee r$
$\sim r$
$\overline{}$
$\therefore q$

Premises

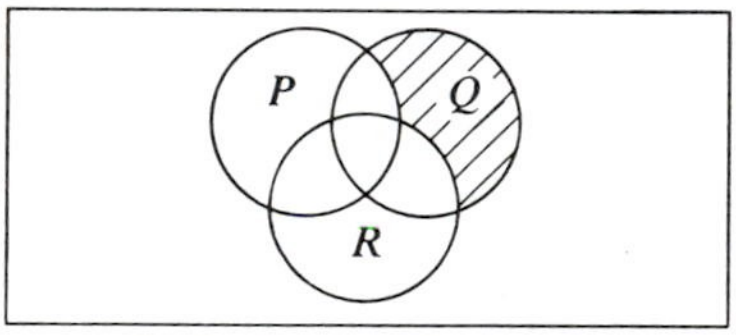

$(P \cup Q) \cap (P' \cup R) \cap R'$

Conclusion

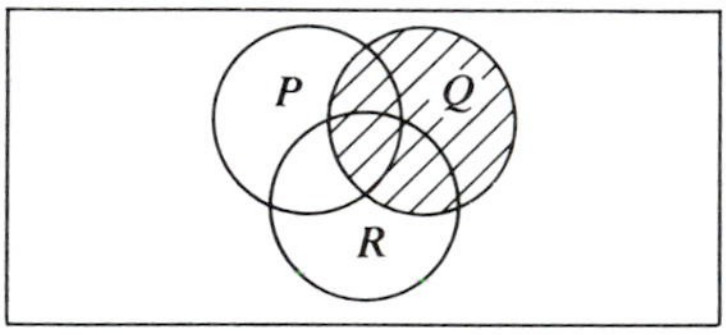

Q

3. Test the validity of the following arguments using truth sets and Venn diagrams:

(a) $q \to \sim p$
$\sim(\sim p)$
$\overline{}$
$\therefore \sim q$

(b) $p \to q$
$\sim p$
$\overline{}$
$\therefore q$

(c) $p \to r$
$\sim(\sim p \wedge q)$
$\sim r$
$\overline{}$
$\therefore \sim q$

(d) $p \to \sim q$
$q \to \sim r$
r
$\overline{}$
$\therefore \sim p$

Venn Diagrams and
Logical Deductive Reasoning

Statements of logic often involve relationships between collections of ideas which use the adjectives "all," "some," and "no." A statement of the form "All A is B" can be symbolized $A \rightarrow B$ and a statement of the form "No A is B" can be symbolized $A \rightarrow \sim B$. These statements and statements of the form "Some A is B" will also be translated using quantifiers later in the chapter.

In this section we will show how Venn diagrams are used to clarify the relationships expressed in these statements. The statement "If a given person is a smoker, then a given person will get cancer" is equivalent to the idea "All smokers get cancer," that is, the set of smokers is a subset of the set of those who get cancer. The diagram below depicts the relationship between the sets.

The statement "If a given person is a smoker, then a given person will get cancer" is equivalent to the idea "All smokers get cancer," that is, the set of smokers is a subset of cancer. The diagram below depicts the relationship between the sets.

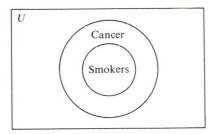

Next consider the idea "Some smokers get cancer." This statement means that there is at least one smoker (*not excluding the possibility of all smokers*) who will get cancer. The Venn diagram below is used to illustrate this relationship:

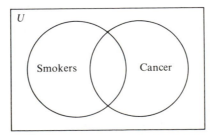

Finally consider the idea: "No smokers get cancer." This is equivalent to the statement "If the given person is a smoker, then the given person will not get cancer," or by using the contrapositive, "If the given person will get cancer, then the given person is not a smoker."

The diagram below of disjoint sets depicts this relationship:

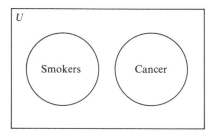

We will now use our knowledge of logic and Venn diagrams to analyze the validity of arguments using deductive reasoning. Deductive reasoning can be described as that kind of reasoning in which the conclusion of an argument must be true whenever the premises of an argument are true or assumed to be true using the rules governing logic.

In order to prove an argument invalid by means of a Venn diagram, we must find a counter-example where the premises of the argument are satisfied and the conclusion is not satisfied using the rules of logic.

Example 1

Draw a Venn diagram and test the validity of the following argument: All guilty people will be arrested. All thieves are guilty people. Therefore, all thieves will be arrested.

This argument is valid since every Venn diagram drawn will satisfy the conditions of both the premises and conclusion using the rules of logic. Note that no counter-example can be constructed.

Example 2

Draw a Venn diagram and test the validity of the following argument: All speeders are dangerous. Some drivers speed. Therefore, some drivers are dangerous.

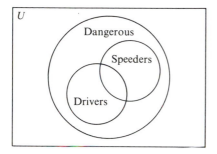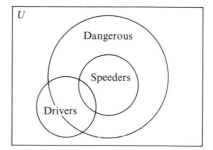

This argument is valid since every Venn diagram drawn will satisfy the conditions of the premises and the condition of the conclusion using the rules of logic. Note that no counter-example can be drawn. Even though one might conclude from the first diagram that all drivers are dangerous, this does not contradict our conclusion that some drivers are dangerous. It was noted above that the word "some" does not exclude the possibility of "all."

Example 3

Draw a Venn diagram and test the validity of the following argument: All boys like sports. All males like sports. Therefore, all boys are males.

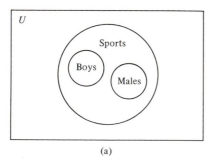
(a)
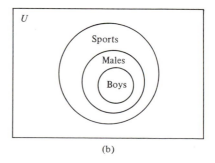
(b)

The Venn diagram (a) presents a counter-example in which the conditions of the conclusion are not satisfied. Therefore, this argument is invalid even though we would observe that the idea of the conclusion is true and that we could draw another diagram (b) that supports the conclusion.

Example 4

Draw a Venn diagram and test the validity of the following argument: Some politicians are honest. All judges are honest. Therefore, no politicians are judges.

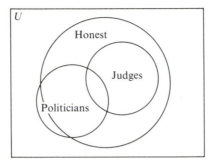

This argument is invalid. The Venn diagram drawn shows a counter-example in which the conditions of the premises are satisfied but the conditions of the conclusion are not satisfied.

Symbolic Translation Using Quantifiers

A more detailed way of symbolizing the statement "All A is B" is to write $(x)(A_x$ is a $B_x)$ or $(x)(A_x \to B_x)$. This is read "For all x, if x is an A, then x is a B." The phrases "for every x" and "for all x" are called universal quantifiers. A statement of the form "No A is B" can be written $(x)(A_x \to \sim B_x)$ and is read "For all x, if x is an A, then x is not a B."

We translate the statement "Some A is B" as $(\exists x)(A_x \wedge B_x)$. This is read "there exists an x such that x is an A and x is a B" or "there is at least one x such that x is an A and x is a B". The phrases "some," "there exist an x such that," and "there is at least one x such that" are called existential quantifiers.

If we translate Example 1 from this section using this system we get;

$$(x)(G_x \to A_x)$$
$$\underline{(x)(T_x \to G_x)}$$
$$\therefore (x)(T_x \to A_x)$$

This argument is valid. It follows the structure of the Hypothetical Syllogism (A_4) from Chapter 3.

Example 2 is translated as follows:

$$(x)(S_x \to D_x) \qquad S = \text{Speeders}$$
$$\underline{(\exists x)(R_x \wedge S_x)} \qquad D = \text{Dangerous}$$
$$(\exists x)(R_x \wedge D_x) \qquad R = \text{Drivers}$$

Here we reason intuitively, that if there is at least one x that is both a driver and a speeder, and all speeders are dangerous, then it is valid to conclude that there is at least one x that is a driver and is dangerous.[1,2]

Equivalences from Chapter 3 also apply to this system. For example: $(x)(A_x \rightarrow B_x)$ is equivalent to $(x)(\sim A_x \vee B_x)$ (Conditional Equivalence E_7).

One very interesting equivalence in this system enables us to change a universal quantification to an existential quantification as follows: $(x)(A_x \rightarrow B_x)$ is equivalent to $\sim(\exists x)(A_x \wedge B_x)$. Intuitively, we know that to say "All A is B" is the same thing as saying "There is no A which is not a B."

Further translations and manipulations of quantified statements are made in more advanced logic courses and will not be covered in our discussion.

Exercises 4.2

Use Venn diagrams to check the validity of each argument:

1. All cats are animals. All animals are living creatures. Therefore, all cats are living creatures.
2. Some grandparents are not living. All living people are proud. Therefore, some grandparents are not proud.
3. All grandparents are proud. Some living people are not proud. Therefore, some living people are not grandparents.
4. All grandparents are proud. Some grandparents are not living. Therefore, some living people are not proud.
5. Some polygons are squares. Some polygons are pentagons. Therefore, some squares are pentagons.
6. All otters are four legged. All alligators are four legged. Therefore, some alligators are otters.
7. Some triangles are scalene. Some triangles are obtuse. Therefore, some obtuse triangles are scalene triangles.
8. Some numbers larger than twelve are divisible by three. Nine is divisible by three. Therefore, nine is larger than twelve.
9. All bananas belong to the fruit family. Some members of the fruit family have pits. Therefore, some bananas have pits.
10–18. Using quantified statements, translate the arguments in problems 1–9 above.

---

[1] If you have one existential premise, you can only have an existential conclusion.

[2] We cannot use truth tables to check the validity of arguments involving quantified statements.

Logic Applied to Sets

In our previous discussion of set theory we saw that the set operations and relations follow certain laws. Some of these laws were proven in our development of Boolean algebra, while others were illustrated by the use of Venn diagrams and membership tables. In this section we shall prove these laws and relations by the use of symbolic logic. To do this we now define the basic notions of set theory in terms of logical statements.

Let A, B, and C be arbitrary sets in a universal set U. We now have the following:

S_1 *Subset* $(A \subseteq B)$ $(x \in A) \rightarrow (x \in B)$

S_2 *Equality* $(A = B)$ $(x \in A) \leftrightarrow (x \in B)$

S_3 *Union* $(A \cup B)$ $[x \in (A \cup B)] \leftrightarrow [(x \in A) \vee (x \in B)]$

S_4 *Intersection* $(A \cap B)$ $[x \in (A \cap B)] \leftrightarrow [(x \in A) \wedge (x \in B)]$

S_5 *Universal Set* (U) $[x \in (U \cup A)] \leftrightarrow (x \in U)$ and

                               $[x \in (U \cap A)] \leftrightarrow (x \in A)$

S_6 *Null Set* $(\emptyset)$ $[x \in (A \cup \emptyset)] \leftrightarrow (x \in A)$ and

                               $[x \in (A \cap \emptyset)] \leftrightarrow (x \in \emptyset)$

S_7 *Complement* (A') $(x \notin A) \leftrightarrow (x \in A')$; $[x \in (A \cup A')] \leftrightarrow (x \in U)$

                         and $[x \in (A \cap A')] \leftrightarrow (x \in \emptyset)$

S_8 *Negation* $(\sim)$ $\sim(x \in A) \leftrightarrow (x \notin A)$

The following are some of the basic laws of sets. Some will be proven and proof of the remainder will be left as exercises.

Theorem 1

$$\text{If } A \subseteq B \text{ and } B \subseteq A, \text{ then } A = B$$

Proof

(1)	$A \subset B$	Premise
(2)	$B \subset A$	Premise
(3)	$(x \in A) \rightarrow (x \in B)$	1, S_1
(4)	$(x \in B) \rightarrow (x \in A)$	2, S_1
(5)	$[(x \in A) \rightarrow (x \in B)] \wedge$	
	$[(x \in B) \rightarrow (x \in A)]$	3, 4, Conjunctive addition
(6)	$(x \in A) \leftrightarrow (x \in B)$	5, Biconditional equivalence
(7)	$A = B$	6, S_2

Theorem 2

$$\text{If } A \subseteq B \text{ and } B \subseteq C, \text{ then } A \subseteq C$$

Theorem 3

$$\text{If } A \subseteq B, \text{ then } A \cap B = A$$

Proof

To show that $A \cap B = A$, we must prove that $[x \in (A \cap B)] \leftrightarrow (x \in A)$. This is a biconditional conclusion and therefore there are two parts to the proof.

(1)	$A \subseteq B$	Premise
(2)	$(x \in A) \to (x \in B)$	1, S_1
(3a)	$x \in (A \cap B)$	Conditional premise
(4a)	$(x \in A) \wedge (x \in B)$	3a, S_4
(5a)	$x \in A$	4a, Conjunctive simplification
(6a)	$[x \in (A \cap B)] \to (x \in A)$	3a, 5a, Conditional form of the conclusion
(3b)	$x \in A$	Conditional premise
(4b)	$x \in B$	2, 3b, Modus ponens
(5b)	$(x \in A) \wedge (x \in B)$	3b, 4b, Conjunctive addition
(6b)	$x \in (A \cap B)$	5b, S_4
(7b)	$(x \in A) \to [x \in (A \cap B)]$	3b, 6b, Conditional form of the conculsion
(3)	$[x \in (A \cap B) \to (x \in A)] \wedge [(x \in A) \to x \in (A \cap B)]$	6a, 7b, Conjunctive addition
(4)	$[x \in (A \cap B)] \leftrightarrow (x \in A)$	3, Biconditional equivalence
(5)	$A \cap B = A$	4, S_2

Theorem 4

$$\text{If } A \subseteq B, \text{ then } A \cup B = B$$

Theorem 5

$$\text{If } A \subseteq B, \text{ then } B' \subset A'$$

Theorem 6

$$(A')' = A$$

Theorem 7

$$A \cup A = A \quad \text{and} \quad A \cap A = A$$

Proof

We shall prove the first of these two statements. The proof of the second is similar. To show that $A \cup A = A$, we must prove the biconditional conclusion

$$[x \in (A \cup A)] \leftrightarrow (x \in A)$$

(1a)	$x \in (A \cup A)$	Conditional premise
(2a)	$(x \in A) \vee (x \in A)$	1a, S_3
(3a)	$x \in A$	2a, Identity equivalence
(4a)	$[x \in (A \cup A)] \rightarrow (x \in A)$	1a, 3a, Conditional form of the conclusion
(1b)	$x \in A$	Conditional premise
(2b)	$(x \in A) \vee (x \in A)$	1b, Disjunctive addition
(3b)	$x \in (A \cup A)$	2b, S_3
(4b)	$(x \in A) \rightarrow [x \in (A \cup A)]$	1b, 3b, Conditional form of the conclusion
(1)	$\{[x \in (A \cup A)] \rightarrow (x \in A)\} \wedge \{(x \in A) \rightarrow [x \in (A \cup A)]\}$	4a, 4b, Conjunctive addition
(2)	$[x \in (A \cup A)] \leftrightarrow (x \in A)$	1, Biconditional equivalence
(3)	$A \cup A = A$	2, S_2

Theorem 8

$$A \cup B = B \cup A \quad \text{and} \quad A \cap B = B \cap A$$

Theorem 9

$$A \cup (B \cup C) = (A \cup B) \cup C \quad \text{and} \quad A \cap (B \cap C) = (A \cap B) \cap C$$

Theorem 10

$$A \cap (B \cup C) = (A \cap B) \cup (A \cap C)$$

and

$$A \cup (B \cap C) = (A \cup B) \cap (A \cup C)$$

Proof

We shall prove the first of these two statements. The proof for the second is similar. To show that $A \cap (B \cup C) = (A \cap B) \cup (A \cap C)$, we must prove that $\{x \in [A \cap (B \cup C)]\} \leftrightarrow \{x \in [(A \cap B) \cup (A \cap C)]\}$.

(1a)	$x \in [A \cap (B \cup C)]$	Conditional premise
(2a)	$(x \in A) \wedge [x \in (B \cup C)]$	1a, S_4
(3a)	$(x \in A) \wedge [(x \in B) \vee (x \in C)]$	2a, S_3
(4a)	$[(x \in A) \wedge (x \in B)] \vee [(x \in A) \wedge (x \in C)]$	3a, Distributive equivalence
(5a)	$[x \in (A \cap B)] \vee [x \in (A \cap C)]$	4a, S_4
(6a)	$x \in [(A \cap B) \cup (A \cap C)]$	5a, S_3
(7a)	$\{x \in [A \cap (B \cup C)]\} \rightarrow \{x \in [(A \cap B) \cup (A \cap C)]\}$	1a, 6a, Conditional form of the conclusion
(1b)	$x \in [(A \cap B) \cup (A \cap C)]$	Conditional premise
(2b)	$[x \in (A \cap B)] \vee [x \in (A \cap C)]$	1b, S_3
(3b)	$[(x \in A) \wedge (x \in B)] \vee [(x \in A) \wedge (x \in C)]$	2b, S_4
(4b)	$(x \in A) \wedge [(x \in B) \vee (x \in C)]$	3b, Distributive equivalence
(5b)	$(x \in A) \wedge [x \in (B \cup C)]$	4b, S_3
(6b)	$x \in [A \cap (B \cup C)]$	5b, S_4
(7b)	$\{x \in [(A \cap B) \cup (A \cap C)]\} \rightarrow \{x \in [A \cap (B \cup C)]\}$	1b, 6b, Conditional form of the conclusion

(1) $(\{x \in [A \cap (B \cup C)]\} \rightarrow \{x \in [(A \cap B) \cup (A \cap C)]\}) \wedge$
      $(\{x \in [(A \cap B) \cup (A \cap C)]\} \rightarrow \{x \in [A \cap (B \cup C)]\})$ 7a, 7b, Conjunc-
                                                                                          tive addition
(2) $\{x \in [A \cap (B \cup C)]\} \leftrightarrow \{x \in [(A \cap B) \cup (A \cap C)]\}$ 1, Biconditional
                                                                                          equivalence
(3) $A \cap (B \cup C) = (A \cap B) \cup (A \cap C)$ 2, S_2

Theorem 11

$$\phi' = U \quad \text{and} \quad U' = \phi$$

Theorem 12

$$(A \cup B)' = A' \cap B' \quad \text{and} \quad (A \cap B)' = A' \cup B'$$

Theorem 13

$$A \cap (A \cup B) = A \quad \text{and} \quad A \cup (A \cap B) = A$$

Theorem 14

$$\text{If } A \subseteq B, \text{ then } A \cup (B \cap A') = B$$

Exercises 4.3

1. The following are proofs of theorems. Fill in the definitions, logically equivalent statements, valid argument forms, and previous step used in each step of the proof:

(a) Theorem 5. If $A \subseteq B$, then $B' \subseteq A'$.

    (1) $A \subseteq B$
    (2) $(x \in A) \rightarrow (x \in B)$
    (3) $\sim (x \in B) \rightarrow \sim (x \in A)$
    (4) $(x \in B') \rightarrow (x \in A')$
    (5) $B' \subseteq A'$

(b) Theorem 13. $A \cap (A \cup B) = A$

    (1a) $x \in [A \cap (A \cup B)]$
    (2a) $(x \in A) \wedge [x \in (A \cup B)]$
    (3a) $x \in A$
    (4a) $\{x \in [A \cap (A \cup B)]\} \rightarrow (x \in A)$
    (1b) $x \in A$
    (2b) $(x \in A) \vee (x \in B)$
    (3b) $x \in (A \cup B)$
    (4b) $(x \in A) \wedge [x \in (A \cup B)]$
    (5b) $x \in [A \cap (A \cup B)]$
    (6b) $(x \in A) \rightarrow \{x \in [A \cap (A \cup B)]\}$
    (7) $\{x \in [A \cap (A \cup B)] \rightarrow (x \in A)\} \wedge \quad \{(x \in A) \rightarrow x \in [A \cap (A \cup B)]\}$
    (8) $\{x \in [A \cap (A \cup B)]\} \leftrightarrow \{x \in A\}$
    (9) $A \cap (A \cup B) = A$

2. Prove the following theorems:

  (a) Theorem 2   (b) Theorem 4   (c) Theorem 6   (d) Theorem 8
  (e) Theorem 9   (f) Theorem 11   (g) Theorem 12

Symbolic Logic Applied to Electrical Switching Circuits

As a final illustration of the use of logic and sets, we shall show how symbolic logic can be used to describe electrical switching circuits, which are an integral component of high-speed electronic computers. Their use in performing the basic arithmetic operations will be shown in the chapter on computers.

Consider an electric wire with a switch A, in which we want current to flow from S to T. For current to flow from S to T, switch A must be on.

The notation $A = 0$ will be used to indicate that switch A is off (open) and $A = 1$ to indicate A is on (closed).

Two or more electrical switches can be arranged in many different and often complex circuits. However, all of these circuits are a combination of two basic circuits.

1. *Series Circuit.*   For current to flow from S to T, both switches A and B must be on. If one of the switches is off, no current will flow. The

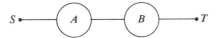

logical connective of conjunction $(A \wedge B)$ will be used to indicate the series arrangement of switches A and B. Therefore, for any arbitrary switch A, we have $A \wedge 0 = 0$ and $A \wedge 1 = A$.

2. *Parallel Circuit.*   For current to flow from S to T, either switch A or B or both must be on. If both switches are off, no current will flow.

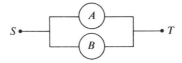

The logical connective of disjunction $(A \vee B)$ will be used to indicate the parallel arrangement of switches A and B. Therefore, for any arbitrary switch A, we have $A \vee 0 = A$ and $A \vee 1 = 1$.

The four possible on–off (1–0) arrangements of switches A and B, the resulting behavior (current flow) of the series circuit $(A \wedge B)$, and the parallel circuit $(A \vee B)$ are given in the following table:

Table 5

A	B	$A \wedge B$	$A \vee B$
1 (on)	1 (on)	1 (current)	1 (current)
1 (on)	0 (off)	0 (no current)	1 (current)
0 (off)	1 (on)	0 (no current)	1 (current)
0 (off)	0 (off)	0 (no current)	0 (no current)

Before we apply these ideas to investigating more complex switching circuits, there is one switch relationship still to be discussed.

3.   *Complementary Switches.*   If two switches are such that when one of them is on, the other is off, and vice versa, we say that the switches are complements of one another. The complement of switch A, denoted by the negation symbol of logic, is $\sim A$. Therefore, for any arbitrary switch A, we have $A \wedge \sim A = 0$ and $A \vee \sim A = 1$.

Table 6

A	$\sim A$
1 (on)	0 (off)
0 (off)	1 (on)

From Tables 5 and 6, we see that the theory of electrical switching circuits is analogous to symbolic logic. If the 1 (on) and 0 (off) were replaced by true and false, respectively, Tables 5 and 6 would be our basic truth tables. Therefore, we are able to apply the laws of symbolic logic in solving switching circuit problems.

Example 1

Construct diagrams for the following switching circuits:

      (a) $A \vee (B \wedge \sim C)$ (b) $[A \wedge (B \vee C)] \vee [\sim A \wedge (B \vee C)]$

(a)   In constructing the diagram for $A \vee [B \wedge \sim C]$, we first construct a diagram indicating that switches B and $\sim C$ are in the following series:

This series arrangement is then put in parallel with switch A.

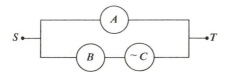

(b)   To construct the diagram for the circuit $[A \wedge (B \vee C)] \vee [\sim A \wedge (B \vee C)]$ we
     construct the circuits for $A \wedge (B \vee C)$ and for $\sim A \wedge (B \vee C)$ and then put
     them in a parallel arrangement. To find the circuit for $A \wedge (B \vee C)$ we first
     construct a circuit indicating that switches B and C are in parallel.

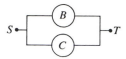

This parallel arrangement is then put in series with switch A.

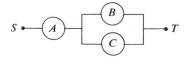

In a similar manner we construct the circuit for $\sim A \wedge (B \vee C)$.

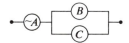

Placing these two circuits in parallel we have the diagram for the circuit

$$[A \wedge (B \vee C)] \vee [\sim A \wedge (B \vee C)]$$

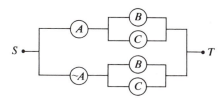

Example 2

Construct the diagram for the switching circuit represented by the logical statement $A \leftrightarrow \sim B$.

To construct the diagram for $A \leftrightarrow \sim B$, we must first write it in terms of series
($\wedge$) and parallel ($\vee$) circuits. We do this by applying the rules for logically
equivalent statements.

$$A \leftrightarrow \sim B$$ Given statement
$$(A \rightarrow \sim B) \wedge (\sim B \rightarrow A)$$ Biconditional equivalence
$$(\sim A \vee \sim B) \wedge (B \vee A)$$ Conditional equivalence

We now can construct the diagram for the circuit $(\sim A \vee \sim B) \wedge (B \vee A)$ since it is equivalent to $A \leftrightarrow \sim B$.

We first construct the two parallel circuits $\sim A \vee \sim B$ and $B \vee A$

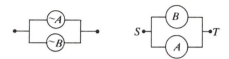

These parallel circuits are then placed in a series arrangement.

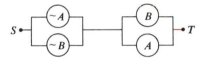

Example 3

Write a symbolic expression for the following circuit:

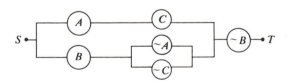

We start with the parallel arrangement of switches $\sim A$ and $\sim C$, symbolized by $\sim A \vee \sim C$. This is in series with B, symbolized by $B \wedge (\sim A \vee \sim C)$. This is in parallel with the series $A \wedge C$, symbolized by $(A \wedge C) \vee [B \wedge (\sim A \vee \sim C)]$. Finally this is in series with the switch $\sim B$, symbolized by

$$\sim B \wedge \{(A \wedge C) \vee [B \wedge (\sim A \vee \sim C)]\}$$

Example 4

What switches must be closed for current to flow through the circuit $A \wedge (B \vee \sim C)$?

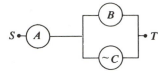

To determine this we construct a switching circuit table for $A \wedge (B \vee \sim C)$.

A	B	C	$A \wedge (B \vee \sim C)$
1	1	1	1
1	1	0	1
1	0	1	0
1	0	0	1
0	1	1	0
0	1	0	0
0	0	1	0
0	0	0	0

We see that current will flow through the circuit when switches A, B, and C are closed; A and B are closed and C is open; and A is closed and B and C are open.

Example 5

Find a simpler circuit to replace the complex circuit $[A \wedge (B \vee C)] \vee [\sim A \wedge (B \vee C)]$.

This is analogous to finding a logically equivalent statement for a given statement. Constructing the switching circuit table for $[A \wedge (B \vee C)] \vee [\sim A \wedge (B \vee C)]$ we have

A	B	C	$[A \wedge (B \vee C)] \vee [\sim A \wedge (B \vee C)]$	$B \vee C$
1	1	1	1	1
1	1	0	1	1
1	0	1	1	1
1	0	0	0	0
0	1	1	1	1
0	1	0	1	1
0	0	1	1	1
0	0	0	0	0

Examining the table we notice that the circuit $B \lor C$ would have the same results as the complex circuit. Therefore, circuit $[A \land (B \lor C)] \lor [\sim A \land (B \lor C)]$ could be replaced by its equivalent circuit $B \lor C$. We could also find a simpler switching circuit for the given circuit by applying the rules for logically equivalent statements.

$[A \land (B \lor C)] \lor [\sim A \land (B \lor C)]$	Given statement
$[(B \lor C) \land A] \lor [(B \lor C) \land \sim A]$	1, Commutative equivalence
$(B \lor C) \land (A \lor \sim A)$	2, Distributive equivalence
$(B \lor C) \land 1$	3, Definition of complement- ary switches
$B \lor C$	4, Definition of series circuit

Exercises 4.4

1. Symbolically describe each of the following switching circuits:

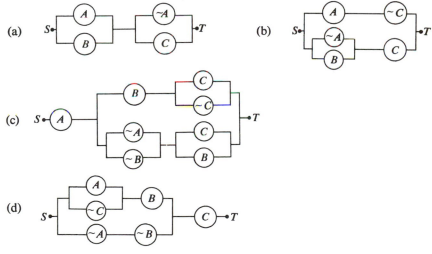

2. Through which switching circuit of Problem 1 will current flow if switches A and B are open and C is closed?

3. Construct switching circuit diagrams for the following statements:

(a) $(\sim A \lor C) \land (B \lor \sim C)$

(b) $[(A \land C) \lor (B \lor \sim C)] \land \sim B$

(c) $[A \land (B \lor C)] \land (B \lor C)$

(d) $\{[A \land (B \lor C)] \land \sim C\} \lor (A \land B)$

4. Through which switching circuits of Problem 3 will current flow if switches *A* and *C* are open and switch *B* is closed?

5. Find simpler switching circuits for the following:

   (a) $[A \vee (B \wedge \sim B)] \wedge [A \wedge (A \vee B)]$

   (b) $[(A \vee C) \wedge (A \vee \sim C)] \vee [\sim A \wedge (\sim A \vee B)]$

   (c) $[A \wedge (B \vee C)] \wedge \{A \vee [A \wedge (B \vee C)]\}$

6. A committee of five members votes on proposals by pressing a button to vote yes; pressing the button closes the switch. Two members of the committee are diametrically opposed; each always votes the opposite of the other. To pass a proposal, three of the five must vote yes. When a proposal passes, current flows through the circuit and a bulb lights. Design a switching circuit and represent it symbolically.

Review Test

1. Test the validity of the following argument by shading the truth sets in the Venn diagrams below:

$$\sim p \to \sim q$$
$$q$$
$$\overline{}$$
$$\therefore p \vee \sim q$$

   (a) Shade in the conjunction of the premises.

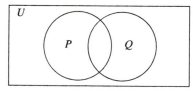

   (b) Shade in the conclusion.

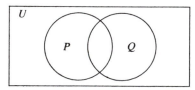

   (c) Is the argument valid?

2. Test the validity of the following argument by shading the truth sets in the Venn diagrams below:

$$\sim p \rightarrow q$$
$$\sim p \lor r$$
$$\sim r$$
$$\therefore p \land q$$

(a) Shade in the conjunction of the premises.

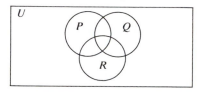

(b) Shade in the conclusion.

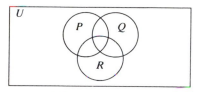

(c) Is the argument valid?

3. Symbolically describe the following switching circuit:

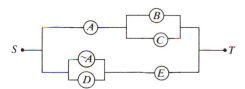

4. A diagram that represents the circuit $[A \lor (B \land C)] \lor (\sim A \lor \sim B)$ is

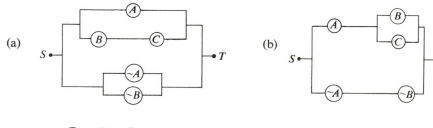

(e) none of these

5. In order for the current to flow through circuit

$(A \lor B) \land (\sim B \land \sim C)$, switches

(a) A, B, and C are closed; (b) A, B, and C are open;

(c) A and B are open and C is closed;

(d) A is closed and B and C are open; (e) none of these.

6. If $x \in (A \cap B)'$, then

(a) $x \in A'$ (b) $x \notin B$ (c) $x \in (A' \cap B')$

(d) $x \notin (A \cup B)$ (e) $x \in (A' \cup B')$

7. If $x \in (A \cup B')$, then

(a) $(x \in A) \lor (x \in B)$ (b) $(x \in A) \lor (x \notin B)$

(c) $(x \in A) \land (x \notin B)$ (d) $(x \notin A) \land (x \notin B)$

(e) $(x \notin A) \lor (x \notin B)$

8. If $x \in (A' \cup B)$, then

(a) $(x \in A') \lor (x \in B)$ (b) $(x \in A') \land (x \in B)$

(c) $(x \notin A) \land (x \notin B)$ (d) $(x \notin A) \lor (x \notin B)$

(e) none of these

9. The statement $(x \in A) \land [(x \in B) \lor (x \in C)]$ is equivalent to

(a) $[(x \in A) \land (x \in B)] \lor [(x \notin A) \land (x \notin C)]$

(b) $[(x \in A) \land (x \in B)] \lor [(x \in A) \land (x \in C)]$

(c) $[(x \notin A) \land (x \notin B)] \lor [(x \notin A) \land (x \notin C)]$

(d) $[(x \notin A) \land (x \notin B)] \lor [(x \in A) \land (x \in C)]$

(e) none of these

10. $A \cap \varnothing =$

(a) A (b) A' (c) $\varnothing$ (d) U (e) none of these

11. $A' \cup (A')' =$

(a) A (b) A' (c) $\varnothing$ (d) U (e) none of these

12. Which is an example of valid reasoning?

(a) All athletes are strong. Jack is an athlete. Therefore, Jack is quick.
(b) All athletes are strong. Joe is strong. Therefore, Joe is an athlete.
(c) All athletes are strong. Jim is not an athlete. Therefore, Jim is not strong.
(d) All athletes are strong. John is an athlete. Therefore, John is strong.

13. The following argument is (a) valid or (b) invalid? All x is y. All z is y. Therefore, all x is z.

5

PROBABILITY

The theory of probability, a relatively new branch of mathematics, developed as seventeenth-century gamblers sought answers to questions dealing with games of chance. They asked, for example, "In a single roll of the dice, what is the chance that the total will equal seven?" "In repeated rolls of the dice what is the chance that the dice will yield a nine before a seven appears?" Questions such as these led the two great mathematicians, Pascal and Fermat, to develop much of the elementary theory of probability that will be discussed in this chapter.

While the history of probability is short, the field has rapidly grown in importance and has extended its applications from the initial concern with gambling to such seemingly unrelated areas as science and psychology. It is now used in the study of the motion of particles, in the theory of heredity, in the determination of insurance rates, and even in the not so exact science of weather prediction.

What is probability? Intuitively we can think of probability as the chance or likelihood of some event occurring. This idea of chance is familiar to most of us, and it is often referred to in our everyday conversations. For example, "Based on his high school record, John has a 50–50 chance of getting into college," "There is only one chance in a hundred that Kay will win the raffle," or, finally, "There is a 20% chance that a male adult will have a heart attack before he reaches his 60th birthday." These statements employ the notion of chance or probability. It is important to note that they are generalizations based on a large number of observations and not just a few cases. For example, the third statement is not based on a study of the medical histories of 10 or 20 adult males but rather on a study of a very large number. Therefore, we can think of probability as the study of the laws of chance dealing with generalizations and not specific cases.

This chapter will deal with the basic concepts on which probability theory is built. We shall primarily use mechanical devices, such as coins, cards, and dice, in our illustrations, while limiting our discussion to problems with finite numbers of outcomes. This will facilitate our work by enabling us to list the possible outcomes of our experiments.

Note: In all of the following material, unless otherwise stated, the devices (coins, dice, cards, etc.) are assumed to be fair or unbiased. For example, a coin is considered to be fair or unbiased if, when it is tossed, the chance of it landing "heads up" is equal to the chance of it landing "tails up."

Finite Sample Spaces, Events, and Probability

Our study of probability will be concerned with answering the following basic question: What is the *probability* or *chance* that one particular event out of a number of possible events will occur?

Definition 1

Sample Space: A sample space is the set of all possible outcomes of an experiment. It is the universal set of the experiment.

Definition 2

Sample Point: A sample point of an experiment is an element of the sample space. Each of the possible outcomes of the experiment is called a sample point or point of the sample space.

Example 1

In the toss of a coin there are two possible outcomes, either heads (h) or tails (t). The sample space for this experiment is $S_1 = \{h, t\}$, where h and t are the sample points of S_1.

Example 2

In the roll of a die there are six possible outcomes, either 1, 2, 3, 4, 5, or 6. The sample space for this experiment is $S_2 = \{1, 2, 3, 4, 5, 6\}$ where the numbers 1, 2, 3, 4, 5, and 6 are the sample points of S_2.

Example 3

A slightly more complex experiment can be performed by combining the experiments of Examples 1 and 2, that is, by tossing a coin and simultaneously rolling a die. If we denote the top face of the coin by c and the number appearing on the top face of the die by d, then an outcome of this experiment can be represented by the ordered pair (c, d). Since there are 2 possible outcomes for c and 6 possible outcomes for d, there are 2×6, or 12, ordered pairs or sample points. Table 1 is a convenient method of listing the points for this experiment.

Table 1

		Outcomes for the Die					
c ╲ d		1	2	3	4	5	6
Outcomes for the Coin	h	$(h, 1)$	$(h, 2)$	$(h, 3)$	$(h, 4)$	$(h, 5)$	$(h, 6)$
	t	$(t, 1)$	$(t, 2)$	$(t, 3)$	$(t, 4)$	$(t, 5)$	$(t, 6)$

The sample space S_3 for this experiment is made up of the twelve ordered pairs. Each ordered pair is a sample point of S_3. To each sample point of a finite sample space we assign a number, which we call its *probability*. This assigned value meets the following three conditions:

(a) The probability assigned to each sample point is nonnegative; that is, it is either positive or zero.

(b) The sum of the probabilities of all the points in a sample space is 1.

(c) Equally likely events have the same probability.

From these three conditions we see that the probability of any sample point x, denoted by $P(x)$ must have the following property:

$$0 \leq P(x) \leq 1$$

To illustrate the above concepts, consider the following three variations of the die-tossing experiment.

Case 1

In this case we assume that the die is unbiased. That is, each of the six points in the sample space $S = \{1, 2, 3, 4, 5, 6\}$ has the *same chance of occurring*. Since there are six points with an equal chance of occurring, each point is assigned a probability of $1/6$. The probability of getting a 2 in a single roll of the die is $1/6$. Symbolically we write $P(2) = 1/6$. The probability of a 5 is also $1/6$; $P(5) = 1/6$. Since each of the six points has a probability of $1/6$, we see that the sum of the probabilities of all the points of the sample space S is 1. That is,

$$P(1) + P(2) + P(3) + P(4) + P(5) + P(6) = 1$$

Case 2

As in Case 1 we assume that the die is unbiased but instead of the six faces being numbered 1–6, five of the faces are colored black (b) with the remaining face colored white (w). The sample space for the roll of this die is $S = \{b, w\}$. It is evident that the chance of a black face and the chance of a white face are not equal. Since five faces are black and only one is white, we assign the probability of $5/6$ to the sample point " black " and $1/6$ to the sample point " white." Symbolically we have $P(b) = 5/6$ and $P(w) = 1/6$. We see that

$$P(b) + P(w) = 5/6 + 1/6 = 1$$

Case 3

In the final case we assume that we have an unbalanced die. That is, the sample space is $S = \{1, 2, 3, 4, 5, 6\}$, but each sample point does not have the same chance of occurring. The die is loaded. In this case we have no reasonable theoretical method of assigning probabilities to the sample points. The only possible way of determining the probabilities of the sample points would be to record the outcomes of repeated rolls of the die. This would enable us to *estimate* the probabilities of the points. These estimates should increase in accuracy as the number of rolls of the die increases.

Suppose through performing an actual experimental situation of repeatedly rolling the die, we estimate the probabilities to be

$$P(1) = P(2) = P(4) = P(5) = 1/8 \quad \text{and} \quad P(3) = P(6) = 1/4$$

Again, we note that the sum of the probabilities assigned to the points of the sample space is 1. That is,

$$P(1) + P(2) + P(3) + P(4) + P(5) + P(6) = 1/8 + 1/8 + 1/4 \\ + 1/8 + 1/8 + 1/4 = 1$$

To summarize, in Case 1 it was clear that each sample point should have the same assigned probability as every other sample point since the die was balanced. In Case 2, we saw that the sample points were not equally likely to occur since the die had five times as many black faces as white. However, since the die was balanced we were able to determine the assigned probabilities. In Case 3, where the probabilities were not known since the die was not balanced, we made inferences about the values of the true probabilities on the basis of experimental results.

We are now ready to discuss the probability of combinations of sample points or events.

Definition 3

An Event: An event E in a sample space S is a subset of the sample space. It is a combination of one or more points of the sample space.

To illustrate, we again consider the die-tossing experiment. The sample space is $S = \{1, 2, 3, 4, 5, 6\}$; $E_1 = \{2, 4, 6\}$ is the event that the die gives an even number; $E_2 = \{1, 3, 5\}$ is the event that the die gives an odd number; $E_3 = \{1\}$ is the event that the die gives a number less than two. A slightly different event E_4 is the event that the die gives a number greater than 6. Since there are no sample points in the sample space that satisfy this condition, the event E_4 would be the null set: $E_4 = \varnothing$.

Definition 4

The Probability of an Event of E: The probability of an event E, denoted by $P(E)$, is the sum of the probabilities of the sample points which make up the event.

Since the probabilities of the sample points are nonnegative numbers and the sum of all the probabilities of all the sample points equals 1, we have for any event E, $0 \leq P(E) \leq 1$.

Note: We assign a probability of zero to the null set of the sample space. That is, $P(\varnothing) = 0$.

Example 4

A balanced die is tossed.   (a) What is the probability that the die gives an even number?   (b) What is the probability that the die gives a number less than 3?

Since the probability of obtaining any face of a die is 1/6 and the probability of an event is the sum of the probabilities of the sample points that make up the event, we have

(a) $E_1 = \{2, 4, 6\}$ represents the event that the die gives an even number and the probability of E_1 is

$$P(E_1) = P(\{2, 4, 6\}) = P(2) + P(4) + P(6) = 1/6 + 1/6 + 1/6 = 3/6 = 1/2$$

(b) $E_2 = \{1, 2\}$ represents the event that the die gives a number less than 3 and the probability of E_2 is

$$P(E_2) = P(\{1, 2\}) = P(1) + P(2) = 1/6 + 1/6 = 2/6 = 1/3$$

Example 5

A die is tossed, but it is observed to be unbalanced. Through repeated tosses the probabilities are estimated to be

$$P(1) = P(2) = P(4) = P(5)$$

and

$$P(3) = P(6) = 2P(1)$$

(a) Find the probabilities assigned to the points of sample space S. (b) What is the probability that the die gives an even number? (c) What is the probability that the die gives a number less than 3?

Since the sum of the probabilities of all the points of a sample space is 1, we have

$$P(1) + P(2) + P(3) + P(4) + P(5) + P(6) = 1$$

By letting $P(1) = x$ and from the estimates of the probabilities of the sample points, we have

$$P(1) = x, \quad P(2) = x, \quad P(3) = 2x, \quad P(4) = x, \quad P(5) = x, \quad \text{and} \quad P(6) = 2x$$

Substituting these values, we have

$$x + x + 2x + x + x + 2x = 1$$
$$8x = 1$$
$$x = 1/8$$

Therefore

(a) $P(1) = 1/8$, $P(2) = 1/8$, $P(3) = 1/4$, $P(4) = 1/8$, $P(5) = 1/8$, and $P(6) = 1/4$.

(b) $P(E_1) = P(\{2, 4, 6\}) = P(2) + P(4) + P(6) = 1/8 + 1/8 + 1/4 = 4/8 = 1/2$.

(c) $P(E_2) = P(\{1, 2\}) = P(1) + P(2) = 1/8 + 1/8 = 2/8 = 1/4$.

Exercises 5.1

1. A coin is tossed. The outcomes are equally likely to occur.

   (a) List the sample space.
   (b) What is the probability that a head will occur?

2. A biased coin is tossed (each outcome does not have the same chance of occurring). By repeatedly tossing the coin, the investigator estimates that a head appears on the top face of the coin 70 times out of each 100 tosses. What is the probability that a tail will occur?

3. Two coins are tossed; one is a penny and the other a nickel.

   (a) List the sample space.
   (b) Find the probability of the following events:
       (1) both coins are heads
       (2) the penny gives a head and the nickel gives a tail
       (3) one coin comes up tails while the other comes up heads
       (4) both coins are tails

4. A coin is tossed three times.

   (a) List the sample space.
   (b) Find the probability of the following events:
       (1) 3 heads
       (2) 2 heads and a tail
       (3) no heads
       (4) exactly 1 head
       (5) at least 1 head
       (6) at most 1 head
       (7) a tail, a head, and a tail, exactly in that order

5. An ordinary deck of playing cards (4 suits—hearts, diamonds, clubs, and spades—with 13 cards in each suit) are shuffled thoroughly. If a single card is drawn, find the probability of the following events:

   (a) an ace of spades
   (b) a black ace
   (c) an ace
   (d) a spade
   (e) a face card (king, queen, or jack)
   (f) a card greater than 6 but less than 9
   (g) a card

6. The 13 spades are taken from a deck of cards and shuffled thoroughly. If a single card is drawn from the 13 spades, find the probability of the following events:

    (a) an ace
    (b) a spade
    (c) a diamond
    (d) a card that is a multiple of 3
    (e) a face card

7. $S = \{a, b, c, d, e\}$ is a sample space of five points with the following probabilities:

$$P(a) = P(b)$$
$$P(c) = P(d) = 2P(a)$$
$$P(e) = 3P(a)$$

    (a) Find the probabilities of the points in the sample space S.
    (b) Find the probabilities of the following events:
        (1) a vowel
        (2) a consonant
        (3) $\{a, c, e\}$

8. $S = \{1, 2, 3, 4, 5, 6\}$ is a sample space of a die with the following assigned probabilities:

$$P(1) = P(2) = P(3)$$
$$P(4) = P(5) = 2P(1)$$
$$P(6) = 3P(1)$$

    (a) Find the probabilities of the points of the sample space S.
    (b) Find the probabilities of the following events:
        (1) an even number
        (2) an odd number
        (3) a number less than 1
        (4) a number greater than 0
        (5) a number greater than 1 but less than 5

9. A committee of three is to be selected at random from a group of five students: Alice, Bruce, Carl, Donna, and Ed. Each student is equally likely to serve on the committee.
    (a) List the sample space S.
    (b) If a committee of three is selected at random, find the probabilities of the following events:
        (1) the committee has 3 boys
        (2) the committee has exactly 1 boy
        (3) the committee has exactly 2 girls
        (4) the committee has at most 1 girl
        (5) Bruce is on the committee
        (6) Bruce is not on the committee

10. One hundred people were surveyed to determine the relationship between sex and political affiliation. In this group, there are a total of 60 males of which 40 are Democrats; 25 of the 40 females are also Democrats. Each person in the survey falls into one of the following categories:

c_1 : male and Democrat
c_2 : male and Republican
c_3 : female and Democrat
c_4 : female and Republican

(a) Complete the following table:

Political Affiliation Sex	Democrat	Republican	Total
Male			
Female			
Total			100

(b) Find the probabilities of the points of the sample space $S = \{c_1, c_2, c_3, c_4\}$.

(c) Find the probability that a person selected at random from this group is a male.

(d) Find the probability that a person selected at random from this group is a Republican.

Combination of Events

In the previous section we introduced the basic ideas of probability. Using these as well as the laws of set theory, we shall investigate the fundamental principles of probability.

Definition 5

The Complement of an Event: The complement of an event E, denoted by E', is the event made up of the sample points in the sample space which are not in the event E.

In the die-tossing experiment, $S = \{1, 2, 3, 4, 5, 6\}$. If E is the event that the die gives a number greater than 4, i.e., $E = \{5, 6\}$, then the complement of E is the event that the die gives a number less than or equal to 4, i.e., $E' = \{1, 2, 3, 4\}$. Note that $P(E) = 2/6$ and $P(E') = 4/6$.

Using this definition, we now state the following theorem.

Theorem 1

Complimentary Theorem. If E is an event of a finite sample space S, then $P(E') = 1 - P(E)$.

Proof

Consider the following Venn diagram:

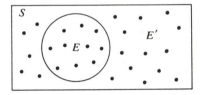

The events E and E' are subsets of the sample space S such that every sample point in S belongs to either E or E', with no points belonging to both. Therefore, the sum of $P(E)$ and $P(E')$ must equal $P(S)$.

$$P(E) + P(E') = P(S)$$

By definition the sum of the probabilities assigned to all the sample points in the sample space S is 1, i.e., $P(S) = 1$. Substituting, we have $P(E) + P(E') = 1$. Therefore,

$$P(E') = 1 - P(E)$$

Example 1

A die is tossed. Find the probability of the event (E_1) that the die gives a number less than 3 and of the event (E_2) that the die gives a number greater than 2.

The probability of E, is:

$$P(E_1) = P(\{1, 2\}) = P(1) + P(2) = 1/6 + 1/6 = 2/6 = 1/3$$

To find the probability of E_2, we can employ the complementary theorem since E_1 and E_2 are complementary events; that is $E_1 = \{1, 2\}$ and $E_2 = \{3, 4, 5, 6\}$. Using Theorem 1, we have

$$P(E_2) = 1 - P(E_1)$$
$$P(E_2) = 1 - 1/3$$
$$P(E_2) = 2/3$$

Theorem 2

The Addition Theorem. If E and F are events in a sample space S, then

$$P(E \cup F) = P(E) + P(F) - P(E \cap F)$$

Proof

Consider the following Venn diagram:

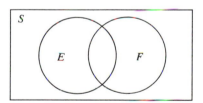

The set $E \cup F$ contains all points of the sample space S which are in set E or in set F or in both. The probability $P(E \cup F)$ is the sum of the probabilities of the sample points in $E \cup F$. Since $P(E) + P(F)$ is the sum of the probabilities assigned to points in set E and set F, respectively, it includes the probabilities of the points in the intersection set, $E \cap F$, twice. Subtracting the probability $P(E \cap F)$ once, we have the addition theorem:

$$P(E \cup F) = P(E) + P(F) - P(E \cap F)$$

Example 2

Suppose we roll a pair of dice, one black and one white. What is the probability that the black die gives a number less than 4 or the white die gives a number less than 3?

If we denote the black die by "b" and the white die by "w", we can represent the outcome of a roll of the dice by the ordered pair (b, w). Since there are 6 possible outcomes for b and 6 for w there are 6×6, or 36, sample points in the sample space of this experiment. The following table is the sample space:

Table 2

		Outcomes of the White Die					
b \ w		1	2	3	4	5	6
Outcomes of the Black Die	1	(1, 1)	(1, 2)	(1, 3)	(1, 4)	(1, 5)	(1, 6)
	2	(2, 1)	(2, 2)	(2, 3)	(2, 4)	(2, 5)	(2, 6)
	3	(3, 1)	(3, 2)	(3, 3)	(3, 4)	(3, 5)	(3, 6)
	4	(4, 1)	(4, 2)	(4, 3)	(4, 4)	(4, 5)	(4, 6)
	5	(5, 1)	(5, 2)	(5, 3)	(5, 4)	(5, 5)	(5, 6)
	6	(6, 1)	(6, 2)	(6, 3)	(6, 4)	(6, 5)	(6, 6)

Since the dice are unbiased, each of the 36 sample points has the same probability. Therefore, the probability of each sample point is $1/36$. Let E be the event that the black die gives a number less than 4 and F be the event that the white die gives a number less than 3. Set E consists of the 18 sample points which are in the first 3 rows of Table 2. Therefore, $P(E) = 18/36$. Set F consists of the 12 sample points which are in the first 2 columns of the table. Therefore, $P(F) = 12/36$. To obtain our solution we cannot just add these two probabilities, because we would be counting the six points, common to both events E and F, twice. These are the points in the set $E \cap F$.

$$E \cap F = \{(1, 1), (1, 2), (2, 1), (2, 2), (3, 1), (3, 2)\}$$

And, therefore, $P(E \cap F) = 6/36$. We must subtract the probability of $E \cap F$ from the sum of the probabilities of E and F in order to obtain the solution. That is,

$$\begin{aligned}
P(E \cup F) &= P(E) + P(F) - P(E \cap F) \\
&= 18/36 + 12/36 - 6/36 \\
&= 24/36 \\
&= 2/3
\end{aligned}$$

Before we state a special case of the addition theorem, we define the following:

Definition 6

Mutually Exclusive Events: If events E and F in a sample space have no sample points in common, i.e., $E \cap F = \emptyset$, they are called mutually exclusive events.

Theorem 3

Addition Theorem for Mutually Exclusive Events. If E and F are mutually exclusive events in a sample space S, then

$$P(E \cup F) = P(E) + P(F)$$

Proof

See problem 9, p. 229.

Example 3

In the two-dice experiment, what is the probability that the sum of the dice is 7 or 11?

Let E be the event that the sum of the dice is 7 and F be the event that the sum of the dice is 11. There are 6 sample points which have a sum of 7, giving $P(E) = 6/36$. There are 2 sample points which have a sum of 11, giving $P(F) = 2/36$. Since E and F are mutually exclusive events, we have

$$P(E \cup F) = P(E) + P(F)$$
$$= 6/36 + 2/36$$
$$= 8/36 = 2/9$$

To conclude, we shall illustrate the concepts of this section by considering the following:

Example 4

In a group of 20 college students, 7 are athletes, 8 are honor students, and 3 are honor athletes. A student is chosen at random from this group. Find the probability that the student selected is

(a) not an honor student

(b) an athlete or an honor student

(c) neither an athlete nor an honor student

(d) an athlete and not an honor student

(e) an athlete or not an honor student

(f) an honor athlete or neither an honor student nor an athlete

Let S be the sample space of the 20 college students, A be the event that an athlete is selected, H be the event that an honor student is selected, and $A \cap H$ be the event that an honor athlete is selected.

Note: Since each student is equally likely to be selected from the group, each sample point has a probability of 1/20. Therefore,

$$P(A) = 7/20, \quad P(H) = 8/20, \quad P(A \cap H) = 3/20$$

We can illustrate the number of sample points in each event with either a Venn diagram or a table.

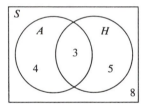

	H	H'	Total
A	3	4	7
A'	5	8	13
Total	8	12	20

(a) $P(\text{not an honor student}) = P(H')$. By the complementary theorem

$$\begin{aligned} P(H') &= 1 - P(H) \\ &= 1 - 8/20 \\ &= 12/20 \\ &= 3/5 \end{aligned}$$

(b) $P(\text{an athlete or an honor student}) = P(A \cup H)$. By the addition theorem

$$P(A \cup H) = P(A) + P(H) - P(A \cap H)$$
$$= 7/20 + 8/20 - 3/20$$
$$= 12/20$$
$$= 3/5$$

(c) P(neither an athlete nor an honor student) $= P[(A \cup H)']$. By the complementary theorem

$$P[(A \cup H)'] = 1 - P(A \cup H)$$
$$= 1 - 3/5$$
$$= 2/5$$

(d) P(an athlete and not an honor student) $= P(A \cap H')$. From the Venn diagram we see that there are 4 sample points in $A \cap H'$. Therefore,

$$P(A \cap H') = 4/20$$
$$= 1/5$$

(e) P(an athlete or not an honor student) $= P(A \cup H')$. By the addition theorem and the solutions to parts (a) and (d)

$$P(A \cup H') = P(A) + P(H') - P(A \cap H')$$
$$= 7/20 + 12/20 - 4/20$$
$$= 15/20$$
$$= 3/4$$

(f) P(an honor athlete or neither an honor student nor an athlete) $= P[(A \cap H) \cup (A \cup H)']$. Since the two events are mutually exclusive, we use the addition theorem for mutually exclusive events and the solution to part (c).

$$P[(A \cap H) \cup (A \cup H)'] = P(A \cap H) + P[(A \cup H)']$$
$$= 3/20 + 2/5$$
$$= 3/20 + 8/20$$
$$= 11/20$$

Exercises 5.2

1. Two coins are tossed; one is a penny and the other a nickel. Find the probability of the following events:

   (a) the penny gives a tail and the nickel a head
   (b) the penny gives a tail or the nickel a head
   (c) both coins do not give a tail
   (d) the penny does not give a tail

2. At the same time two individuals A and B show their right hand to each other with 1, 2, or 3 fingers extended. Each player is equally likely to extend 1, 2, or 3 fingers.

   (a) List a sample space for the game.

   Find the following probabilities:

   (b) both players do not extend 3 fingers
   (c) one player extends 3 fingers and the other extends less than 3 fingers
   (d) one player extends 3 fingers or the other extends less than 3 fingers
   (e) the total number of fingers extended is 3 or 2

3. An ordinary deck of playing cards is shuffled thoroughly. Find the probability that the card selected is

   (a) not an ace of spades
   (b) a face card and a diamond
   (c) a face card or a diamond
   (d) a black ace or a diamond
   (e) a card greater than 6 and less than 9, or a diamond

4. $S = \{a, b, c, d, e\}$ is a sample space of five points with assigned probabilities

$$P(a) = P(b)$$
$$P(c) = P(d) = 2P(a)$$
$$P(e) = 3P(a)$$

   If $A = \{a, c, e\}$ and $B = \{d, e\}$, find the following probabilities:

   (a) $P(A')$
   (b) $P(A \cap B)$
   (c) $P(A \cup B)$
   (d) $P[(A \cap B)']$
   (e) $P[(A \cup B)']$

5. In a class of 50 students, 25 take German, 15 take French, and 10 take both German and French.

   (a) Illustrate the problem with both a Venn diagram and a table.

   Find the probability that a student chosen at random

   (b) does not take German
   (c) takes German or French
   (d) takes neither language
   (e) takes German and not French
   (f) takes German or does not take French
   (g) takes both languages or takes neither language

6. From a group of 40 stockholders, 10 own stock A, 15 own stock B, and 20 own neither stock.

   (a) Illustrate the problem with both a Venn diagram and table.

Find the probability that a stockholder selected at random

(b)   does not own stock A
(c)   owns stock A or stock B
(d)   owns neither stock
(e)   owns stock A and does not own stock B
(f)   owns stock A or does not own stock B

7. In a group of 50 girls, 25 of the girls have brown eyes and the others have blue eyes; 30 girls are brunettes while the rest are blondes; there are 5 blue-eyed blondes.

(a)   Illustrate the problem with both a Venn diagram and a table.

Find the probability that a girl chosen at random for a date by a boy is

(b)   a brown-eyed blonde
(c)   a brown-eyed blonde or a blue-eyed blonde
(d)   brown-eyed or a blonde
(e)   a blue-eyed brunette
(f)   a blue-eyed brunette or a brown-eyed blonde

8. A single card is drawn from an ordinary deck of playing cards. What is the probability that the card is an ace, a king, a queen, or a jack?

9. Prove the addition theorem for mutually exclusive events: If E and F are mutually exclusive events ($E \cap F = \varnothing$) in a sample space S, then $P(E \cup F) = P(E) + P(F)$.

Conditional Probability
and Independent Events

Consider again our experiment with the two dice. If we want to find the probability that the black die gives a 2, we calculate the probability by examining the sample space for the experiment (see Table 2). There are 6 sample points for which the black die is 2, each with a probability of 1/36. Therefore, if E is the event that the black die is 2, we have $P(E) = 6/36 = 1/6$.

Now, suppose that we are given additional information. That is, suppose we want to find the probability that the black die is 2, given the fact that the sum of the dice is less than 4. While this problem appears similar to the first, it is, in fact, an entirely new problem. The sample space no longer consists of 36 sample points, but is instead made up of the 3 sample points whose sum is less than 4. This new sample space is represented by S_1.

$$S_1 = \{(1,1),\ (1,2),\ (2,1)\}$$

We now assign probabilities to the 3 sample points in our new sample space. Since the probabilities of the 3 points were equal in the original sample space S, we assign equal probabilities of $1/3$ to each of the 3 sample in S_1. In general, we assign probabilities to points in the new sample space in direct proportion to their probabilities in the original sample space.

In S_1 there is only one point $(2,1)$ for which the black die is 2. Therefore, if E_1 is the event that the black die is 2, given that the sum of the dice is less than 4, we have $P(E_1) = 1/3$.

We can now see the difference between the two problems. In the first, which had *no conditions* placed on it, the probability that the black die is 2 was $1/6$. However, in the second problem, the probability that the black die gives a 2 had to be computed on the new sample space, which was determined by the added *condition* that the sum of the dice be less than 4. In this case, the probability was $1/3$. This second problem illustrates the concept of conditional probability, which we define as follows.

Definition 7

Conditional Probability: Let E and F be events in a finite sample space S, with $P(F) \neq 0$. The conditional probability of event E occurring, given that event F has occurred, denoted by $P(E|F)$, is

$$P(E|F) = \frac{P(E \cap F)}{P(F)}$$

In other words, the probability of event E occurring, given that event F has occurred, is the ratio of the probability of E and F to the probability of F alone. The events E and F may or may not be mutually exclusive.

We can now apply the formula of Definition 7 to find the probability that the black die yields a 2, given that the sum of the dice is less than 4. Let E be the event that the black die is 2, F the event that the sum of the dice is less than 4, and $E \cap F$ the event that the black die is 2 and the sum is less than 4. Referring to the sample space of the two-dice experiment (Table 2), we see that there are 6 points in E, 3 points in F, and 1 point in $E \cap F$. Therefore, $P(F) = 3/36$ and $P(E \cap F) = 1/36$. Applying the formula for conditional probability, we have

$$P(E|F) = \frac{P(E \cap F)}{P(F)} = \frac{1/36}{3/36} = 1/3$$

This could also be written $P(b = 2 \,|\, b + w < 4) = 1/3$.

Example 1

In the two-dice experiment, what is the probability that the sum of the dice is greater than 8, given that the white die is 6?

Let E be the event that the sum of the dice is greater than 8, F the event that the white die is 6, and $E \cap F$ the event that the sum of the dice is greater than 8 and the white dice is 6. Referring to the sample space of the two-dice experiment (Table 2) we see that there are 10 points in E, 6 points in F, and 4 points in $E \cap F$. Therefore, $P(F) = 6/36$ and $P(E \cap F) = 4/36$. Applying the definition of conditional probability, we have

$$P(E|F) = \frac{P(E \cap F)}{P(F)} = \frac{4/36}{6/36} = 2/3$$

Solving intuitively, we see that the condition that the white die is 6 determines a new sample space $S_1 = \{(1, 6), (2,6), (3, 6), (4, 6), (5, 6), (6, 6)\}$. Each sample point of S_1 is assigned a new probability of $1/6$. There are 4 points in S_1 which have a sum greater than 8; therefore, the probability that the sum of the dice is greater than 8, given the condition that the white die is 6, is $4/6$ or $2/3$. This could also be written $P(b + w > 8 | w = 6) = 2/3$.

We shall now examine what happens to conditional probability when the additional information or condition has no effect on the probability of the event being considered.

Example 2

In the two-dice experiment, what is the probability that the black die gives a number less than 4, given that the white die is less than 3?

Let E be the event that the black die is less than 4, F the event that the white die is less than 3, and $E \cap F$ the event that the black die is less than 4 and the white die is less than 3. Referring to the sample space of the two-dice experiment (Table 2), we see that there are 18 points in E, 12 points in F, and 6 points in $E \cap F$. Therefore, $P(E) = 18/36 = 1/2$, $P(F) = 12/36 = 1/3$, and $P(E \cap F) = 6/36 = 1/6$. Applying the definition of conditional probability, we have

$$P(E|F) = \frac{P(E \cap F)}{P(F)} = \frac{6/36}{12/36} = 1/2$$

Note that $P(E|F) = 1/2 = P(E)$. It appears that the additional information or the condition of event F occurring has no effect on the probability of event E. Logically we see that the outcome of one die will have no effect on the other, just as the toss of one coin will not effect the toss of another. When this occurs, we say that the events are independent of one another.

Definition 8

Independent Events: If E and F are events in sample space S, such that $P(E|F) = P(E)$, then the event E is said to be independent of event F.

Using this definition, we can now prove the following theorem:

Theorem 4

Independent Theorem. If E and F are independent events in a sample space S, then

$$P(E \cap F) = P(E) \cdot P(F)$$

Proof

There are two distinct parts to the proof. One for the condition that $P(F) \neq 0$ and one for $P(F) = 0$.

(1)   Assume $P(F) \neq 0$. By definition of conditional probability we have

$$P(E|F) = \frac{P(E \cap F)}{P(F)}$$

or

$$P(E \cap F) = P(E|F) \cdot P(F)$$

Since events E and F are given as independent, we have, from the definition of independent events, $P(E|F) = P(E)$. Therefore, by substitution we have $P(E \cap F) = P(E) \cdot P(F)$.

(2)   Assume $P(F) = 0$. Since $E \cap F$ is a subset of F, the probability $P(E \cap F)$ cannot be greater than the probability of F. Therefore, since a probability must be nonnegative, we have $P(E \cap F) = 0$.

Therefore, from 1 and 2 we see that for two independent events E and F we have $P(E \cap F) = P(E) \cdot P(F)$.

Note: If $P(E \cap F) \neq P(E) \cdot P(F)$, then the events E and F are not independent. This statement is true since it is the contrapositive of the independent theorem.

Example 3

In the two-dice experiment, what is the probability that the black die gives a number less than 5 and the white die a number greater than 4?

Let E be the event that the black die gives a number less than 5 and F the event that the white die is greater than 4. Referring to the sample space of the two-dice experiment (Table 2), we see that there are 24 points in E and 12 points in F. Therefore, $P(E) = 24/36$ and $P(F) = 12/36$. Since E and F are independent events we have, by Theorem 4,

$$\begin{aligned} P(E \cap F) &= P(E) \cdot P(F) \\ &= 24/36 \cdot 12/36 \\ &= 2/3 \cdot 1/3 \\ &= 2/9 \end{aligned}$$

The result can be verified, using Table 2, by counting the number of points in $E \cap F$. Since there are 8 points in $E \cap F$ and each has a probability of $1/36$, we have $P(E \cap F) = 8/36 = 2/9$.

The independent theorem can be extended to find the probability of any finite number of independent events. We shall state the extended theorem without proving it.

Theorem 5

Extended Independent Theorem. If $E_1, E_2, \ldots, E_n$ are pairwise independent events in a sample space S, then

$$P(E_1 \cap E_2 \cap \cdots \cap E_n) = P(E_1) \cdot P(E_2) \cdots P(E_n)$$

It also holds, by the contrapositive of this theorem, that if

$$P(E_1 \cap E_2 \cap \cdots \cap E_n) \neq P(E_1) \cdot P(E_2) \cdots P(E_n),$$

the events are *not independent*.

Example 4

A perfectly balanced die is tossed three times. Find the probability that the first two tosses are odd and the third toss yields a number less than 3.

If E represents the event that the first toss is odd, F the event that the second toss is odd, and G the event that the third toss yields a number less than 3, then $P(E) = 3/6$, $P(F) = 3/6$, and $P(G) = 2/6$. Since E, F, and G are independent, we have by Theorem 5,

$$
\begin{aligned}
P(E \cap F \cap G) &= P(E) \cdot P(F) \cdot P(G) \\
&= 3/6 \cdot 3/6 \cdot 2/6 \\
&= 1/2 \cdot 1/2 \cdot 1/3 \\
&= 1/12
\end{aligned}
$$

Example 5

In a certain school there are 50 students in the senior class. There are a total of 25 studying mathematics, 15 studying English, and 10 studying history. Of these students there are 10 studying both mathematics and English, 3 studying both mathematics and history, and 4 studying both English and history. Of these students, 2 are studying all three subjects.

(a) Find the probability that a student selected at random is studying just mathematics.

(b) Find the probability that a student selected at random is not studying any one of the three subjects.

(c) Are the events not studying mathematics, not studying English, and not studying history independent?

First, we illustrate the above example with the following Venn diagram:

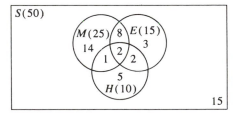

The number of students in each of the 8 disjoint subsets was calculated from the information given in the problem. For example, since there are 2 students taking all three subjects and a total of 10 students taking mathematics and English, there must be 8 students taking just mathematics and English. The other values were computed in a similar manner.

Let M be the event that the student is taking mathematics, E the event that the student is taking English, and H the event that the student is taking history. The probabilities of these events and their complements are

$$P(M) = 25/50 = 1/2, \quad P(E) = 15/50 = 3/10, \quad P(H) = 10/50 = 1/5$$

$P(M') = 25/50 = 1/2, \quad P(E') = 35/50 = 7/10, \quad P(H') = 40/50 = 4/5$

(a) The event that a student is taking just mathematics can be represented by the set $M \cap E' \cap H'$ which contains 14 sample points. Therefore, $P(M \cap E' \cap H') = 14/50 = 7/25$.

(b) The event that a student is taking none of the three courses can be represented by the set $M' \cap E' \cap H'$, which contains 15 sample points. Therefore, $P(M' \cap E' \cap H') = 15/50 = 3/10$.

(c) To determine whether M', E', and H' are independent events, we apply the extended independent theorem:

$$P(M' \cap E' \cap H') = P(M') \cdot P(E') \cdot P(H')$$
$$3/10 = 1/2 \cdot 7/10 \cdot 4/5$$
$$3/10 \neq 7/25$$

Since $P(M' \cap E' \cap H') \neq P(M') \cdot P(E') \cdot P(H')$, the events $M', E',$ and H' are not independent.

Examining events M, E, and H of Example 5, we see that they also are not independent. This is because

$$P(M \cap E \cap H) \neq P(M) \cdot P(E) \cdot P(H)$$
$$2/50 \neq 1/2 \cdot 3/10 \cdot 1/5$$
$$1/25 \neq 3/100$$

This observation leads us to suspect that there is a relationship between the independence of events and the independence of their complements. This relationship is shown in the following theorem.

Theorem 6

If E and F are independent events, then E' and F, E and F', and E' and F' are also independent.

Proof

The following table lists the probabilities for events E, E', F, and F':

Events	F	F'	Totals
E	$P(E \cap F)$	$P(E \cap F')$	$P(E)$
E'	$P(E' \cap F)$	$P(E' \cap F')$	$P(E')$
Totals	$P(F)$	$P(F')$	1

From the table we see that

$$P(E \cap F) + P(E \cap F') = P(E)$$

or

$$P(E \cap F') = P(E) - P(E \cap F)$$

Since E and F are independent, we have

$$P(E \cap F) = P(E) \cdot P(F)$$

Substituting this in the second equation, we have

$$P(E \cap F') = P(E) - P(E) \cdot P(F)$$

$$P(E \cap F') = P(E) \cdot [1 - P(F)]$$

However, since $1 - P(F) = P(F')$, we have

$$P(E \cap F') = P(E) \cdot P(F')$$

This shows that E and F' are independent events when E and F are independent. In a similar manner we could also show that events E' and F, and E' and F', are also independent (see Problem 11).

Exercises 5.3

1. A card is drawn from a shuffled deck of ordinary playing cards and a die is tossed. Find the probability of the following events:

   (a) an ace is selected from the deck and a 5 appears on the top of the die
   (b) an ace is selected from the deck, given that a 5 appears on the top face of the die
   (c) a spade is selected from the deck and the top face of the die is less than 3

    (d)   a spade is selected from the deck, given that the top face of the die is less than 3

2. We toss two coins, a penny and a nickel.

    (a)   List the sample space.

Find the probability that

    (b)   at least one head appears
    (c)   the nickel is a tail
    (d)   the nickel is a tail, given that the penny is a tail
    (e)   both coins are tails
    (f)   both coins are tails, given that both coins are alike

3. We toss two biased coins, one a penny and the other a nickel. We estimate the probability of the penny landing heads to be 0.2 and the nickel landing heads to be 0.6.

    (a)   List the sample space.

Find the probability that

    (b)   at least one head appears
    (c)   the nickel is a tail
    (d)   the nickel is a tail, given the penny is a tail
    (e)   both coins are tails
    (f)   both coins are tails, given that both coins are alike

4. At the same time, two players A and B show their right hand with one, two, three, or four fingers extended. Each player is equally likely to extend one, two, three, or four fingers.

    (a)   List a sample space for this game.

Find the probability that

    (b)   the sum of the fingers extended is 4 or 5
    (c)   the sum of the fingers extended is neither 4 nor 5
    (d)   the sum of the fingers extended is less than 6
    (e)   the sum of the fingers extended is less than 4
    (f)   the sum of the fingers extended is 4, given that the sum of the fingers extended is less than 6
    (g)   the number of fingers extended by player A is less than 3
    (h)   the number of fingers extended by player B is less than 3
    (i)   that only one of the players extended less than 3 fingers

5. In a group of 40 college students enrolled in both mathematics and science, 5 failed mathematics, 10 failed science, and 3 failed both.

    (a)   Illustrate this by means of a table or a Venn diagram.

Find the probability that a student selected at random

    (b)   passed both mathematics and science

(c) passed mathematics, given that he failed science
(d) failed mathematics, given that he passed science

6. Twenty-four mixed colored chips are in a box. The chips are either large or small and are either black or white. There are twice as many large chips as small ones and there are three times as many black chips as white ones. There are exactly four small white chips.

   (a) Illustrate this by either a table or a Venn diagram.
   (b) Find the probability that
        (1) a large chip is chosen
        (2) a small chip is chosen
        (3) a large black chip is chosen, given that a black chip is chosen
        (4) a large black chip is chosen, given that a large chip is chosen
   (c) Are the events black chip and large chip independent in this experiment? Why?

7. In a group of 50 college athletes, 10 are on the basketball team and 35 are on the football team. If independence is assumed between the two teams, estimate the number of students who play both sports.

8. If the probability that Joan will pass mathematics is 8/9 and the probability that Judy will pass mathematics is 9/10, what is the probability that both will pass mathematics?

9. A die is tossed four times. Find the probability that

   (a) all four tosses yield a 3
   (b) the first two tosses are even and the last two tosses are odd
   (c) at least one toss yields a 3
   (d) none of the tosses are even
   (e) exactly one toss is even

10. In a group of 100 students, 50 students are studying algebra, 30 are studying geometry, 20 are studying trigonometry, 10 are studying both algebra and trigonometry, 5 are studying both geometry and trigonometry, 17 are studying both algebra and geometry, and 3 are studying all three subjects.

    (a) Find the probability that a student selected at random is studying only algebra.
    (b) Find the probability that a student selected at random is studying algebra and trigonometry but not geometry.
    (c) Are the events studying algebra, studying geometry, and studying trigonometry independent?
    (d) Find the probability that a student selected at random is studying geometry, given that he is already studying algebra and trigonometry.

11. Complete the proof of Theorem 6: If E and F are independent events, then E' and F, and E' and F', are also independent. (*Hint*: First show that E' and F are independent and then show that E' and F' are independent.)

Expected Value

To conclude this chapter, we shall show an application of probability that enables us to calculate the expected outcome (gain or loss) when we perform experiments that involve chance.

Example 1

Suppose an individual takes out a $10,000.00, one-year-term life insurance policy which has an annual premium of $100.00. The probability that an individual of his age and in his general physical condition will die during the year is estimated to be 0.005. What is the insurance company's expected gain?

This problem consists of two mutually exclusive events. The event E_1 that he lives throughout the year has a probability of 0.995. The event E_2 that he dies during the coming year has a probability of 0.005. If E_1 occurs the insurance company gains $100.00, while if E_2 occurs it will lose $9900.00. This information can be summarized in the following table:

Event (E)	$P(E)$	Gain or Loss (y)	$y \cdot P(E)$
E_1 (he lives)	0.995	+ $100.00	+$99.50
E_2 (he dies)	0.005	− $9900.00	− $49.50

Expected Value $= +$50.00$

In the last column of the table we calculated the individual expected gain ($+$99.50) and the expected loss ($-$49.50). Summing the expected gain and loss, we see that for each policy of this type the insurance company can expect to make $50.00 profit before expenses (salesman's commissions, operation costs, etc.).

Definition 9

Expected Value: Given a sample space S, with mutually exclusive events $E_1, E_2, E_3, \ldots, E_n$ and with associated values $x_1, x_2, \ldots, x_n$, then the expected value of the x's, denoted by $E(X)$, is defined by the following:

$$E(X) = x_1 \cdot P(E_1) + x_2 \cdot P(E_2) + x_3 \cdot P(E_3) + \cdots + x_n \cdot P(E_n)$$

The expected value, $E(X)$, is also called the mean, or average, value.

Example 2

A biased coin is tossed twice. The probability of it coming up heads has been estimated to be 0.7. Find the expected number of heads in the two tosses of the coin.

The sample space and associated probabilities for this experiment are listed in the following table:

Event (E)		$P(E)$	x	Number of Heads	$x \cdot P(E)$
E_1	(H,H)	0.49	x_1	2	0.98
E_2	(H,T)	0.21	x_2	1	0.21
E_3	(T,H)	0.21	x_3	1	0.21
E_4	(T,T)	0.09	x_4	0	0

$$E(x) = 2(0.49) + 1(0.21) + 1(0.21) + 0(0.09)$$
$$= 0.98 + 0.21 + 0.21 + 0$$
$$= 1.40 \text{ heads}$$

This result provides a good illustration of exactly what we mean by the expected value. We know that it is impossible to get 1.40 heads on a single trial of this experiment. The expected value of 1.40 heads indicates that if we perform this experiment a large number of times, the number of heads per trial will average out to be 1.40 heads.

Exercises 5.4

1. A die is rolled once. Find the expected number of dots appearing on the top face of the die.

2. A coin is tossed twice. Find the expected number of heads.

3. An unbalanced die is tossed once. Through repeated tosses the probabilities are estimated to be $P(1) = P(2) = P(4) = P(5)$ and $P(3) = P(6) = 2P(1)$. Find the expected number of dots appearing on the top face of the die.

4. In a certain country the probability of a newborn child being a male is 0.4. Find the expected number of males in families with three children. (Assume the children are born separately and the births are independent of one another.)

5. A gambler pays $2.00 to play a game in which two dice are rolled. He receives $6.00 if a 2 or a 12 appears; $2.00 if a 3 or an 11 appears; and $1.00 if a 4, 5, 9, or 10 appears; otherwise he receives nothing. What is his expected gain or loss?

6. A certain item is mass produced. If it is nondefective, the company makes a profit of $2.00 per item. If when it comes off the assembly line it needs minor adjustments, the profit is reduced to $1.00 per item. If it needs major adjustments the company loses $1.00 per item. Finally, if it cannot be repaired and must be scrapped the company loses $10.00 per item. What is the company's expected gain per item if 90% of the items are nondefective, 5% need minor repairs, 3% need major repairs, and the remainder must be scrapped.

7. How large a premium should an insurance company charge a man for a $1000.00, one-year-term life insurance policy, if it wants to make a $10.00 per policy profit before expenses. The probability that the man will live for another year is 0.990.

Review Test

1. For the finite sample space $S = \{a, b, c, d\}$, which of the following satisfies the conditions for the probabilities of the sample points?

(a) $P(a) = 1/2$ $P(b) = 1/3$ $P(c) = 1/4$ $P(d) = 1/5$
(b) $P(a) = 1/2$ $P(b) = 1/4$ $P(c) = 1/8$ $P(d) = 1/8$
(c) $P(a) = 1/2$ $P(b) = 1/4$ $P(c) = -1/4$ $P(d) = 1/2$
(d) $P(a) = 1/2$ $P(b) = 1$ $P(c) = 0$ $P(d) = -1/2$
(e) none of these

Complete problems 2–4 with respect to the following information:

A committee consists of 8 Democrats, 5 Republicans, 4 Liberals, and 3 Conservatives. A member of the committee is randomly selected to be chairman of the committee.

2. What is the probability that a Conservative is selected?

(a) 1/4   (b) 3/17   (c) 3/20   (d) 17/20   (e) none of these

3. What is the probability that a Liberal or Conservative is selected?

(a) 1/2   (b) 7/20   (c) 13/20   (d) 3/20   (e) none of these

4. What is the probability that neither a Liberal nor Conservative is selected?

   (a) 13/20   (b) 7/20   (c) 3/20   (d) 1/2   (e) none of these

5. The probability that an event will occur is p. The probability that this event will not occur is

   (a) $1/p$ (b) $p - 1$ (c) $1 - p$ (d) p (e) none of these

6. Three horses A, B, and C are in a race. The probability that A will win is 2/5 and the probability that B will win is 1/5. The probability that C will win is

   (a) 1/5   (b) 2/5   (c) 3/5   (d) 4/5   (e) none of these

7. Let A and B be mutually exclusive events with $P(A) = 1/4$ and $P(A \cup B) = 1/3$. Therefore, $P(B)$ is

   (a) 1/12   (b) 1/6   (c) 1/3   (d) 1/4   (e) none of these

8. A fair die is tossed. If the number is odd, what is the probability that it is less than 4?

   (a) 1/3   (b) 2/3   (c) 1/2   (d) 1/4   (e) none of these

9. If A is a subset of B with $P(A) \neq 0$, then $P(B/A)$ is

   (a) 0   (b) 1   (c) 1/2   (d) 1/4   (e) none of these

10. Let A and B be independent events with $P(B) \neq 0$. If $P(A) = 1/2$, then $P(A/B)$ is

   (a) 1/8   (b) 1/4   (c) 1/2   (d) 1   (e) none of these

Complete problems 11–15 with respect to the following information:

If A and B are events with $P(A) = 1/2$, $P(A \cup B) = 3/4$, and $P(B) = 3/8$, then

11. $P(A \cap B)$ is

   (a) 2/3   (b) 1/4   (c) 7/8   (d) 1/8   (e) none of these

12. $P(A' \cap B')$ is

   (a) 2/3   (b) 1/4   (c) 7/8   (d) 1/8   (e) none of these

13. $P(A' \cup B')$ is

   (a) 2/3   (b) 1/4   (c) 7/8   (d) 1/8   (e) none of these

14. $P(B \cap A')$ is

   (a) 2/3   (b) 1/4   (c) 7/8   (d) 1/8   (e) none of these

15. $P(A'/B')$ is

(a) 2/3   (b) 1/4   (c) 7/8   (d) 1/8   (e) none of these

Complete problems 16–19 with respect to the following information:

The probability that the U.S. men's swimming team will take first place in the Olympic games is 1/4 and the probability that the U.S. women's swimming team will take first place in the Olympic games is 1/3.

16. The probability that both teams will take first place in the Olympics is

(a) 1/12   (b) 1/6   (c) 1/3   (d) 1/4   (e) none of these

17. The probability that at least one team will take first place in the Olympics is

(a) 1/12   (b) 1/6   (c) 1/2   (d) 1/4   (e) none of these

18. The probability that neither team will take first place in the Olympics is

(a) 1/12   (b) 1/3   (c) 1/2   (d) 2/3   (e) none of these

19. The probability that only the U.S. women's swimming team will take first place in the Olympics is

(a) 1/12   (b) 3/4   (c) 1/4   (d) 1/2   (e) none of these

20. Find the expected number of heads in three tosses of a biased coin if $P(H) = 2/3$.

(a) 1.75   (b) 1.50   (c) 2.00   (d) 1.85   (e) none of these

6

THE COMPUTER

Since the beginning of time, man has sought to make his work easier and, in the process, has developed many ingenious time-saving devices. One of the latest in man's long list of creations is the high-speed electronic computer. Most of the great scientific and medical achievements of the past few years are due, in a substantial part, to the speed and efficiency of the so-called electronic brain. Computers now touch on almost every facet of our daily lives. For example, they print and process most pay, social security, and other checks. Industry uses them to maintain inventory, control assembly lines, keep personal records, and assist in the testing of new products. They are used to compile weather forecasts, monitor the critical body functions of patients in intensive-care units, guide astronauts through space, and they are now being used by the Internal Revenue Service to audit the millions of yearly tax returns. In fact, the day may fast be approaching when each persons' detailed life history and financial record will be stored on a piece of metalic tape in some computer. It would probably be more appropriate to call our era the Age of the Computer rather than the Atomic Age.

In this chapter we shall give an overview of what a computer is and how it functions. We shall discuss the number systems that computers use in performing their calculations, show how symbolic logic is used in designing the computer circuits which perform these calculations, and finally discuss the planning that must be done by man in order for a computer to operate.

A computer is an exceptionally fast and versatile device for processing data. It can accept the data to be processed and the instructions on how to process it. It can store this information in its memory unit. It can perform the fundamental arithmetic operations of addition, subtraction, multiplication, and division as well as the simple logical operations of distinguishing zero from nonzero and plus from minus. It can store and later use the intermediate results obtained in the solution of a problem. Finally, it can emit the results. Therefore, a computer is a device that will accept data, process it, and give the results.

There are many different types of computers, ranging from a slide rule or desk calculator to the multimillion-dollar, high-speed electronic computer. In our discussion, when we use the term computer, we shall be referring to the high-speed electronic computer.

Computers are generally divided into two broad classifications: analog and digital. An analog computer operates by converting a physical quantity such as speed into a measurable quantity such as miles. An automobile speedometer is an example of a simple analog computer. It transforms the physical quantity speed into the measurable quantity distance. A clock or watch is another example of an analog computer since it converts time into the measurable quantity of the distance between the markings on the face of the clock. A general-purpose analog computer, such as the type found in an industrial complex, is a good deal more sophisticated.

While analog computers are used in industry, the digital computer is used extensively in business, science, and administrative work. For this reason, we shall concern ourselves only with digital computers.

The Digital Computer

A digital computer is a device which operates with numbers rather than with physical quantities. While this makes it less useful for studying physical problems, it does have definite advantages over the analog computer. The accuracy of a digital computer can be increased by simply adding more digits. It can deal with the letters of the alphabet as well as numbers and, therefore, can be used with more types of problems. It can be directed to make basic decisions and, most important of all, it is more easily programmed than the analog computer.

Digital computers come in varied sizes with different capabilities, but they all have the same basic areas of operation: input, control, arithmetic–logic, memory, and output. It is interesting to note that these components are not necessarily distinct physical units and are often interrelated in the design of the computer. However, we shall discuss them as separate entities since they perform different tasks. The areas of operation of these five units can be briefly described as follows:

(a) *Input.* The input unit provides the means of entering data or instructions into the computer's memory. There are several input devices and media for entering information: cards, paper tape, magnetic tape, or manual entry by means of the computer's typewriter.

(b) *Control.* The control unit is the work supervisor. It takes one instruction at a time from the list of instructions stored in the memory unit. It then interprets the instruction and carries it out. It takes the next instruction and continues to follow this procedure until all of the stored instructions are executed. In carrying out the instructions, the control unit directs the operation of the other four units: input, memory, arithmetic–logic, and output. In brief, the control unit coordinates the operation of the components of the computer.

(c) *Arithmetic–Logic.* The arithmetic–logic unit performs the arithmetic operations of addition, subtraction, multiplication, and division as well as the simple logical operations of determining whether or not a number is zero and whether or not a number is positive or negative. It is these two simple logical operations that enable a computer and its programmer to deal with conditional instructions and to make " decisions."

(d) *Memory.* The memory unit is the storage area of the computer. The input data and instructions, as well as the intermediate steps in a problem, can be stored in the memory unit of the computer. This information can be taken from the memory unit and used without being destroyed. That is, even though the data has been used, it can remain in the memory unit for use at a later time.

(e) *Output.* The output unit enables the computer to relay any information stored in the computer to the operator. There are several output devices and media, some of which are similar to the input media, namely, electric typewriter or high-speed printer, cards, paper tape, magnetic tape, and other special devices.

These components of a digital computer differ from one computer to another and depend on the computer's size and use. The relationship of these components is represented in the following diagram:

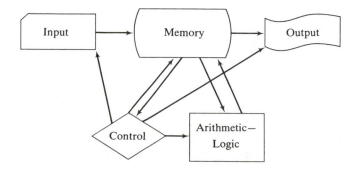

Computer Numeration Systems

In describing a digital computer, we stated that it is a device that works with numbers. Therefore, we shall now examine the various number systems that computers use in performing their operations.

We shall first turn our attention to the basic concepts and rules of our everyday system—the decimal system. This will help us to better understand how the computer number systems function. The decimal system, which is based on powers of 10 (i.e., a base 10 system), is a positional number system with the basic symbols 0, 1, 2, 3, 4, 5, 6, 7, 8, and 9. Before we discuss the representation of numbers in this system, we shall list the powers of 10 to assist us in our future work:

Power of 10	Units	Description
$10^4 = (10)(10)(10)(10)$	10,000	Ten thousand
$10^3 = (10)(10)(10)$	1,000	One thousand
$10^2 = (10)(10)$	100	One hundred
$10^1 = 10$	10	Ten
$10^0 = 1$	1	One
$10^{-1} = 1/10$	1/10	One-tenth
$10^{-2} = 1/(10)(10)$	1/100	One-hundredth
$10^{-3} = 1/(10)(10)(10)$	1/1,000	One-thousandth
$10^{-4} = 1/(10)(10)(10)(10)$	1/10,000	One-ten–thousandth

These are the most frequently used powers of 10 and the list could be extended indefinitely in either direction.

Numbers are now written by combining the basic symbols on a horizontal line as illustrated below. (*Note:* The number of positions to the left and right of the decimal may be extended as far as necessary.)

10^3	10^2	10^1	10^0	.	10^{-1}	10^{-2}	10^{-3}
1000	100	10	1	Decimal point	$\frac{1}{10}$	$\frac{1}{100}$	$\frac{1}{1000}$
Thousands	Hundreds	Tens	Units		Tenths	Hundredths	Thousandths

The symbols that are used and their position on this line determine the numbers that they represent. The decimal point divides the number into two distinct parts. The positions to the left of the decimal indicate the whole units of the number, while the positions to the right represent the fractional part of one unit. When the number to be represented contains no fractional parts of a unit the decimal point may be omitted. In this case the last symbol on the right is taken to be the units place.

Of all the decimal–system number representations, the one that we are most familiar with and interested in is that of dollars and cents. This is a base 10 system with the dollar representing the basic unit. Examining the amount $472.53, we see that we have four 100 dollars plus seven 10 dollars plus two 1 dollars plus five dimes (tenths of a dollar) plus three cents (hundredths of a dollar). The 53 (fifty-three cents) could also be interpreted as being 53 of the 100 parts that make up one dollar.

To conclude our discussion on the decimal representation of numbers, we shall write two numbers in expanded notation.

$$40.35 \text{ (forty and thirty-five hundredths)}$$
$$= 4(10) + 0(1) + 3(1/10) + 5(1/100)$$
$$= 4(10^1) + 0(10^0) + 3(10^{-1}) + 5(10^{-2})$$

$$637.079 \text{ (six hundred thirty-seven and seventy-nine thousandths)}$$
$$= 6(100) + 3(10) + 7(1) + 0(1/10) + 7(1/100) + 9(1/1000)$$
$$= 6(10^2) + 3(10^1) + 7(10^0) + 0(10^{-1}) + 7(10^{-2}) + 9(10^{-3})$$

The base 10 or decimal system is not the only positional numeration system in use today. The digital computer uses the base 2 or binary system. Programmers have also found other numeration systems that can be easily converted to the binary system, namely the base 8 (octal system) and the base 16 (hexadecimal system).

A logical explanation for man's use of the decimal system is that he has ten fingers on which to count. Therefore, it seems reasonable that an electronic digital computer would count best in the base 2 system since the two symbols 0 and 1 can be used to represent an "on" or "off" electric switch. In the base 2 system, numbers are written as in the decimal system except that the positions on the horizontal line now represent powers of 2 instead of powers of 10.

$\cdots$	2^3	2^2	2^1	2^0	$\cdot$	2^{-1}	2^{-2}	2^{-3}	$\cdots$
	8	4	2	0	Binary point	$\frac{1}{2}$	$\frac{1}{4}$	$\frac{1}{8}$	
	Eights	Fours	Twos	Units		Halves	Fourths	Eighths	

Examining the expanded form of the base 2 numbers 101.1_2 and 1110.1001_2, we see that

$$101.1_2 \quad = 1(2^2) + 0(2^1) + 1(2^0) + 1(2^{-1})$$

$$1110.1001_2 = 1(2^3) + 1(2^2) + 1(2^1) + 0(2^0) + 1(2^{-1}) + 0(2^{-2}) + 0(2^{-3}) \\ + 1(2^{-4})$$

Note: When a base other than 10 is used, that base will be written as a subscript immediately following the number.

We shall now show how numbers are represented in the base 8 or octal system. This system has eight symbols: 0, 1, 2, 3, 4, 5, 6, and 7. The positions of octal numbers are shown in the following diagram:

$\cdots$ $\overline{8^3}$	$\overline{8^2}$	$\overline{8^1}$	$\overline{8^0}$	$\bullet$	$\overline{8^{-1}}$	$\overline{8^{-2}}$	$\overline{8^{-3}}$ $\cdots$
512	64	8	1	Octal point	$\frac{1}{8}$	$\frac{1}{64}$	$\frac{1}{512}$
Five-hundred-twelves	Sixty-fours	Eights	Units		Eighths	Sixty-fourths	Five-hundred-twelfths

For example, the expanded forms of the octal numbers 2376_8 and 402.34_8 are

$$2376_8 = 2(8^3) + 3(8^2) + 7(8^1) + 6(8^0)$$

$$402.34_8 = 4(8^2) + 0(8^1) + 2(8^0) + 3(8^{-1}) + 4(8^{-2})$$

Conversion to Base 10.   It is now relatively simple to convert a number written in a base b to base 10. This conversion is accomplished by writing the base b number in expanded form and then carrying out the indicated operations in base 10. To illustrate we shall convert several of the previous examples to base 10.

$$
\begin{aligned}
1110.1001_2 &= 1(2^3) + 1(2^2) + 1(2^1) + 0(2^0) + 1(2^{-1}) + 0(2^{-2}) + 0(2^{-3}) \\
&\quad + 1(2^{-4}) \\
&= 1(8) + 1(4) + 1(2) + 0(1) + 1(1/2) + 0(1/4) + 0(1/8) \\
&\quad + 1(1/16) \\
&= 8 + 4 + 2 + 1/2 + 1/16 = 14 + 9/16 \\
&= 14 + .5625 \\
&= 14.5625
\end{aligned}
$$

$$402.34_8 = 4(8^2) + 0(8^1) + 2(8^0) + 3(8^{-1}) + 4(8^{-2})$$
$$= 4(64) + 0(8) + 2(1) + 3(1/8) + 4(1/64)$$
$$= 256 + 2 + 3/8 + 4/64 = 258 + 7/16$$
$$= 258 + .4375$$
$$= 258.4375$$

Conversion from Base 10 to Base b. To change a number from base 10 to base b we must convert the base 10 number into powers of b. We shall examine two of the several procedures that are used to convert from base 10 to base b. Consider the problem of converting 447 to a base 2 representation. To express 447 in base 2 we must write it as powers of 2. We shall first list the following powers of 2, and then determine how many of these are in our base 10 number:

$$2^0 = 1, \quad 2^1 = 2, \quad 2^2 = 4, \quad 2^3 = 8, \quad 2^4 = 16, \quad 2^5 = 32, \quad 2^6 = 64,$$
$$2^7 = 128, \quad 2^8 = 256, \quad 2^9 = 512$$

The number 2^9, or 512, is the last power of 2 that we list because it exceeds the number that we wish to convert. That is, 512 is greater than 447; therefore, there are no 2^9 in 447.

However there is one 2^8 in 447.	447
Subtracting $1(2^8) = 256$ from 447	$-256 = 1(2^8)$
leaves a remainder of 191; 191 contains one 2^7.	191
Subtracting $1(2^7) = 128$ from 191	$-128 = 1(2^7)$
leaves a remainder of 63; 63 contains no 2^6.	63
Subtracting $1(2^5) = 32$ from 63	$-32 = 1(2^5)$
leaves a remainder of 31; 31 contains one 2^4.	31
Subtracting $1(2^4) = 16$ from 31	$-16 = 1(2^4)$
leaves a remainder of 15; 15 contains one 2^3.	15
Subtracting $1(2^3) = 8$ from 15	$-8 = 1(2^3)$
leaves a remainder of 7; 7 contains one 2^2.	7
Subtracting $1(2^2) = 4$ from 7	$-4 = 1(2^2)$
leaves a remainder of 3; 3 contains one 2^1.	3
Subtracting $1(2^1) = 2$ from 3	$-2 = 1(2^1)$
leaves a remainder of 1.	1
Subtracting $1(2^0) = 1$ from 1	$-1 = 1(2^0)$
leaves a remainder of 0.	0

The number 447 expressed as powers of 2 is

$$447 = 1(2^8) + 1(2^7) + 0(2^6) + 1(2^5) + 1(2^4) + 1(2^3) + 1(2^2) + 1(2^1)$$
$$+ 1(2^0)$$
$$= 110111111_2$$

We shall now express 447 in base 8. The powers of 8 are as follows:

$$8^0 = 1, \quad 8^1 = 8, \quad 8^2 = 64, \quad 8^3 = 512$$

We follow the same procedure as we did in converting from base 10 to base 2.

$$
\begin{array}{r}
447 \\
-384 = 6(8^2) \\
\hline
63 \\
-56 = 7(8^1) \\
\hline
7 \\
-7 = 7(8^0) \\
\hline
0
\end{array}
$$

Therefore,

$$447 = 6(8^2) + 7(8^1) + 7(8^0)$$
$$= 677_8$$

There is another, shorter procedure that can be used to convert base 10 numbers to other bases. To obtain the number expressed in the new base we perform successive division by the desired base, and the remainders which are written after each division are read in reverse order to obtain the desired number. For example, converting 447 to base 2 and to base 8,

$2\overline{)447}$	*Remainders*		$8\overline{)447}$	*Remainders*
$2\overline{)223}$	1		$8\overline{)\ 55}$	7
$2\overline{)111}$	1		$8\overline{)\ \ 6}$	7
$2\overline{)\ 55}$	1		0	6
$2\overline{)\ 27}$	1			
$2\overline{)\ 13}$	1			
$2\overline{)\ \ 6}$	1			
$2\overline{)\ \ 3}$	0			
$2\overline{)\ \ 1}$	1			
0	1			

Reading the remainders from bottom to top we have

$$447 = 110111111_2 = 677_8$$

These procedures have been applied only to converting whole numbers to another base. If the number to be converted has a fractional part, the conversion of the whole and fractional parts must be carried out separately. The whole portion of the number is converted as above and the fractional part is converted by the following procedure.

In writing numbers in positional notation the basimal point is used to separate the number into its whole and fractional parts. The powers of the base to the right of the basimal point are negative and in converting base 10 decimals to other bases, we must express the new number in terms of these negative powers of the new base. *Note:* A *basimal point* is the general term used to describe the point that separates a number into its whole and fractional parts. In the base 10 system it is called a decimal point.

We shall describe the procedure for converting decimal fractions to a different base by means of the following illustrations.

Example 1

Convert .75 to base 2

To convert .75 to base 2 we multiply the decimal fraction (.75) by (2/2) in order to determine how many (1/2) or (2^{-1}) there are in .75; that is,

$$.75 = (.75)(2/2) = 1.5/2 = 1/2 + .5/2$$

This shows that there is one (2^{-1}) in (.75). Next we multiply (.5/2) by (2/2) in order to determine how many (1/4) or (2^{-2}) there are in (.5/2); that is,

$$(.5/2) = (.5/2)(2/2) = 1.0/4 = 1/4 + 0/4$$

This shows that there is one (2^{-2}) in (.5/2). Since the remainder (0/4) is zero we stop. However, if the remaining portion were not zero we would multiply it by (2/2) to determine the number of (1/8) or (2^{-3}). This process is repeated until there is no fractional part or until the fraction repeats.

For the above example we have:

$$.75 = 1(1/2) + 1(1/4) = 1 \ (2^{-1}) + 1(2^{-2}) = .11_2$$

The above process can be written in the following form.

$$.75 = (.75)(2/2) = (1.5)/2 = \underline{1/2} + .5/2$$

$$= \underline{1/2} + (.5/2)(2/2) = \underline{1/2} + 1.0/4 = \underline{1/2} + \underline{1/4} + 0/4$$

$$= 1(2^{-1}) + 1(2^{-2}) = .11_2$$

(*Note:* Multiplying by (2/2) = 1 does not change the value of the number.)

Example 2

Convert .75 to base 8.

In converting a decimal fraction to base 8, we multiply repeatedly by (8/8).

$$.75 = (.75)(8/8) = 6.00/8 = \underline{6/8} + .0/8 = 6(1/8) = 6(8^{-1}) = .6_8$$

Example 3

Convert .1875 to base 2 and to base 8.

$$.1875 = (.1875) \times (2/2) = 0.3750/2 = \underline{0/2} + .375/2 = \underline{0/2} + (.375/2) \times (2/2)$$

$$= \underline{0/2} + 0.75/4 = \underline{0/2} + \underline{0/4} + .75/4 = \underline{0/2} + \underline{0/4} + (.75/4) \times (2/2)$$

$$= \underline{0/2} + \underline{0/4} + 1.5/8 = \underline{0/2} + \underline{0/4} + \underline{1/8} + .5/8 = \underline{0/2} + \underline{0/4} + \underline{1/8}$$

$$+ (.5/8) \times (2/2)$$

$$= \underline{0/2} + \underline{0/4} + \underline{1/8} + \underline{1/16} = \underline{0}(1/2) + \underline{0}(1/4) + \underline{1}(1/8) + \underline{1}(1/16)$$

$$= \underline{0}(2^{-1}) + \underline{0}(2^{-2}) + \underline{1}(2^{-3}) + \underline{1}(2^{-4}) = \underline{.0011_2}$$

$$.1875 = (.1875) \times (8/8) = 1.5000/8 = \underline{1/8} + .5/8 = \underline{1/8} + (.5/8) \times (8/8)$$

$$= \underline{1/8} + \underline{4/64} = \underline{1}(1/8) + \underline{4}(1/64) = \underline{1}(8^{-1}) + \underline{4}(8^{-2}) = \underline{.14_8}$$

Example 4

Convert 231.625 to base 2.

We first convert 231 by means of repeated division:

	Remainder
2)231	
2)115	1
2) 57	1
2) 28	1
2) 14	0
2) 7	0
2) 3	1
2) 1	1
0	1

We have $231 = 11100111_2$.

Next we convert .625 to base 2:

$$.625 = (.625/1) \times (2/2) = 1.250/2 = \underline{1/2} + (.25/2 \times 2/2)$$

$$= \underline{1/2} + .50/4 = \underline{1/2} + \underline{0/4} + (.5/4 \times 2/2) = \underline{1/2} + \underline{0/4} + \underline{1/8}$$

$$= 1(1/2) + 0(1/4) + 1(1/8) = 1(2^{-1}) + 0(2^{-2}) + 1(2^{-3})$$

$$= .101_2$$

Therefore, combining we have $231.625 = 11100111.101_2$.

Conversion from One Base to Another.   Conversion of numbers from one base to another can be accomplished by first translating the number from the first numeration system to the decimal system and then from the decimal system to the desired numeration system. However, the conversion from the binary system to the octal system and from the octal system to the binary system can be performed directly since $2^3 = 8^1$. That is, powers of 8 are also powers of 2. Each group of three binary digits is equivalent to one octal digit.

Example 1

Convert 10110111.1_2 to an octal number.

Group the base 2 number in groups of three beginning at the binary point. When necessary, add one or two zeros to the left of the whole part or to the right of the fractional part.

$$
\begin{array}{llllll}
10110111.1_2 = 0 & 1 & 0 & 1 & 1 & 0 & 1 & 1 & 1 . 1 & 0 & 0 \\
= 0 + 2 + 0 & & 4 + 2 + 0 & & 4 + 2 + 1 . 4 + 0 + 0 \\
= \quad 2 & & \quad 6 & & \quad 7 \quad . \quad 4_8 \\
= 267.4_8
\end{array}
$$

Example 2

Convert 3.14_8 to base 2.

Write the three-digit binary number for each octal number.

$$
\begin{array}{cc}
3 \ . \ 1 & 4_8 \\
= 011 \ . \ 001 & 100 \\
= \ 11 \ . \ 0011_2
\end{array}
$$

Before we proceed to a discussion of the fundamental arithmetic operations in different bases, it is worthwhile to observe how the counting numbers are represented in the bases that we have discussed.

Base 10 (Decimal)	Base 2 (Binary)	Base 8 (Octal)
1	1_2	1_8
2	10_2	2_8
3	11_2	3_8
4	100_2	4_8
5	101_2	5_8
6	110_2	6_8
7	111_2	7_8
8	1000_2	10_8
9	1001_2	11_8
10	1010_2	12_8
11	1011_2	13_8
12	1100_2	14_8
13	1101_2	15_8
14	1110_2	16_8
15	1111_2	17_8
16	10000_2	20_8
17	10001_2	21_8
18	10010_2	22_8
19	10011_2	23_8
20	10100_2	24_8
.	.	.
.	.	.
.	.	.
30	11110_2	36_8
.	.	.
.	.	.
.	.	.
50	110010_2	62_8
.	.	.
.	.	.
.	.	.
100	1100100_2	144_8
.	.	.
.	.	.
.	.	.

Fundamental Operations

The operations of addition, multiplication, subtraction, and division in other bases are performed in a manner similar to the decimal system. The difficulty in performing these operations in systems other than base 10 arises from our lack of familiarity with the addition and multiplication tables for these different bases. To facilitate the computations in the different systems we shall construct the addition and multiplication tables. For the base 2 (binary) system we have the following tables:

+	0	1
0	0	1
1	1	10

×	0	1
0	0	0
1	0	1

These tables are constructed by performing the indicated operations and then writing the results in base 2 notation.

Example 1

Find the sum of 111_2 and 101_2.

$$
\begin{array}{llll}
& \overset{1}{} & & \overset{1}{} & & \overset{1}{} & \\
(a) & \begin{array}{r} 0\,1\,1\,1_2 \\ +\,1\,0\,1_2 \\ \hline 0_2 \end{array} &
(b) & \begin{array}{r} 0\,1\,1\,1_2 \\ +\,1\,0\,1_2 \\ \hline 0\,0_2 \end{array} &
(c) & \begin{array}{r} 0\,1\,1\,1_2 \\ +\,1\,0\,1_2 \\ \hline 1\,0\,0_2 \end{array} &
(d) & \begin{array}{r} 0\,1\,1\,1_2 \\ +\,1\,0\,1_2 \\ \hline 1\,1\,0\,0_2 \end{array}
\end{array}
$$

Example 2

Subtract 1011_2 from 1101_2.

$$
\begin{array}{lll}
(a) & \begin{array}{r} 1\,1\,0\,1_2 \\ -\,1\,0\,1\,1_2 \\ \hline 0_2 \end{array} &
(b) \quad \begin{array}{r} 0\ 10\ \text{Borrow} \\ 1\,1\,0\,1_2 \\ -\,1\,0\,1\,1_2 \\ \hline 1\,0_2 \end{array} \quad
\begin{array}{r} \text{Check:}\ 1\,0\,1\,1_2 \\ +\quad 1\,0_2 \\ \hline 1\,1\,0\,1_2 \end{array}
\end{array}
$$

Example 3

Multiply 1011_2 by 110_2.

$$
\begin{array}{r}
1\ 0\ 1\ 1_2 \\
\times 1\ 1\ 0_2 \\
\hline
0\ 0\ 0\ 0 \\
1\ 0\ 1\ 1 \\
1\ 0\ 1\ 1 \\
\hline
1\ 0\ 0\ 0\ 0\ 1\ 0_2
\end{array}
$$

Example 4

Divide 11110_2 by 101_2.

$$
\begin{array}{r}
1\ 1\ 0_2 \\
1\ 0\ 1_2 \overline{)1\ 1\ 1\ 1\ 0_2} \\
1\ 0\ 1 \\
\hline
1\ 0\ 1 \\
1\ 0\ 1 \\
\hline
0
\end{array}
\qquad
\begin{array}{r}
\text{Check:}\quad 1\ 1\ 0_2 \\
\times 1\ 0\ 1_2 \\
\hline
1\ 1\ 0 \\
0\ 0\ 0 \\
1\ 1\ 0 \\
\hline
1\ 1\ 1\ 1\ 0_2
\end{array}
$$

From these examples we can see the advantages of binary multiplication and division. In each step of multiplication we are multiplying by either 0 or 1. Since 0 times a number equals 0 and 1 times a number equals the number, the multiplication is relatively simple. Multiplication in the binary system actually reduces to an addition problem. With regard to division, the divisor is either contained in the portion of the dividend being tested (quotient of 1) or it is not (quotient of 0). Therefore, division in the binary system reduces to several problems in subtraction.

One major disadvantage of the binary system is the number of digits that are needed to represent large numbers. For example, the number $283_{10} = 100011011_2$ requires only three places in base 10 and requires nine places in base 2.

For the base 8 (octal) system we have the following addition and multiplication tables:

+	0	1	2	3	4	5	6	7
0	0	1	2	3	4	5	6	7
1	1	2	3	4	5	6	7	10
2	2	3	4	5	6	7	10	11
3	3	4	5	6	7	10	11	12
4	4	5	6	7	10	11	12	13
5	5	6	7	10	11	12	13	14
6	6	7	10	11	12	13	14	15
7	7	10	11	12	13	14	15	16

×	0	1	2	3	4	5	6	7
0	0	0	0	0	0	0	0	0
1	0	1	2	3	4	5	6	7
2	0	2	4	6	10	12	14	16
3	0	3	6	11	14	17	22	25
4	0	4	10	14	20	24	30	34
5	0	5	12	17	24	31	36	43
6	0	6	14	22	30	36	44	52
7	0	7	16	25	34	43	52	61

Example 5

Add 43.65_8, 7.32_8, 106.1_8, and 77.63_8.

Carry 1_8

(a)
$$
\begin{array}{r}
4\ 3\ .\ 6\ 5_8 \\
7\ .\ 3\ 2_8 \\
1\ 0\ 6\ .\ 1\ 0_8 \\
7\ 7\ .\ 6\ 3_8 \\
\hline
2_8
\end{array}
$$

Carry 2_8

(b)
$$
\begin{array}{r}
4\ 3\ .\ 6\ 5_8 \\
7\ .\ 3\ 2_8 \\
1\ 0\ 6\ .\ 1\ 0_8 \\
7\ 7\ .\ 6\ 3_8 \\
\hline
.\ 1\ 2_8
\end{array}
$$

Carry 3_8

(c)
$$
\begin{array}{r}
4\ 3\ .\ 6\ 5_8 \\
7\ .\ 3\ 2_8 \\
1\ 0\ 6\ .\ 1\ 0_8 \\
7\ 7\ .\ 1\ 2_8 \\
\hline
1\ .\ 1\ 2_8
\end{array}
$$

Carry 1_8

(d)
$$
\begin{array}{r}
4\ 3\ .\ 6\ 5_8 \\
7\ .\ 3\ 2_8 \\
1\ 0\ 6\ .\ 1\ 0_8 \\
7\ 7\ .\ 6\ 3_8 \\
\hline
6\ 1\ .\ 1\ 2_8
\end{array}
$$

(e)
$$
\begin{array}{r}
4\ 3\ .\ 6\ 5_8 \\
7\ .\ 3\ 2_8 \\
1\ 0\ 6\ .\ 1\ 0_8 \\
7\ 7\ .\ 6\ 3_8 \\
\hline
2\ 6\ 1\ .\ 1\ 2_8
\end{array}
$$

Example 6

Subtract 36.74_8 from 43.25_8.

(a)
2 12 Borrow
$$
\begin{array}{r}
4\ 3\ .\ 2\ 5_8 \\
-3\ 6\ .\ 7\ 4_8 \\
\hline
.\ 3\ \ 1_8
\end{array}
$$

(b)
12 Borrow
3 2 12
$$
\begin{array}{r}
4\ 3\ .\ 2\ 5_8 \\
-3\ 6\ .\ 7\ 4_8 \\
\hline
4\ .\ 3\ 1_8
\end{array}
$$

Example 7

Multiply 27.5_8 by 6.43_8.

$$
\begin{array}{r}
2\,7\,.\,5\,0_8 \\
\times 6\,.\,4\,3_8 \\
\hline
1\;0\;6\;7 \\
1\;3\;6\;4 \\
2\;1\;5\;6 \\
\hline
2\;3\;2\,.\,5\,2\,7_8
\end{array}
$$

Example 8

Divide 2733.5_8 by 53_8.

$\begin{array}{r}42.7_8\\[2pt]53_8\overline{)2733.5_8}\\254\\\hline173\\126\\\hline455\\455\\\hline0\end{array}$	*Trial Divisors* $1 \times 53_8 = 53_8$ $2 \times 53_8 = 126_8$ $3 \times 53_8 = 201_8$ $4 \times 53_8 = 254_8$ $5 \times 53_8 = 327_8$ $6 \times 53_8 = 402_8$ $7 \times 53_8 = 455_8$

To assist us in performing the operation of division in bases other than 10, it is often useful to list the trial divisors. When we divide in base 10, this is not necessary since we are familiar with the multiplication tables and can therefore mentally construct the list of trial divisors.

Exercises 6.1

1. Convert the following to numerals in the base indicated:

    (a)   384 to base 2   (f)   188 to base 8
    (b)   83 to base 8   (g)   13.5 to base 2
    (c)   25 to base 2   (h)   13.5 to base 8
    (d)   396 to base 8   (i)   25.625 to base 2
    (e)   159 to base 2   (j)   25.625 to base 8

2. Convert the following to numerals in the decimal system:

(a) 101011_2 (f) 7.54_8
(b) 327_8 (g) 1101.01_2
(c) 1011_2 (h) 31.2_8
(d) 670_8 (i) 111.1_2
(e) 101.11_2 (j) 43.6_8

3. Convert the following to the octal system:

(a) 10111.11_2 (c) 110.1_2
(b) 101.01_2 (d) 10101.0111_2

4. Convert the following to the binary system:

(a) 241.3_8 (c) 26.17_8
(b) 2.54_8 (d) 76.52_8

5. Perform the following operations and give the results in the given base unless otherwise stated:

(a) $11100.01_2 + 1111.11_2 + 1010.1_2$
(b) $374.6_8 + 72.56_8 + 721.4_8$
(c) $1110.1_2 + 25.6_8 + 23$; give the result in the decimal system
(d) $100001_2 - 10000_2$
(e) $1001.1_2 - 100_2$
(f) $367.2_8 - 64.35_8$
(g) $30.4_8 - 110.1_2$; give the result in the octal system
(h) $10.1_2 \times 11.011_2$
(i) $1101.101_2 \times 1.001_2$
(j) $60.72_8 \times 30.6_8$
(k) $23.5_8 \times 101.1_2$; give the result in the binary system
(l) $110001_2 \div 111_2$
(m) $110111_2 \div 1011_2$
(n) $54_8 \div 13_8$
(o) $100.6_8 \div 7_8$
(p) $100100_2 \div 11_8$; give the result in the binary system

6. The digits for base 16, or the hexadecimal system, are given as follows:

$$1, 2, 3, 4, 5, 6, 7, 8, 9, a, b, c, d, e, \text{ and } f \quad \text{where} \quad 10 = a_{16},$$
$$11 = b_{16}, \quad 12 = c_{16}, \quad 13 = d_{16}, \quad 14 = e_{16}, \quad \text{and} \quad 15 = f_{16}$$

Construct addition and multiplication tables.

7. Convert the following to numerals in the hexadecimal system:

(a) 176   (c) 12.5
(b) 255   (d) 27.25

8. Convert the following to numerals in the decimal system:

(a) $3db_{16}$ (c) $12e_{16}$
(b) def_{16} (d) $af.8_{16}$

9. Perform the following operations:

(a) Change $3e_{16}$ to the binary system. (*Hint*: $2^4 = 16^1$.)
(b) Change 1011001_2 to the hexadecimal system.
(c) $1f5_{16} + e0b_{16}$
(d) $1db_{16} + 39a_{16}$
(e) $91d_{16} - 1cd_{16}$
(f) $fde_{16} - ce_{16}$
(g) $2d_{16} \times 12_{16}$
(h) $3f1_{16} \times 3a_{16}$
(i) $20_{16} \div 8_{16}$
(j) $30_{16} \div 6_{16}$

Calculations by Computer*

In the previous section we examined the various number systems that are used in computers. In this section we shall show how the computer uses one of these systems, the binary number system, in performing the basic arithmetic operation of addition. Our discussion will be limited to the operations of addition and subtraction, since an investigation of how the computer performs all of its operations is beyond the scope of this book.

To begin with, it is important to remember that computers cannot think. The only way they can perform operations such as addition is by being specially wired to do so. We shall now show how the principles of symbolic logic and switching circuits are used in wiring a computer to perform the operation of addition.

Let us recall that there are two basic switching circuits, the series circuit and the parallel circuit, which can be represented by the logical connectives of "and" and "or," respectively. In both circuits, A and B are simple

Series Circuit   Parallel Circuit

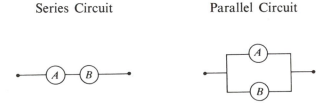

---

*Before studying this section, a student should be familiar with the binary number system and the theory of electrical switching circuits.

switches that can be turned *on* or *off*. We represent a switch being " on "
by either the number 1 or the logical value " *true.* " " *Off* " is represented
by the number 0 or the value " *false.* "

A computer is designed to operate using a base 2, or binary system, for
the simple reason that the binary number system has only two digits,
0 and 1, which can be represented inside the machine by " off " and " on "
switches. When we enter a number into the machine, it is immediately
converted into base 2. If we enter a 5, it is converted by the machine to
101_2. If we enter a 9, it is converted to 1001_2. In other words we, as the
persons using the machine, still use the base 10 number system. The
machine converts the numbers into binary numbers as we enter them and
after it performs its calculations, the binary numbers are converted back
into base 10 numbers before they are reported to us. We never actually
see the base 2 numbers. The computer, on the other hand, is primarily
interested in base 2 numbers since these are the numbers with which it
can perform its calculations. Let us now investigate how a computer is
wired to do addition.

The Half Adder

The simplest addition problem involves the adding of two single-digit
binary numbers A and B. There are only four possible combinations of
this problem:

$A = 0$	$A = 0$	$A = 1$	$A = 1$
$B = 0$	$B = 1$	$B = 0$	$B = 1$
Sum $= 0$	Sum $= 1$	Sum $= 1$	Sum $= 0$
Carry $= 0$	Carry $= 0$	Carry $= 0$	Carry $= 1$

Since the numbers are in base 2, we have only these four possibilities.
Note that we actually get two answers, the sum and the carry, each is a
single digit, and each is either a 0 or a 1. We can represent the entire
problem and answer by the familiar truth table.

A	B	Sum	Carry
0	0	0	0
0	1	1	0
1	0	1	0
1	1	0	1

In the section on symbolic logic we saw that it is possible to write an expression or compound statement for some given set of truth values. This is not always easy to do, especially when the truth tables are long. Examining the above truth table, we see that the logical expression for the sum is $A \leftrightarrow \sim B$ and for the carry is $A \wedge B$. That is,

				Sum		Carry	
A		B		$A \leftrightarrow \sim B$		$A \wedge B$	
0	F	0	F	0	F	0	F
0	F	1	T	1	T	0	F
1	T	0	F	1	T	0	F
1	T	1	T	0	F	1	T

Having written the correct expressions for the sum and carry digits, we can draw the circuits which correspond to these expressions. The carry digit is represented by the simple series circuit.

It is a more difficult task to represent the sum digit, since its logical expression is more complicated. We cannot build a circuit until all conditional and biconditional symbols have been converted to conjunctions and/or disjunctions. We do this by applying the biconditional equivalence. That is, $A \leftrightarrow \sim B$ is equivalent to the expression $(A \vee B) \wedge (\sim A \vee \sim B)$. This is true since

$$(A \leftrightarrow \sim B) \leftrightarrow [(A \to \sim B) \wedge (\sim B \to A)]$$
$$\leftrightarrow [(\sim A \vee \sim B) \wedge (B \vee A)] \leftrightarrow [(A \vee B) \wedge (\sim A \vee \sim B)]$$

Therefore, the circuit for $A \leftrightarrow \sim B$ is

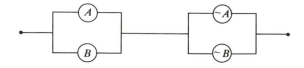

Now let us take these two circuits and place them in a box. The two ends marked "A" and "B" represent a means of turning on the switches A and B.

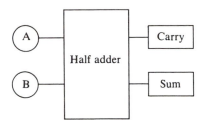

Let us now examine how the computer would add 1 plus 1. That is, suppose that $A = 1$ and $B = 1$. The computer would send a signal into the box at terminal "A" and at terminal "B." Since A and B both equal 1, this signal would turn on all switches marked "A" and all switches marked "B". It would also turn off all switches marked "$\sim A$" and all switches marked "$\sim B$". The answer to the problem would then appear at the sum and carry terminals. For this case the sum would be 0 (no electrical current at the sum terminal) and carry would be 1 (electrical current at the carry terminal).

If we now change the problem to $A = 0$ and $B = 1$, the computer would signal only the "B" terminal to turn on all the switches marked "B" and turn off all of those marked "$\sim B$". The switches marked "A" would remain off, the switches marked "$\sim A$" would remain on. The answer would come out sum = 1 and carry = 0.

The solution of the problem thus becomes a very simple one. The computer has merely to look at the numbers it wants to add and send the signals to the correct terminals. It then sends out the answer from the output terminals.

The Full Adder

We could not expect much from a computer if it could only add single-digit numbers. Let us now see how we can wire the machine to add numbers several digits long.

To add two-, three-, or four-digit numbers, we can follow the procedure that we follow in ordinary addition. If we were adding the numbers 011_2 and 001_2, our work would appear as follows:

(a) Add the last two digits.

$$
\begin{array}{ccc}
0 & 1 & 1 \\
0 & 0 & 1 \\
\hline
 & & 0 \\
 & & \text{Carry } 1
\end{array}
$$

(b) Now add the next pair of digits, and then add the carry digit held over from the last addition.

$$
\begin{array}{cccl}
 & 1 & & \text{Previous carry} \\
0 & 1 & 1 & \\
0 & 0 & 1 & \\
\hline
 & 0 & 0 & \\
\end{array}
$$

New carry $= 1$

(c) Repeat step (b) using the new carry.

$$
\begin{array}{cccl}
1 & & & \text{Previous carry} \\
0 & 1 & 1 & \\
0 & 0 & 1 & \\
\hline
1 & 0 & 0 & \\
\end{array}
$$

New carry $= 0$

We have now completed the addition. We stop here because we are dealing strictly with three-digit numbers. If there were more digits, we would continue until we had added the last set of digits.

It is clear that to completely solve an addition problem, we need more than half adders. We need a circuit that is capable of adding three digits together. It must add the two given digits and the carry from the previous addition. A half adder cannot do this since it is only designed to add two digits at a time.

To build a full adder, a device which considers the previous carry in its addition, we begin as we did with the half adder. We write down all of the possible combinations of the addition of 3 one-digit binary numbers. There are eight possibilities.

(a) Previous carry $= 0$ (b) Previous carry $= 0$

$$
\begin{array}{r}
A = 0 \\
+ B = 0 \\
\hline
\text{Sum} = 0 \\
\text{New carry} = 0
\end{array}
\qquad
\begin{array}{r}
+ A = 0 \\
+ B = 1 \\
\hline
\text{Sum} = 1 \\
\text{New carry} = 0
\end{array}
$$

(c) Previous carry $= 0$

$+ A = 1$

$+ B = 0$

$\overline{}$

Sum $= 1$

New carry $= 0$

(d) Previous carry $= 0$

$+ A = 1$

$+ B = 1$

$\overline{}$

Sum $= 0$

New carry $= 1$

(e) Previous carry $= 1$

$A = 0$

$+ B = 0$

$\overline{}$

Sum $= 1$

New carry $= 0$

(f) Previous carry $= 1$

$+ A = 0$

$+ B = 1$

$\overline{}$

Sum $= 0$

New carry $= 1$

(g) Previous carry $= 1$

$+ A = 1$

$+ B = 0$

$\overline{}$

Sum $= 0$

New carry $= 1$

(h) Previous carry $= 1$

$+ A = 1$

$+ B = 1$

$\overline{}$

Sum $= 1$

New carry $= 1$

For each of the eight possibilities we get two answers, the sum and the new carry, each of which is either 0 or 1. We can represent the eight addition problems by the following truth table:

P	A	B	Sum	New Carry
0	0	0	0	0
0	0	1	1	0
0	1	0	1	0
0	1	1	0	1
1	0	0	1	0
1	0	1	0	1
1	1	0	0	1
1	1	1	1	1

Note: The letter P represents the carry from the previous addition.

We now write the logical expressions that represent the sum and new carry. As might be expected, the expressions are rather complicated since there are three component statements, P, A, and B. The statements and circuits for the sum and new carry are as follows:

Sum

$$(\sim P \wedge \sim A \wedge B) \vee (\sim P \wedge A \wedge \sim B) \vee (P \wedge \sim A \wedge \sim B) \vee (P \wedge A \wedge B)$$

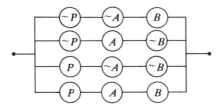

Carry

$$(P \wedge A) \vee (P \wedge B) \vee (A \wedge B)$$

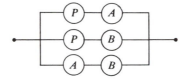

Once again, we imagine these circuits placed in a box.

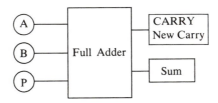

Now that we have seen how a half adder and a full adder are built, the problem of adding two numbers of any size is solved by constructing a circuit consisting of one half adder and a series of full adders wired as follows:

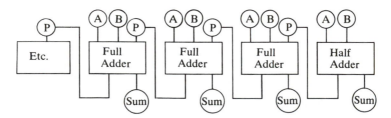

For example, to add the two 4-digit numbers 1101_2 and 0101_2 the network would appear as follows:

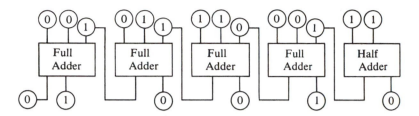

In other words, we feed the digits of the first number into the B terminals and the second number into the A terminals. The answer will come out of the output terminals. In this case, the correct answer is 10010_2, a five-digit number. Each digit comes out of one of the adders in the correct order.

It is also possible to wire a computer to subtract, multiply, and divide. The same principles are involved, the only difference being the level of difficulty. The circuits for multiplication and division are exceedingly complicated and will not be considered here. We shall instead turn to the problem of programming a computer to solve more complex problems after the circuits have been built.

Exercises 6.2

1. Design a half-subtractor. You are given two digits, A and B, and you want to compute $D = A - B$. The answer will be two digits long (D is the difference and P is the borrow instead of the carry). There are four cases.

$$
\begin{array}{cccc}
A = 0 & A = 1 & A = 0 & A = 1 \\
-B = 0 & -B = 0 & -B = 1 & -B = 1 \\
\hline
D = 0 & D = 1 & D = 1 & D = 0 \\
P = 0 & P = 0 & P = 1 & P = 0
\end{array}
$$

2. A young computer designer, frustrated in his attempts to design a full adder, suddenly comes to the following realization. "I don't have to design a new circuit after all. If I have to add three digits, I can add the first two and then add the third. I'll bet I can build the circuit with two half adders (which I know how to build)." Actually, the designer has forgotten about the carry. His problem is not as simple as it sounds. He is quite right, however, in saying that a full adder can be built using ready-built half adders. See if you can do this. How many half adders do you need? Why isn't this method ordinarily used?

Computer Programming

In a general sense, programming is using a computer to solve a problem. Since the computer cannot think for itself or make unplanned decisions, the programmer must direct it in solving the problem. That is, the programmer decides how to solve the problem and uses the computer to perform the time-consuming calculations. Three basic steps are taken in programming a computer to solve a problem. First, the programmer carefully analyzes the problem and develops a detailed plan of how to solve it. That is, he must list *all* the steps that the computer must perform in order to solve the problem. A computer will do exactly what it is designed and programmed to do—nothing more or nothing less. Second, these steps are coded or written in a language that the machine can "understand" (accept). There are many different computer languages and most of the larger computers will accept more than one language. Three of the more popular computer languages are FORTRAN, COBOL, and PL/I. Finally, these coded instructions and data are put on an input medium of the computer and processed.

In our discussion of computer programming we shall concentrate our attention on the first step, namely analyzing the problem and developing a detailed plan of attack for its solution. With regard to step two, coding the instructions, we shall only briefly mention the basic operational instructions. A practical discussion of the use of any one of the computer languages would require a course in itself. Since there are many types of digital computers, each with its own distinct control system, we shall not discuss the operation of a computer.

Communicating with the Computer

When we speak of communicating with a computer, we actually mean communicating with the control unit, since it is this unit that supervises the work of the input, output, arithmetic–logic, and memory units. We communicate with the control unit by means of a program.

A *program* is a series of coded instructions, written in a computer language that the machine can accept. These instructions direct the control unit as to what steps to take in order to solve the problem. Every step needed to solve the problem must be specified in the program. There are five basic types of instruction that can be given to the control unit:

(a) *Input Instructions.* These instructions direct the control unit to activate an input device such as a card reader. When we want the machine to receive data from the typewriter, we instruct it to activate the typewriter and we then type in the data.

(b) *Output Instructions.* These instructions are similar to input instructions except that they direct the control unit to print out certain data through various devices such as a typewriter, a printer, or a card punch.

(c) *Arithmetic Instructions.* These instructions cause the arithmetic–logic unit to perform the basic operations of addition, subtraction, multiplication, and division.

(d) *Logical Instructions.* These instructions cause the arithmetic–logic unit to make decisions.

(e) *Store Instructions.* Whenever we arrive at some imtermediate results in a problem which we may need later on, we must instruct the control unit to store this result in the memory unit of the machine. The computer does not automatically remember its results.

Flow Charts

To assist in developing their plan of attack, programmers use a convenient device called a flow chart. A *flow chart* is a graphic or pictorial representation of the steps that are necessary in order to solve a problem. It shows precisely what is to be done and the order in which it is to be done.

In constructing a flow chart, various symbols such as rectangles, squares, and circles are used to represent the operations that are to be performed. The following is a list of basic symbols and the operations they represent:

○   Start–Stop   ◇   Decision

▱   Input–Output   ⟶| Flow

▭   Process($+$, $\times$, $-$, $\div$)–Store

These symbols may vary from one programmer to another. One may use a circle to represent START or STOP while another may use it to represent a decision. However, it is important that each programmer be consistent in his use of symbols. This will help to avoid unnecessary errors and assist in the detection of flow charting errors.

We shall conclude our discussion of computers by illustrating how we construct a flow chart for the solution of a problem.

Example 1

The XYZ Company plans to bill
its customers by mail and wants
to use a computer to calculate
and print the bill. There is a 5%
sales tax on any item that costs
$1.00 or more. Construct a flow
chart for the program that will
print the customer's name and
address as well as the cost of
the item, the tax, and the total
cost.

Enter a customer's name,
address, and the cost of the
item into the memory unit
of the computer.

The cost of the item is
compared to $1.00. If it
is less than $1.00, there is
no tax. If it is $1.00 or
more, there is a tax equal
to 5% of the cost of the
item.

The tax is added to the cost
in order to obtain the total
cost.

The bill is printed.

If this is the last customer, the
computer stops. If not, the
computer reads the information
on the next customer and then
repeats the process.

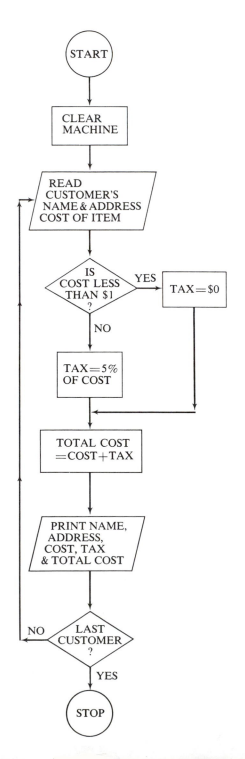

Example 2

Construct the flow chart for the program that will print out the whole numbers from 1 to 1000.

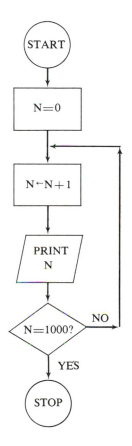

The number is initially set equal to zero.

The number is increased by one. The number in the memory unit is replaced by a new number, which is equal to the old number plus one.

The number is printed.

If the number printed is 1000, the computer stops. If not, the computer is instructed to increase the number by one and repeat the process.

Example 3

There are 100 men employed by the XYZ Company. The manager wishes to calculate the average salary. Construct a flow chart for the solution of this problem. (*Note*: The average salary is computed by dividing the sum of the 100 salaries by the number of employees.)

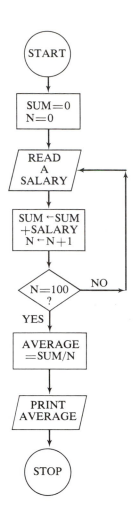

The sum and number of salaries is initially set equal to zero.

After a salary is read it is added to the sum already in the computer. In other words, the sum is replaced by the new sum, which is found by adding the salary just read to the previous sum. The number of salaries read is increased by one.

If the salary that has just been read is the hundredth salary, the average is computed. If not, the next salary is read and the process repeated.

Example 4

At the end of each week, the name, the number of hours worked, and the rate per hour for each employee of the XYZ Company are recorded. The company wants to use its computer to calculate the salary and print the pay checks for each of its employees. Construct a flow chart for the solution of this problem based on the following conditions. Each employee

(a)   receives time and one-half for all work over 40 hours per week

(b)   pays 10% federal tax if his total weekly pay is less than $100.00, 13% if $100.00 or more but less than $200.00, and 16% if more than $200.00

(c)   pays 1% of his salary for union dues

(d)   pays $2.00 per week for health insurance

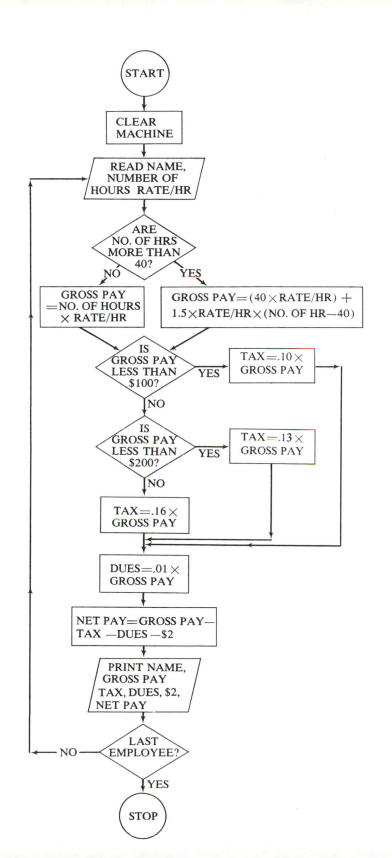

Exercises 6.3

1. In Example 1, suppose there is a delivery charge which is 1% of the cost of the item. Alter the flow chart to calculate and include this charge in the bill.

2. In Example 4, the XYZ Company, wishes to calculate the average net pay of its employees. Make the necessary adjustments in the flow chart so that the average salary will be computed and printed out.

3. The State Park Commission wants to use a computer to calculate and print the pay checks for the part-time summer employees. The information on the check is to include the employee's name, number of hours worked, rate per hour, gross pay, tax, insurance, and net pay. Construct the flow chart for the solution of this problem based on the following conditions. Each employee

    (a) earns $2.50 per hour
    (b) pays 5% federal tax if he earns $50.00 or less, and 8% if more than $50.00
    (c) pays $1.00 insurance

4. Construct the flow chart for finding the area and length of the diagonal for 150 rectangles given the height and width of each rectangle. Print out the height, width, area, and length of the diagonal for each rectangle.

5. Construct a flow chart for the program that will compute and print out the squares and square roots of the whole numbers from 1 to 100. (*Note*: Computers can calculate square roots.)

6. Construct a flow chart for the program that will print out the multiplication tables for the whole numbers from 1 to 12.

7. There are 200 students enrolled in a special ecological summer program at Local College. All students take the same three courses for which they will receive grades based on a 4-point marking system. The courses, number of credits, and grading system are as follows:

Course	Credits	Grade	Quality Points
Environmental Science	4	A	4
Society and Pollution	3	B	3
Health Problems	2	C	2
Total	9	D	1
		F	0

Construct a flow chart for the program that will print the name, courses grades, and grade-point average for each student.

Review Test

1. Which unit of the computer directs the operation of the other units?

   (a) input   (b) logic   (c) control   (d) memory
   (e) none of these

2. Which unit of the computer provides the means of entering data or instructions into the computer's memory?

   (a) arithmetic   (b) control   (c) logic   (d) output
   (e) none of these

3. Which of the following diagrams shows the relationship between the memory, control, and arithmetic–logic units of the computer?

(a)
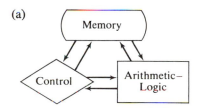

(b)
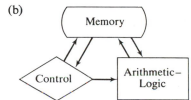

(c)
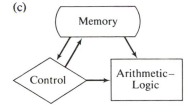

(d)
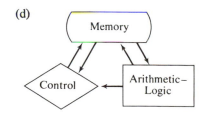

   (e) none of these

4. Which of the following diagrams shows the functional relationship between the input, memory, and output units of the computer?

(a)
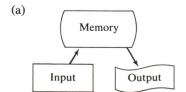

(b)
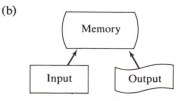

(c) 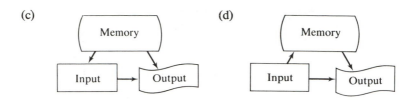 (d)

(e)   none of these

5. $10^{-4} =$

(a)   .001   (b)   .0004   (c)   .0001   (d)   .4
(e)   none of these

6. Which positional number system uses the symbols 0, 1, 2, 3, 4, 5, 6, 7?

(a)   binary   (b)   decimal   (c)   hexadecimal   (d)   septal
(e)   none of these

7. The base 10 representation of 101.01_2 is

(a)   2.1   (b)   5.25   (c)   5.1   (d)   5.01   (e)   none of these

8. The base 10 representation of 127.4_8 is

(a)   10.8   (b)   12.74   (c)   78.5   (d)   87.5   (e)   none of these

9. The base 2 representation of 27.25 is

(a) 11011.1_2 (b) 11011_2 (c) 1011.5_2 (d) 11011.01_2
(e)   none of these

10. The base 8 representation of 431 is

(a) 657_8 (b) 756_8 (c) 675_8 (d) 567_8 (e)   none of these

11. The base 8 representation of 101.011_2 is

(a) 5.4_8 (b) 5.3_8 (c) 3.5_8 (d) 53_8 (e)   none of these

12. The sum of $1011_2 + 1101_2 + 111_2$ is

(a) 11101_2 (b) 11110_2 (c) 10111_2 (d) 1111_2
(e)   none of these

13. The sum of $427_8 + 367_8 + 155_8 + 67_8$ is

(a) 1016_8 (b) 1162_8 (c) 1062_8 (d) 1262_8
(e)   none of these

14. The difference of 2735_8 minus 1473_8 is

   (a) 1262_8 (b) 1242_8 (c) 1062_8 (d) 1042_8
   (e) none of these

15. The product of 1011_2 times 101_2 is

   (a) 110111_2 (b) 11011_2 (c) 10111_2 (d) 10000_2
   (e) none of these

16. The product of 627_8 times 53_8 is

   (a) 42135_8 (b) 41235_8 (c) 32135_8 (d) 51235_8
   (e) none of these

17. The quotient of 100101011_2 divided by 1101_2 is

   (a) 10111_2 (b) 11011_2 (c) 11101_2 (d) 1011_2
   (e) none of these

18. The circuit representing a series arrangement of switches *A*, *B*, and *C* is

(a)
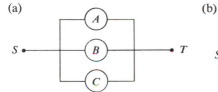
(b)
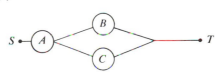

(c)

(d)
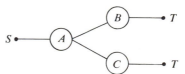

   (e) none of these

19. Which binary numbers is the half adder capable of adding?

   (a) one-digit   (b) 2 one-digit   (c) two-digit   (d) two
   (e) none of these

20. Which binary numbers is the full adder capable of adding?

   (a) 2 one-digit   (b) 3 one-digit   (c) three-digit   (d) two
   (e) none of these

21. The circuit

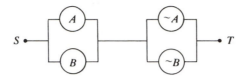

for the sum digit of the half adder is represented by the logical statement

(a) $(A \wedge B) \vee (\sim A \wedge \sim B)$ (b) $(A \vee B) \wedge (\sim A \vee \sim B)$
(c) $(A \rightarrow B) \wedge (\sim A \rightarrow \sim B)$ (d) $(A \wedge \sim A) \vee (B \wedge \sim B)$
(e) none of these

22. The logical statement $(P \wedge A) \vee (P \wedge B) \vee (A \wedge B)$ for the carry digit of the full adder represents which circuit?

(a)   (b)

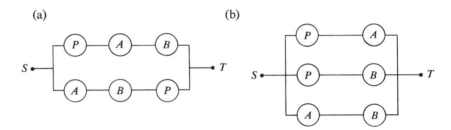

(c)   (d)

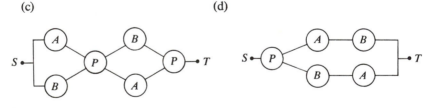

(e) none of these

23. The pictorial representation of the steps that are necessary to solve a problem is a

(a) program   (b) control   (c) instruction   (d) flow chart
(e) none of these

24. The series of coded instructions, written in a form that the computer can accept, is a

(a) flow chart   (b) binary system   (c) program   (d) list
(e) none of these

25. The flow charting symbol for a decision is

(a)   (b)

(c)   (d)

(e)   none of these

26. Which instruction is represented by the flow charting symbol?

(a)   process   (b)   decision   (c)   read   (d)   store
(e)   none of these

27. What does the following flow charting symbol mean?

$N \leftarrow N+1$

(a)   store the number $N+1$ and add N
(b)   store the number N and add $N+1$
(c)   replace the number N by the number $N+1$
(d)   add $N+1$ to N
(e)   none of these

28. What does the following flow charting symbol mean?

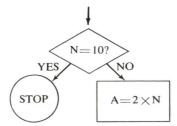

(a)   if N = 10, stop; if N ≠ 10, repeat the process
(b)   if N = 10, multiply N by 2; if N ≠ 10, stop
(c)   set N = 10; either stop or multiply N by 2
(d)   if N = 10, stop; multiply N by 2
(e)   none of these

29. The flow chart for the program to compute the sum of the salaries of the
    100 employees of the XYZ Company is

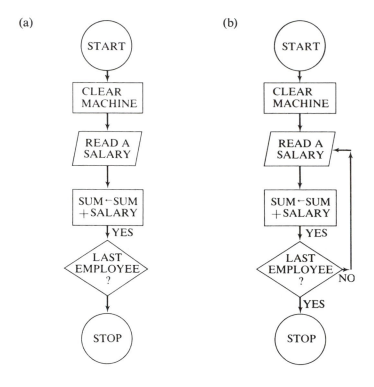

(c)   (d)

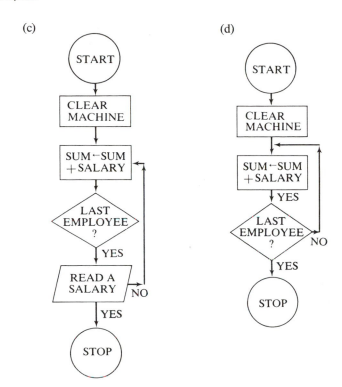

(e)   none of these

30. The accompanying flow chart is for the program that reads a list of names, as well as the age and sex of each person, and prints a list of (a)   19-year-olds   (b)   19-year-old females   (c)   19-year-old males   (d)   males   (e)   none of these

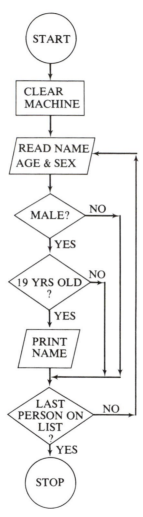

7

NUMBER SYSTEMS

If asked the question "What is a number?" a person would probably respond, "You know, a number is something like 2, -3, or 1/2." If we examine this answer, we would agree that these examples are indeed numbers but that listing examples does not constitute a definition of "number." This idea of number, which we understand intuitively, presents a problem when we try to give a satisfactory definition.

Historically the concept of number begins with man, for he alone among the animals seems to have a "number sense." His first ideas of the property of numbers were probably those of "more than," "same as," and "less than," which were used in comparing the number of warriors in his tribe to those of his enemy or in some other similar comparison. Eventually this concept was refined to a point where he could compare or tally the amount of objects in two groups. He realized that if he put a feather in his headdress for each enemy slain, the collection of feathers and enemies slain had a common property. This sameness or the comparison of the two sets was the beginning of the number concept as we know it today.

Words and symbols were developed to represent this property which enabled man to indicate specific degrees of sameness or difference between sets. The words that were first used to stand for different numbers usually related to objects. For example, the word for five in many languages can be traced to the word for hand. Based on these ideas we now give the following definition of number.

Definition 1

Number: A number is an abstract property of a set which indicates quantity or position.

From this definition we see that a number is a mental abstraction; that is, it does not exist in a physical sense. It is a property that man attributes to sets. We must be careful not to confuse the concept of number with the symbols that are used to represent it. These symbols, which we call numerals, are only the physical representation of the idea.

Definition 2

Numeral: A numeral is a symbol used to represent a number.

If we examine the set of players on a basketball team we see that this set has a number property. We say that there are five players and we use the numeral 5 to represent this number. The number five does not exist by itself; it is a property of the basketball team. When, as children, we started to learn numbers, we studied the following type of illustrations:

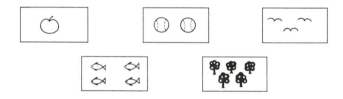

We examined sets of objects and gave number properties to the sets; one apple, two balls, three birds, etc. It probably would have been impossible for us to learn the concept of number if we did not associate it with sets of objects.

When, as in the above illustrations, numbers are used to tell *how many* objects there are in a set, or *how much*, they are referred to as cardinal numbers.

Definition 3

Cardinal Number: A cardinal number is a number that indicates quantity.

The other basic type of numbers are those used to indicate the *relative position* of the elements of a set when the elements are ordered in a certain way. These are the ordinal numbers: first, second, third, etc.

Definition 4

Ordinal Number: An ordinal number is a number that indicates position.

These numbers are used to tell the rank or relative position of an element in a set. For example, Hawaii was the 49th state admitted to the Union and Alaska was the 50th.

During the last 5000 years many systems have been developed for representing numbers. Some of these can be seen by examining the different ways the number four was represented by past civilizations. It is important to remember that while these symbols are different they all represent the same idea or number.

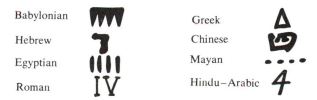

Babylonian	Greek
Hebrew	Chinese
Egyptian	Mayan
Roman	Hindu–Arabic

Although these ancient numerals seem strange and their use in performing the basic arithmetic operations appears difficult to us, people trained in them were able to solve many complex problems. It is our lack of familiarity with these systems that make them appear more complicated than they really are.

The Decimal Numeration System

The system that is almost universally used today is the familiar decimal system. Technically the decimal system is a base 10 positional system with the basic symbols 0, 1, 2, 3, 4, 5, 6, 7, 8, 9. Many people have never heard our numbers described as they are above. Most of us take our numbers for granted and think that they and the related operations have always been and will always be just as they are today. The idea is far from the truth. Our system of representing numbers has gone through a long and often controversial development and even today there is a small group trying to change our system to the base 12, or duodecimal system.

It was probably around the year 500 A.D. that the Hindus developed a positional notation for the decimal system. The symbols for one to nine were developed first and it was not until sometime afterward that the zero concept was introduced. Before the introduction of zero as a place holder, the Hindus stated each multiple of ten. For example, the number 2035 was two one thousands, three tens, and five. When zero was introduced to fill the vacant powers of ten, it became possible to write numbers by using the nine symbols and zero.

These symbols were introduced into Europe by the Arabs sometime between 700 and 1000 A.D. and from then until the year 1500 A.D. a struggle between the proponents of the Roman and Hindu–Arabic systems ensued. The decimal system finally won out and the sixteenth century Europeans used a decimal system whose symbols are basically the same as they are today.

As we already stated, the decimal system is a base ten positional numeration system with the basic symbols 0, 1, 2, 3, 4, 5, 6, 7, 8, and 9. Numbers are now written by combining the basic symbols on the horizontal line.

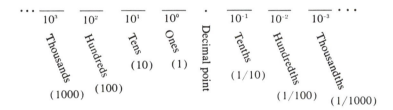

The symbols that are used and their position on this line with respect to the decimal point determine the numbers that they represent. Intuitively, we can think of the decimal point as dividing a number into two parts with the part of the number to the left of the decimal indicating the number of whole units and the part to the right, the fractional part of one unit that is contained in the number. For example, if the distance from town A to town B is two and one-half miles (2 1/2) and from A to D is four and three-quarter miles (4 3/4) we could write these distances in the following decimal form. The distance from A to B equals 2.5 miles. The distance from A to C equals 4.75 miles. In both cases the portion of the number to the left of the decimal indicates the number of whole miles while the portion to the right of the decimal indicates the fractional part of one mile.

When the number to be represented contains no fractional part of a unit the decimal point may be omitted. In this case the last symbol on the right is taken to be the units place.

Of all the decimal system number representations, the one that we are usually most familiar with and interested in is that of dollars and cents. This is a base ten system with the dollar representing the basic unit. Examining the amount \$472.53 (four hundred seventy-two dollars and fifty-three cents), we see that we have four one hundred dollars plus seven ten dollars plus two one dollars plus five dimes (five tenths of a dollar) plus three cents (three hundredths of a dollar). The .53 (fifty-three cents) could also be interpreted as being 53 of the 100 equal parts that make up one dollar.

To conclude our discussion on the decimal representation of numbers, we will write the following numbers in what is commonly known as expanded notation:

40.35 (forty and thirty-five hundredths)

$4(10) + 0(1) + 3(1/10) + 5(1/100)$

$4(10^1) + 0(10^0) + 3(10^{-1}) + 5(10^{-2})$

637.079 (six hundred thirty-seven and seventy-nine thousandths)

$6(100) + 3(10) + 7(1) + 0(1/10) + 7(1/100) + 9(1/1000)$

$6(10^2) + 3(10^1) + 7(10^0) + 0(10^{-1}) + 7(10^{-2}) + 9(10^{-3})$

In Chapter 6 we studied the decimal system of representing numbers. We shall now turn our attention to examining these numbers, the operations defined on these numbers, and their properties. To do this, we shall group the numbers into sets. To assist us in our study we shall assume that each number consists of three distinct parts; the digits to the left of the decimal point, the digits to the right of the decimal point, and the sign (plus or minus).

Note: With the exception of zero, every number is either positive ($+$) or negative ($-$). Zero is a special number that is neither positive nor negative. Therefore, with regards to sign, we have three type of numbers; the positive numbers, the negative numbers, and the number zero.

Sets of Numbers

Since in the decimal system the number of positions to the left and right of the decimal can be extended indefinitely, there are infinitely many numbers thus represented. All of these possible numbers (positive, negative and zero) form a set of numbers called the *real numbers*. Within this set of numbers there are many different subsets, some of which we shall now discuss. We begin our investigation of the real numbers by dividing them into two disjoint subsets; one set consists of all real numbers for which the decimal portion of the number is infinite repeating (including repeating zeroes), and the other set consists of all numbers for which the decimal portion of the number is infinite nonrepeating. The real numbers that have infinite nonrepeating decimals are called *irrational numbers*. The real numbers that have infinite repeating decimals are called *rational numbers*.

Note: The decimal portion of every real number is considered to have an infinite number of places or to extend indefinitely to the right of the decimal point.

Consider the following examples of rational and irrational numbers:

Rational Numbers	*Irrational Numbers*
(a) $-23 = -23.\bar{0}$	(h) $\sqrt{2} = 1.414214\ldots$
(b) $0 = 0.\bar{0}$	(i) $\sqrt{3} = 1.732051\ldots$
(c) $2\ 1/2 = 2.5\bar{0}$	(j) $e = 2.718281\ldots$
(d) $-4\ 5/8 = -4.625\bar{0}$	(k) $\pi = 3.141592\ldots$
(e) $12\ 2/3 = 12.\bar{6}$	(l) $1/\sqrt{2} = .707107\ldots$
(f) $-5\ 4/11 = -5.\overline{36}$	(m) $1/\pi = .318309\ldots$
(g) $1/7 = .\overline{142857}$	(n) $\pi^2 = 9.869604\ldots$

Note: The bars above the numerals in examples a–g are used to indicate the portion of the decimal which repeats.

In examining the examples of rational numbers we see that examples a–d have repeated zeroes. The repeated zeroes are not normally indicated. They are used here to stress the fact that the zeroes do repeat. These numbers are normally written -23, 0, 2.5, and -4.625. Examples e–g are illustrations of numbers which have repeating digits other than zeroes. In example e, the 6's repeat, in example f the sequence 36 repeats, and in example g the sequence 142857 repeats.

In the examples of irrational numbers (h–n) only the first six decimal places of each are given. If these numbers were expanded indefinitely there would be no repeating sequence. There are proofs to show that these numbers are indeed irrational. The proof of the irrationality of $\sqrt{2}$ will be presented in the next section.

Within the set of rational numbers there is a subset whose elements are numbers that have only repeated zeroes to the right of the decimal point (i.e., they are whole numbers). These numbers are called *integers*. For example, the numbers -4, 0, and 11 are integers. They can be represented in their decimal form as follows:

$$-4 = -4.\bar{0}$$
$$0 = 0.\bar{0}$$
$$11 = 11.\bar{0}$$

There is a relationship between the set of rational numbers and its proper subset of integers. This relationship will be stated without proof.

A number is a rational number if and only if it can be written in the form of a fraction P/Q, where P and Q are integers and $Q \neq 0$.

In other words, every infinite repeating decimal can be written as a fraction with an integral numerator and a nonzero integral denominator and every fraction with an integral numerator and a nonzero integral denominator can be written as an infinite repeating decimal (this includes repeated zeroes).

Example 1

Show that the fractions (a) 7/4, (b) 3/8, (c) 12/11, and (d) 1/13 can be written as infinite repeating decimals.

(a) $7/4 = 4 \overline{)7.000000} \quad \dfrac{1.750000}{} \cdots = 1.75\bar{0}$

(b) $3/8 = 8 \overline{)3.0000000} \quad \dfrac{.3750000}{} \cdots = .375\bar{0}$

(c) $12/11 = 11 \overline{)12.000000} \quad \dfrac{1.090909}{} \cdots = 1.09\overline{09}$

(d) $1/13 = 13 \overline{)1.000000000000} \quad \dfrac{.076923076923}{} \cdots = .\overline{076923}$

Example 2

Show that the repeating decimals (a) $3.764\bar{0}$, (b) $\overline{43}.$, and (c) $2.0\bar{3}$ can be written as fractions with integral numerators and nonzero integral denominators.

(a) When only zeroes repeat it is an easy matter to convert the number to a fraction. Since in the number $3.764\bar{0}$ the decimal portion .764 means seven-hundred-sixty-four thousandths, we have

$$3.764\bar{0} = \frac{3764}{1000}$$

When digits other than zero repeat, as in b and c, we use the following procedure:

(b)   Step 1 $N = .\overline{43} = .434343 \ldots$

Step 2   Since the repeating portion of the decimal consists of two digits we multiply the number by 100. If there were three digits in the repeating portion of the decimal, we would multiply by 1000, etc.

$$100N = 43.434343 \ldots$$

Step 3   We now subtract N from $100N$.

$$
\begin{aligned}
100N &= 43.434343 \ldots \\
-N &= -.434343 \ldots \\
\hline
99N &= 43.000000 \ldots
\end{aligned}
$$

or

$$99N = 43$$

Step 4   Solving for N we have $N = 43/99$. Therefore, since $N = .\overline{43}$ we have $.\overline{43} = 43/99$.

(c)   Step 1 $N = 2.0\overline{3} = 2.0333 \ldots$

Step 2   Multiply the number by 10.

$$10N = 20.333 \ldots$$

Step 3   We now subtract N from $10N$.

$$
\begin{aligned}
10N &= 20.3333 \ldots \\
-N &= -2.0333 \ldots \\
\hline
9N &= 18.30000
\end{aligned}
$$

or

$$9N = 18.3$$

Step 4   Solving for N we have $N = 18.3/9$. We note that the numerator 18.3 is not an integer. To write the fraction with an integral numerator and denominator, we multiply both the numerator and denominator by 10; that is,

$$N = \frac{18.3}{9} = \frac{18.3 \times 10}{9 \times 10} = \frac{183}{90}$$

Therefore, since $N = 2.0\overline{3}$ we have

$$2.0\overline{3} = 183/90$$

From our discussion we can see that a real number must belong to either the set of rational numbers or the set of irrational numbers and that every rational number can be written as a fraction with an integral numerator and a nonzero integral denominator. Using these facts, we shall now prove that $\sqrt{2}$ is an irrational number. (*Note:* A square root of a number is that number which when multiplied by itself gives the number under the radical sign; i.e., $\sqrt{9} = 3$, since $3 \times 3 = 9$. In our discussion we will deal only with the positive square roots.)

First, we can show that there exists a number ($\sqrt{2}$) which when multiplied by itself equals 2. Consider a square, each side of which is one unit in length. It follows from the Pythagorean theorem ($a^2 + b^2 = c^2$)* that the length of the diagonal of this square is $\sqrt{2}$.

$$1^2 + 1^2 = c^2$$
$$1 \ + 1 = c^2$$
$$2 = c^2$$
$$\sqrt{2} = c$$

Now that we have shown that $\sqrt{2}$ does exist, we shall use an indirect method to prove that $\sqrt{2}$ is irrational. There are two possibilities:

(a)   There is an integral fraction a/b, such that $a/b = \sqrt{2}$ (i.e., $\sqrt{2}$ is rational).

(b)   There is no such fraction (i.e., $\sqrt{2}$ is irrational).

Proof

(1)   Assume an integral fraction a/b exists such that $\sqrt{2} = a/b$. A theorem exists that states that every fraction can be reduced to lowest terms. Therefore,

---

* The Pythagorean theorem, named after the Greek philosopher Pythagoras, states that in a right triangle, the sum of the squares of the lengths of the sides is equal to the square of the length of the hypotenuse ($a^2 + b^2 = c^2$).

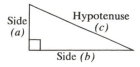

we shall also assume that the fraction a/b is in lowest terms. Squaring both sides of the equality,

$$\sqrt{2} = a/b$$

we have

$$2 = a^2/b^2$$

which yields

$$2b^2 = a^2$$

Since a^2 is a multiple of 2 it is an even number. Furthermore, since if the square of a number is even that number is also even, we see that a is an even number.

(2) Since a is an even number it can be written as two times some other integer, that is; $a = 2p$. Therefore,

$$2b^2 = (2p)^2$$
$$2b^2 = 4p^2$$
$$b^2 = 2p^2$$

(3) By the same chain of reasoning we used for a, we see that b is also even. Setting $b = 2q$ we have $\sqrt{2} = a/b = 2p/2q$. This contradicts our assumption that a/b is in lowest terms. Therefore, the assumption that $\sqrt{2} = a/b$ is false, and the alternative hypothesis that no integral fraction for $\sqrt{2}$ exists is true.

Note: The statement "if the square of a number is even, that number is also even" can be proven but it is not within the range of the material in this book.

The set of integers contains a subset of numbers that is the most fundamental or basic set of all numbers. This is the subset of the counting numbers $\{1, 2, 3, 4, 5, \ldots\}$. This subset of integers is called the set of *natural numbers*. We note that zero is not a natural number.

To conclude our discussion of sets of numbers we shall discuss a subset of the natural numbers that have fascinated mathematicians for centuries. These are the natural numbers greater than one that are exactly divisible only by themselves and one. They are called *prime numbers*. The first few primes are 2, 3, 5, 7, 11, 13, 17, 19, 23, 29, 31, 37, and 41. Examining these numbers we see that 2 is the only even prime number. If a natural number greater than one is not a prime number, it is called a *composite number*. That is, a composite number is a number that is exactly divisible by at least one number other than itself and one. The first few composite numbers are 4, 6, 8, 9, 10, 12, 14, 15, 16, 18, and 20.

Mathematicians have sought to find a pattern for the primes and a formula for generating all the prime numbers but so far have been unsuccessful. No pattern appears to exist. However, it has been proven that there is an infinite number of primes. The proof is again indirect.

Proof

Assume that there is a finite number of primes, 2, 3, 5, 7, ..., p with p the largest prime. We now form the number,

$$N = (2 \times 3 \times 5 \times 7 \times \cdots \times p) + 1$$

There are two possibilities for N. It is either a prime number or it is composite. If N is prime, our assumption that p is the largest prime would be contradicted since N is larger than p. If N is a composite number it has a factor other than itself or one. None of the primes 2, 3, 5, 7, ..., p can be factors of N since N divided by any one of them would yield a remainder of 1. Therefore, there must be a prime number q which is a factor of N but not one of the primes 2, 3, 5, 7, ..., p. This contradicts our assumption that there is a finite number of primes. Therefore, since both possibilities for N contradict the assumption that the set of primes is finite the other alternative that the set of prime numbers must have an infinite number of elements is true.

With regard to the prime numbers, there are many unsolved or unproven conjectures. The eighteenth-century Russian mathematician Goldbach stated that "every even number greater than 2 can be written as the sum of two primes." For example,

$4 = 2 + 2$	$10 = 3 + 7$
$6 = 3 + 3$	$12 = 5 + 7$
$8 = 3 + 5$	$14 = 7 + 7$

Another unproven conjecture about the prime numbers is that there is an infinite number of twin primes. Twin primes are consecutive prime numbers that differ by 2. For example,

3 and 5	29 and 31
5 and 7	41 and 43
11 and 13	59 and 61
17 and 19	71 and 73

These are just two of the many intriguing problems of our number system and even though they are still unsolved, many new ideas have developed out of the many attempts to solve them.

The Number Line

We can now provide a physical representation for the real numbers. We take a line of infinite length and arbitrarily select a point, which we designate as zero:

We now select a point some distance to the right of 0. This point represents 1 and the distance from 0 to 1 establishes our basic unit length:

Now using the basic unit of length, we mark off the points that represent the other integers:

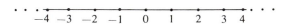

The points on the line between the integers represent all of the other rational numbers as well as the irrational numbers.

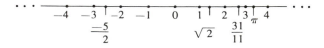

Therefore, the set of real numbers can be represented by the points on a line. That is, for each point on the line there is a real number and for each real number there is a point on the line.

To conclude this section we shall summarize the relationships of the various sets of numbers by the following chart.

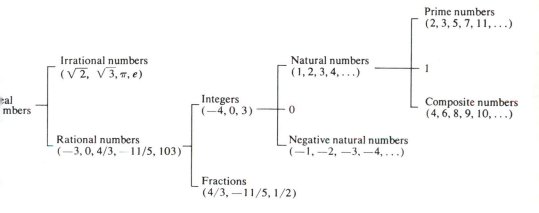

Exercises 7.1

1. Write the following numbers in decimal notation:

   (a) $(4 \times 10^2) + (1 \times 10^1) + (3 \times 10^0) + (6 \times 10^{-1}) + (5 \times 10^{-2})$
   (b) $(7 \times 10^2) + (0 \times 10^1) + (9 \times 10^0) + (0 \times 10^{-1}) + (2 \times 10^{-2})$
   (c) $(2 \times 10^3) + (0 \times 10^2) + (0 \times 10^1) + (1 \times 10^0)$
   (d) $(5 \times 10^{-1}) + (2 \times 10^{-2})$
   (e) $(6 \times 10^0) + (8 \times 10^{-1}) + (7 \times 10^{-2})$

2. Write the following numbers in expanded notation:

   (a) 402.508
   (b) 30.0034
   (c) -4.52
   (d) 0.0017
   (e) 690.

3. Write the following integral fractions as decimals:

   (a) 5/6   (b) 1/3   (c) 2/9   (d) 1/7   (e) 1/11

4. Change the following repeating decimals to integral fractions:

   (a) $.\overline{6}$ (b) $.\overline{81}$ (c) $4.\overline{12}$ (d) $-12.\overline{072}$ (e) $-5.0\overline{6}$

5. Assume the following to be true. *If the square root of a positive integer is not a positive integer, then the square root is an irrational number.* Determine which of the following numbers are rational and which are irrational:

(a) $\sqrt{50}$ (b) $\sqrt{16}$ (c) $\sqrt{7}$

(d) $\sqrt{225}$ (e) $\sqrt{6}$ (f) $\sqrt{200}$

(g) $\sqrt{9}$ (h) $\sqrt{145}$ (i) $\sqrt{49}$

6. Given that

$$U = \{x \,|\, x \text{ is a real number}\}$$
$$R = \{x \,|\, x \text{ is a rational number}\}$$
$$I = \{x \,|\, x \text{ is an integer}\}$$
$$N = \{x \,|\, x \text{ is a natural number}\}$$

(a) Define set R'.
(b) Sketch the Venn diagram showing the relationship of the sets U, R, I, and N.
(c) Indicate which of the following statements are true and which are false:

(a) $R \cup R' = U$ (b) $R \cap R' = \phi$

(c) $I \subset N$ (d) $I \subset R'$

(e) $N \subset R'$ (f) $R' \subset R$

7. Complete the following table by classifying each number. Use $\in$ to show that the number belongs to the set of numbers defined at the top of each column and $\notin$ to show that the number does not belong to the set of numbers defined at the top of each column.

Numbers	Real Numbers	Irrational Numbers	Rational Numbers	Integers	Non-negative Integers	Natural Numbers
347	$\in$	$\notin$	$\in$	$\in$	$\in$	$\in$
-28						
$\sqrt{3}$						
0						
3.1415926...						
$-\sqrt{2}$						
$\left(\dfrac{7.63}{.00\overline{123}}\right)$						
$\left(\dfrac{17}{23}\right)$						
$-\sqrt{16}$						
$1.\overline{4}$						
2^3						
$\sqrt[3]{8}$						
$\left(\dfrac{\sqrt{4}}{5}\right)$						

8. The *density property of the rational numbers* states that given any two rational numbers a and b there is a third rational number c between them such that $c = (1/2) \times (a + b)$. In order to determine a rational number between 5/8 and 6/8, we compute the arithmetic mean as follows:

$$c = \frac{1}{2} \times \left(\frac{5}{8} + \frac{6}{8}\right) = \frac{1}{2} \times \left(\frac{11}{8}\right) = \frac{11}{16}$$

Determine the arithmetic mean of each of the following pairs of rational numbers:

(a) $\frac{1}{4}, \frac{1}{5}$ (b) $\frac{3}{5}, \frac{4}{5}$ (c) $\frac{4}{5}, \frac{7}{9}$

(d) Is the set of integers dense? Explain.

Now that we have discussed the various sets of numbers, we shall investigate what happens when the operations of addition and multiplication are performed on these sets. We shall study the real number system by stating and proving some of the principles of arithmetic.

Mathematical Systems

A mathematical system is a set of elements together with an operation or operations defined for these elements. The basic properties of the elements and the operations are listed, and from these, other properties are derived. There are many mathematical systems, some more complex than others. Before we study the real number system, we shall study one of the simpler systems in order to help us understand the basic properties of the real numbers.

Modular Arithmetic

We shall introduce modular arithmetic by studying one particular modular arithmetic system. Suppose we have a five-minute clock. The numbers on the face of the clock form the set of elements {0, 1, 2, 3, 4}

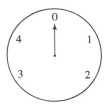

of a system, which is called a modular 5 (or mod 5) system. The operations of addition and multiplication are defined on the set and are given by Tables 1 and 2.

Table 1

+	0	1	2	3	4
0	0	1	2	3	4
1	1	2	3	4	0
2	2	3	4	0	1
3	3	4	0	1	2
4	4	0	1	2	3

Table 2

×	0	1	2	3	4
0	0	0	0	0	0
1	0	1	2	3	4
2	0	2	4	1	3
3	0	3	1	4	2
4	0	4	3	2	1

The sum $1 + 2$ can be thought of as moving the clock hand one place from its initial point, 0, to 1, and then two places from 1 to 3. For the sum $2 + 4$ the clock hand (clock a) is moved two places from 0 to 2 and then four places from 2, past 0, to 1. The product 3×2 can be thought of as moving the clock hand (clock b) from 0, six places, to 1, that is, three moves of two places each, from 0 around to 1.

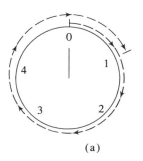

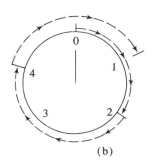

(a)   (b)

A simpler and more direct method of obtaining the results of addition and multiplication in this system is possible since modular arithmetic is really an arithmetic of remainders. That is, the results of adding or multiplying any two numbers in an arbitrary modulus, m, is the remainder that is obtained when the ordinary sum or product is divided by m. For example, to add $2 + 3 + 4$ in mod 5, we take the sum of ordinary addition, 9, and divide by the modulus 5. The remainder 4 is the sum, i.e., $2 + 3 + 4 = 4$ (mod 5).

In general, there are as many remainders as the size of the modulus. In modulus m, the remainders are $0, 1, 2, 3, \ldots, (m-1)$. Modulus 5 has five remainders: 0, 1, 2, 3, and 4. For example,

(a) $(2 \times 3) \times 4$ $2 \times 3 \times 4 = 24$
     $= (1) \times 4 = 4$ (mod 5) $24/5 = 4 +$ remainder of $\underline{4}$

(b) $(3 \times 4) + 4$ $(3 \times 4) + 4 = 16$
     $= (2) + 4 = 1$ (mod 5) $16/5 = 3 +$ remainder of $\underline{1}$

(c) $(4 + 3) \times 2$ $(4 + 3) \times 2 = 14$
     $= (2) \times 2 = 4$ (mod 5) $14/5 = 2 +$ remainder of $\underline{4}$

(d) $[(3 \times 3) + 1] + 4$ $[(3 \times 3) + 1] \times 4 = 40$
     $= [(4) + 1] \times 4$ $40/5 = 8 +$ remainder of $\underline{0}$
     $= [\quad 0 \quad] \times 4 = 0$ (mod 5)

Now that we have examined how the operations of addition and multiplication are performed in modulo 5 we can investigate the properties of these two operations with respect to the mod 5 system.

Examining the addition and multiplication tables (Tables 1 and 2), we can see that when we add or multiply in mod 5 the result is always one of the five remainders. That is, the set of elements $\{0, 1, 2, 3, 4\}$ is closed with respect to addition and multiplication modulo 5.

Definition 5

Closure: A set S is closed with respect to an operation if, whenever we perform the operation on any two elements of the set, the result is also an element of the set.

In our previous discussions of the logical connectives of conjunction ($\wedge$) and disjunction ($\vee$), and the set operations of union ($\cup$) and intersection ($\cap$), we observed that they possess certain properties which we called commutative, associative, and distributive. The operations of addition and multiplication modulo 5 also possess these properties.

Commutative Laws.   The operations of addition and multiplication, defined on a set S, are said to be commutative if for any two elements a and b of S we have

$$a + b = b + a \qquad \text{(commutative law of addition)}$$

$$a \times b = b \times a \qquad \text{(commutative law of multiplication)}$$

That is, the order in which the operations of addition and multiplication are performed on a pair of elements is unimportant. For example, in mod 5 we have

$$
\begin{aligned}
4 + 2 &= 2 + 4 \\
1 &= 1 \qquad (\text{mod } 5) \\
2 \times 4 &= 4 \times 2 \\
3 &= 3 \qquad (\text{mod } 5)
\end{aligned}
$$

Associative Laws.   The operations of addition and multiplication, defined on a set S, are said to be associative if for any three elements a, b, and c of S we have

$$a + (b + c) = (a + b) + c \qquad \text{(associative law of addition)}$$

$$a \times (b \times c) = (a \times b) \times c \qquad \text{(associative law of multiplication)}$$

That is, the sum or product of several elements is the same regardless of the manner in which the terms are grouped. For example, in mod 5 we have

$$
\begin{aligned}
2 + (3 + 4) &= (2 + 3) + 4 \\
2 + (\quad 2 \quad) &= (\quad 0 \quad) + 4 \\
4 &= 4 \qquad (\text{mod } 5) \\
3 \times (2 \times 1) &= (3 \times 2) \times 1 \\
3 \times (\quad 2 \quad) &= (\quad 1 \quad) \times 1 \\
1 &= 1 \qquad (\text{mod } 5)
\end{aligned}
$$

Distributive Law.   The operation of multiplication is said to be distributive over the operation of addition if for any three elements a, b, and c of the set S, on which the operations are defined, we have

$$a \times (b + c) = (a \times b) + (a \times c)$$

For example in modulo 5, we have

$$3 \times (2 + 4) = (3 \times 2) + (3 \times 4)$$
$$3 \times (\quad 1 \quad) = (\quad 1 \quad) + (\quad 2 \quad)$$
$$3 = 3 \qquad \qquad (\text{mod } 5)$$

Note: Addition is not distributive over multiplication. That is, in general,

$$a + (b \times c) \neq (a + b) \times (a + c) \ (\text{mod } 5)$$

For example,

$$4 + (3 \times 2) \overset{?}{=} (4 + 3) \times (4 + 2)$$
$$4 + (\quad 1 \quad) \overset{?}{=} (\quad 2 \quad) \times (\quad 1 \quad)$$
$$0 \neq 2$$

We cannot assume that the operations of addition and multiplication mod 5 have the above properties and it is not enough to show that they are true for a few particular cases. We must show that they are true for all of the possible arrangements of the elements of the set. However, we can show that a property does not apply by showing one case for which it does not hold true, as we did with addition distributed over multiplication. That is, we find a counterexample.

+	0	1	2	3	4
0	0	1	2	3	4
1	1	2	3	4	0
2	2	3	4	0	1
3	3	4	0	1	2
4	4	0	1	2	3

There is a short method of determining whether an operation is commutative. If the table for the operation is symmetric with respect to the diagonal from the upper left to lower right of the table, the operation is commutative. The table for addition modulo 5 is symmetric with respect to the diagonal and the operation of addition is therefore commutative mod 5. There are no short methods of testing the associative and distributive properties.

In our study of mathematical systems there are certain elements that deserve special attention. Examining the addition and multiplication tables modulo 5, we see that whenever an element is added to 0, the result is that same element. Also, when we multiply an element by 1, the result is that same element. That is

$$
\begin{array}{ll}
0 + 0 = 0 & 0 \times 1 = 0 \\
1 + 0 = 1 & 1 \times 1 = 1 \\
2 + 0 = 2 & 2 \times 1 = 2 \\
3 + 0 = 3 & 3 \times 1 = 3 \\
4 + 0 = 4 & 4 \times 1 = 4
\end{array}
$$

For this reason we give these elements special names. The element 0 is called the identity element of addition or the additive identity and the element 1 is called the identity element of multiplication or the multiplicative identity.

Definition 6

Identity Element: An element of set S is said to be an identity element with respect to an operation if when operated on by any element of S the identity element yields that element.

For example, for any element of S (mod 5) we have

$$
\begin{array}{ll}
a + 0 = a & \text{(additive identity)} \\
a \times 1 = a & \text{(multiplicative identity)}
\end{array}
$$

Again examining the addition and multiplication tables modulo 5, we see that for each element of the set it is possible to find an element so that the sum of the two elements yields the additive identity (0), and that for each element (except 0) it is possible to find an element so that the product of the two elements yields the multiplicative identity (1). That is,

$$
\begin{array}{ll}
0 + \underline{0} = 0 & 0 \times \underline{?} = 1 \text{ (none exists)} \\
1 + \underline{4} = 0 & 1 \times \underline{1} = 1 \\
2 + \underline{3} = 0 & 2 \times \underline{3} = 1 \\
3 + \underline{2} = 0 & 3 \times \underline{2} = 1 \\
4 + \underline{1} = 0 & 4 \times \underline{4} = 1
\end{array}
$$

For this reason, we give these pairs of elements special names. Two elements that when added together yield the additive identity (0) are called additive inverses of one another, and two elements that when multiplied together yield the multiplicative identity (1) are called multiplicative inverses of one another.

Definition 7

Inverse Elements: An element of a set S is said to be an inverse of an element with respect to an operation if, when one operates on the other, they yield the identity element with respect to that operation.

For example, for the element 4 (mod 5) the additive inverse is 1 and the multiplicative inverse is 4, itself, That is,

$$4 + \underline{1} = 0 \qquad \text{and} \qquad 4 \times \underline{4} = 1$$

We see that for modulo 5, each element has an additive inverse and that each element, except 0, has a multiplicative inverse. Now using the inverses we can define the operations of subtraction and division modulo 5.

Definition 8

Subtraction: To subtract one element from another we add the additive inverse of the number to be subtracted.

For example in mod 5, since the additive inverse of 2 is 3 and the additive inverse of 4 is 1, we have

$$4 - 2 = 4 + (-2) = 4 + (3) = 2$$
$$2 - 4 = 2 + (-4) = 2 + (1) = 3 \qquad \text{or} \qquad 2 - 4 = -2 = 3$$

Definition 9

Division: To divide one element by another, we multiply the first element by the multiplicative inverse of the divisor.

Since 2 and 3 are multiplicative inverses of each other and 0 has no multiplicative inverse, we have

$$4 \div 2 = 4 \times (2^{-1}) = 4 \times (3) = 2$$
$$2 \div 3 = 2 \times (3^{-1}) = 2 \times (2) = 4$$
$$3 \div 0 \quad \text{(this operation cannot be performed)}$$

To simplify our work with subtraction and division, additive inverses will be denoted by a minus sign before the element. For example, the additive inverse of 2 is denoted by (-2). The multiplicative inverse will be denoted by writing the element to the minus one power. For example, the multiplicative inverse of 2 is denoted 2^{-1}. The modulo 5 inverses are therefore

Additive Inverses	Multiplicative Inverses
$-0 = 0$	0^{-1} = does not exist
$-1 = 4$	$1^{-1} = 1$
$-2 = 3$	$2^{-1} = 3$
$-3 = 2$	$3^{-1} = 2$
$-4 = 1$	$4^{-1} = 4$

We see that subtraction and division are really not independent operations but are the inverse operations of addition and multiplication. Subtraction is equivalent to adding the additive inverse and division is equivalent to multiplying by the multiplicative inverse. If an element has a multiplicative inverse, it is possible to divide by that element. Since zero has no multiplicative inverse, it is not possible to divide by zero. To illustrate these ideas, consider the following examples:

(a) $[2 - 3] - 4$
$= [2 + (-3)] + (-4)$
$= [2 + (2)] \quad + (1)$
$= [\quad 4 \quad] + 1$
$= 0 \pmod 5$

(b) $1/4$
$= 1 \div 4$
$= 1 \times 4^{-1}$
$= 1 \times 4$
$= 4 \pmod 5$

(c) $4 \times (3 \div 2)$
$= 4 \times (3 \times 2^{-1})$
$= 4 \times (3 \times 3)$
$= 4 \times 4$
$= 1 \pmod 5$

(d) $1/2 + 2/4$
$= (1 \div 2) + (2 \div 4)$
$= (1 \times 2^{-1}) + (2 \times 4^{-1})$
$= (1 \times 3) + (2 \times 4)$
$= (3) + (3)$
$= 1 \pmod 5$

$$(e) \frac{2 \times [3 - 2]}{2}$$
$$= \{2 \times [3 + (-2)]\} \div 2$$
$$= \{2 \times [3 + 3]\} \times 2^{-1}$$
$$= \{2 \times 1\} \times 3$$
$$= 2 \times 3$$
$$= 1 \ (\text{mod } 5)$$

Now that we have studied the modulo 5 system, we shall examine the modulo 6 system to see if the same properties apply. The addition and multiplication tables for modulo 6 are as follows:

Table 3

+	0	1	2	3	4	5
0	0	1	2	3	4	5
1	1	2	3	4	5	0
2	2	3	4	5	0	1
3	3	4	5	0	1	2
4	4	5	0	1	2	3
5	5	0	1	2	3	4

Table 4

×	0	1	2	3	4	5
0	0	0	0	0	0	0
1	0	1	2	3	4	5
2	0	2	4	0	2	4
3	0	3	0	3	0	3
4	0	4	2	0	4	2
5	0	5	4	3	2	1

Examining the tables we see that since the result of adding or multiplying any two elements of the mod 6 system is one of the elements {0, 1, 2, 3, 4, 5}, the set is closed with respect to both operations. The operations are also commutative since the tables are symmetric with respect to the diagonal. If we were to test all the possible arrangements of the elements we would see that the operations are associative and that multiplication is distributive over addition. For example,

$$2 + (3 + 5) = (2 + 3) + 5 \qquad 2 \times (3 \times 5) = (2 \times 3) \times 5$$
$$2 + \quad 2 \quad = \quad 5 \quad + 5 \qquad 2 \times \quad 3 \quad = \quad 0 \quad \times 5$$
$$4 \quad = \quad 4 \ (\text{mod } 6) \qquad 0 \quad = \quad 0 \ (\text{mod } 6)$$

$$2 \times (3 + 5) = (2 \times 3) + (2 \times 5)$$
$$2 \times \quad 2 \quad = \quad 0 \quad + \quad 4$$
$$4 \quad = \quad 4 \ (\text{mod } 6)$$

The tables also show that 0 is the additive identity and that 1 is the multiplicative identity. In determining the additive and multiplicative inverses

Additive Inverses		Multiplicative Inverses	
$0 + \underline{0} = 0$	$-0 = 0$	$0 \times \underline{?} = 1$	0^{-1} does not exist
$1 + \underline{5} = 0$	$-1 = 5$	$1 \times \underline{1} = 1$	$1^{-1} = 1$
$2 + \underline{4} = 0$	$-2 = 4$	$2 \times \underline{?} = 1$	2^{-1} does not exist
$3 + \underline{3} = 0$	$-3 = 3$	$3 \times \underline{?} = 1$	3^{-1} does not exist
$4 + \underline{2} = 0$	$-4 = 2$	$4 \times \underline{?} = 1$	4^{-1} does not exist
$5 + \underline{1} = 0$	$-5 = 1$	$5 \times \underline{5} = 1$	$5^{-1} = 5$

we see that every element has an additive inverse but only 1 and 5 have multiplicative inverses. Therefore, we can subtract any element but divide only by 1 or 5 (mod 6). Consider the following examples:

(a) $(4 + 3) - 5$

$= (4 + 3) + (-5)$
$= 1 + 1$
$= 2 \pmod 6$

(b) $\dfrac{(5 - 3) - 2}{5}$

$= \{[5 + (-3)] + (-2)\} \div 5$
$= \{[5 + 3] + 4\} \times 5^{-1}$
$= \{2 + 4\} \times 5$
$= 0 \times 5$
$= 0 \pmod 6$

(c) $4 + 2/3 =$
$4 + (2 \times 3^{-1}) =$
cannot be completed since 3^{-1} does not exist.

   To conclude this section we shall consider a system that is not a modulo system. Suppose we assume an operation $*$ on the set $S = \{a, b, c, d\}$, the table for this operation being as follows:

$*$	a	b	c	d
a	a	a	c	d
b	a	b	c	d
c	d	c	b	c
d	c	d	a	b

Examining the table we see that the operation $*$ is closed on S. The operation is not commutative since the table is not symmetric with respect to the upper-left to lower-right diagonal; that is,

$$d * c \neq c * d$$
$$a \neq c$$

In testing to determine if the operation $*$ is associative we will examine several cases. (In testing to determine if the associative and distributive properties apply, we should test all the possible arrangements of the elements. However, this is exceedingly time consuming. Therefore, we shall select three cases at random and if the property holds for the three cases we shall *assume* the property holds for all the possible arrangements.)

1. $c * (d * b) \stackrel{?}{=} (c * d) * b$
$\quad c * (d) \stackrel{?}{=} (c) * b$
$\quad\quad\quad c = c$

2. $(c * d) * c \stackrel{?}{=} c * (d * c)$
$\quad (c) * c \stackrel{?}{=} c * (a)$
$\quad\quad\quad b \neq d$

In the first case, the associative property holds but in the second case it does not. Therefore, the operation $*$ is not associative. We cannot test for a distributive property since we have only one operation and the distributive property requires two.

The identity element for operation $*$ is b and all of the elements except a have inverses. That is,

Identity Element	*Inverse Element*	
$a * \underline{b} = a$	$a * \underline{?} = b$	$a^{-1} = ?$
$b * \underline{b} = b$	$b * \underline{b} = b$	$b^{-1} = b$
$c * \underline{b} = c$	$c * \underline{c} = b$	$c^{-1} = c$
$d * \underline{b} = d$	$d * \underline{d} = b$	$d^{-1} = d$

With the exception of a, which has no inverse, each element is its own inverse.

We have examined these various systems to help us better understand the properties of our own numbers, which we shall investigate in the next section.

Exercises 7.2

1. If possible solve the following in modulo 5.

   (a) $2 - 4$ (f) $(3 \times 1) + (3 \times 3)$
   (b) $1 - 4$ (g) $2 \times (1 - 4)$
   (c) $(2 + 3) - 2$ (h) $3 \div 4$
   (d) $2 + (3 - 2)$ (i) $4 \div 3$
   (e) $3 \times (1 + 3)$ (j) $[(2 + 3) \times (4 - 2)] \div 3$

2. If possible solve the following in modulo 6.

   (a) $3 \times (3 + 1)$ (f) $3 \div 5$
   (b) $(3 \times 3) + (3 \times 1)$ (g) $(4 \div 2) \div 5$
   (c) $(2 - 5) + (1 - 3)$ (h) $3 \div 4$
   (d) $3 \times (2 - 4)$ (i) $[(2 + 3) \times (5 - 2)] \div 1$
   (e) $(3 \times 2) - (3 \times 4)$ (j) $[(0 + 1) \times (4 - 0)] \div 3$

3. Construct the addition and multiplication tables for modulo 3.

   (a) Is the set closed under addition?
   (b) Is the set closed under multiplication?
   (c) Is addition commutative?
   (d) Is multiplication commutative?
   (e) Is addition associative?
   (f) Is multiplication associative?
   (g) What is the additive identity?
   (h) What is the multiplicative identity?
   (i) What are the additive inverses?
   (j) What are the multiplicative inverses?
   (k) Is multiplication distributive over addition?

   If possible solve the following in modulo 3.

   (l) $2 - 1$
   (m) $(2 + 1) \times 2$
   (n) $(2 \times 2) + (1 \times 2)$
   (o) $2 \div 1$
   (p) $1 \div 0$

4. Construct the addition and multiplication tables for modulo 4. Answer parts
   a–k of problem 3.
   If possible solve the following in modulo 4.

   (l) $(1 + 2) \times 3$
   (m) $(1 \times 3) + (2 \times 3)$
   (n) $(0 - 1) \times 3$
   (o) $(0 \times 3) - (1 \times 3)$
   (p) $2 \div 2$

5. Consider a set of elements $\{s, e, t\}$ together with the operation $*$, defined as follows:

$*$	s	e	t
s	t	s	e
e	s	e	t
t	e	t	s

(a) Is the set closed under the operation $*$?
(b) Is operation $*$ commutative?
(c) Is operation $*$ associative?
(d) What is the identity element for the operation $*$?
(e) What are the inverses of the operation $*$ for the elements s, e, t?
(f) Is there a distributive property?

6. Suppose we define a system on the set $S = \{T, F\}$ and two operations, conjunction ($\wedge$) and disjunction ($\vee$). The operations are defined as follows:

$\wedge$	T	F
T	T	F
F	F	F

$\vee$	T	F
T	T	T
F	T	F

(a) Is the set closed under conjunction?
(b) Is the set closed under disjunction?
(c) Is conjunction commutative?
(d) Is disjunction commutative?
(e) Is conjunction associative?
(f) Is disjunction associative?
(g) What is the conjunctive identity?
(h) What is the disjunctive identity?
(i) What are the conjunctive inverses for the elements T, F?
(j) What are the disjunctive inverses for the elements T, F?
(k) Is conjunction distributive over disjunction?
(l) Is disjunction distributive over conjunction?

7. Complete the following table. Use "T" to show that the set of numbers defined at the top of each column of the table has the property listed to the left and "F" to show that the set of numbers at the top of each column of the table does not have the property listed to the left.

Property	Real Numbers	Irrational Numbers	Rational Numbers	Integers	Nonnegative Integers	Natural Numbers
Closed under addition						
Commutative addition						
Associative addition						
The additive identity (0)						
Inverse for addition (negatives)						
Closed under multiplication						
Commutative multiplication						
Associative multiplication						
The multiplicative identity (1)						
Inverses for multiplication (reciprocals except for 0)						
Multiplication distributive over addition						

The Structure of Algebra

In the study of beginning algebra students are given rules and procedures such as the following:

(a) $(+2) \times (-3) = -6$
(b) $(-2) \times (-3) = +6$
(c) If $x + 3 = 5$, then $x = 5 - 3 = 2$
(d) To solve $x^2 - 5x + 4 = 0$, we factor the expression and set the factors equal to zero; that is,

$$x^2 - 5x + 4 = 0$$
$$(x - 4) \cdot (x - 1) = 0$$
$$(x - 4) = 0 \quad \text{or} \quad (x - 1) = 0$$
$$\therefore x = 4 \quad \text{and} \quad x = 1 \quad \text{are the solutions}$$

These and many similar procedures are generally stated without proof. The student is told that these rules always apply but that their proofs are beyond the scope of an elementary or intermediate course in algebra. In this section we shall prove many of these fundamental laws of our number system and give the rationale for some of the various procedures used in the solution of equations.

In proving or developing the laws of a mathematical system we begin with some undefined terms and some basic assumptions or postulates about the system. Using these as the foundation on which to build our system, we develop and prove new laws or theorems.

We begin with a set S, or elements $\{a, b, c, \ldots\}$, with the operations of $+$ and $\times$ defined on the set. The basic postulates of equality apply to the members of set S.

Postulates of Equality.   For any elements a, b, and c of set S we have the following postulates:

P_1 Reflexive Postulate: $a = a$.

P_2 Symmetric Postulate: If $a = b$, then $b = a$.

P_3 Transitive Postulate: If $a = b$ and $b = c$, then $a = c$.

P_4 Substitution Postulate: If $a = b$, then b can be substituted for a in any statement without changing the value of that statement.

These four postulates of equality seem so obvious that it is almost superfluous to state them. However, since we are developing a formal system, it is necessary to state all of the basic assumptions. The following are the postulates that relate the elements of set S and the operations of $+$ and $\times$ defined on S.

Postulates of the System (Set S with $+$ and $\times$).   For any elements a, b, and c of set S we have the following postulates:

P_5 Closure Postulate: The sum $a + b$ and the product $a \times b$ are unique elements of set S. (*Note:* By unique we mean the sum $a + b$ or product $a \times b$ is one and only one element of set S.)

P_6 Commutative Postulate: $a + b = b + a$ and $a \times b = b \times a$.

P_7 Associative Postulate: $a + (b + c) = (a + b) + c$ and $a \times (b \times c) = (a \times b) \times c$.

P_8 Distributive Postulate: $a \times (b + c) = (a \times b) + (a \times c)$.

P_9 Identity Postulate: There are unique elements of set S, 0 and 1, called, respectively, the additive and multiplication identities such that $a + 0 = a$ and $a \times 1 = a$. (Since $+$ and $\times$ are commutative we have $0 + a = a$ and $1 \times a = a$.)

P_{10} Inverse Postulate: For each element a in set S, there exists a unique element $(-a)$ called the additive inverse of a, such that $a + (-a) = 0$; also, for each element a in set S (except 0) there exists a unique element (a^{-1}) called the multiplicative inverse of a, such that $a \times (a^{-1}) = 1$. (Since $+$ and $\times$ are commutative we have $(-a) + a = 0$ and also $(a^{-1}) \times a = 1$.)

We have already discussed the last six postulates of the system in the previous section and have seen that they apply to the real and rational numbers as well as to the modulo 5 system. Therefore, any theorems or laws that we prove based on these assumptions will hold true for these systems as well as any other system to which these postulates apply.

Note: Any system in which the above postulates hold true is called a *field*. Therefore, the sets of rational and real numbers with the operations of $+$ and $\times$ form a field.

We shall now prove some of the fundamental theorems of a field.

Theorem 1A

If a, b, and c are elements of set S, and $a = b$, then $a + c = b + c$

Proof

(1)	$(a + c) \in S$	Closure
(2)	$(a + c) = (a + c)$	Reflexive
(3)	$a = b$	Given
(4)	$a + c = b + c$	P_4 (Substitution, 3 in 2)

Theorem 1B

If a, b, and c are elements of set S, and $a = b$, then $a \times c = b \times c$

Proof

Exercise 2

Theorem 2A

If a, b, c, and d are elements of set S, and $a = b$ and $c = d$, then $a + c = b + d$

Proof

(1)	$a = b$	Given
(2)	$a + c = b + c$	1, theorem $1A$
(3)	$c = d$	Given
(4)	$a + c = b + d$	P_4 (Substitution, 3 in 2)

Theorem 2B   Exercise 3.

Theorem 3A

If a, b, and c are elements of set S and $a + c = b + c$, then $a = b$

Proof

(1)	$(a + c) \in S$ and $(b + c) \in S$	Closure
(2)	$(a + c) = (b + c)$	Given
(3)	$(-c) \in S$	Inverse
(4)	$[a + c] + (-c) = [b + c] + (-c)$	2, 3, Theorem 1A
(5)	$a + [c + (-c)] = b + [c + (-c)]$	4, Associative
(6)	$[c + (-c)] = 0$	Inverse
(7)	$a + 0 = b + 0$	Substitution, 6 in 5
(8)	$a + 0 = a$ and $b + 0 = b$	Identity
(9)	$a = b$	Substitution, 8 in 7

Theorem 3B

If a, b, and c are elements of set S, and $a \times c = b \times c$ and $c \neq 0$, then $a = b$

Proof

Exercise 4

Theorem 4

If a, b, and c are elements of set S, then $(a + b) \times c = (a \times c) + (b \times c)$.

Proof

This theorem is similar to the distributive postulate, which states that multiplication on the left is distributive over addition. Here we shall prove that multiplication on the right is also distributive over addition (see Exercise 5).

(1) $(a + b) \times c \in S$	?
(2) $(a + b) \times c = c \times (a + b)$	Commutative
(3) $c \times (a + b) = (c \times a) + (c \times b)$	?
(4) $(c \times a) = (a \times c)$ and $(c \times b) = (b \times c)$	?
(5) $c \times (a + b) = (a \times c) + (b \times c)$	Substitution, 4 in 3
(6) $(a + b) \times c = (a \times c) + (b \times c)$	?

Theorem 5A

If a is an element of set S, then $-(-a) = a$

Proof

(1) $-a \in S$ and $-(-a) \in S$	Inverse
(2) $-(-a) + (-a) = 0$	Inverse
(3) $[-(-a) + (-a)] + a = 0 + a$	2, Theorem 1A
(4) $-(-a) + [(-a) + a] = 0 + a$	3, Associative
(5) $-(-a) + 0 = 0 + a$	4, Inverse
(6) $-(-a) = a$	5, Identity

Theorem 5B

If a is an element of set S, and $a \neq 0$, then $(a^{-1})^{-1} = a$

Proof

Exercise 6

Theorem 6A

For the additive identity 0 of set S, we have $(-0) = 0$

Proof

(1) $(-0) \in S$	Inverse
(2) $(-0) + 0 = 0$	Inverse
(3) $0 + 0 = 0$	Identity
(4) $(-0) + 0 = 0 + 0$	Substitution, 3 in 2
(5) $(-0) = 0$	4, Theorem 3A

Theorem 6B

For the multiplicative identity 1 of set S we have $(1)^{-1} = 1$

Proof

Exercise 7

Theorem 7A

If a and b are elements of set S and $a = b$, then $-a = -b$

Proof

(See Exercise 8)

$$
\begin{array}{lll}
(1) & -a \in S \ \text{ and }\ -b \in S & ? \\
(2) & a = b & \text{Given} \\
(3) & b = a & ? \\
(4) & b + (-a) = a + (-a) & ? \\
(5) & b + (-a) = 0 & ? \\
(6) & -b = -b & ? \\
(7) & (-b) + [b + (-a)] = (-b) + 0 & ? \\
(8) & [(-b) + b] + (-a) = (-b) + 0 & ? \\
(9) & 0 + (-a) = (-b) + 0 & ? \\
(10) & (-a) = (-b) & ?
\end{array}
$$

Theorem 7B

If a and b are elements of set S, and $a = b$, $a \neq 0$ and $b \neq 0$, then $a^{-1} = b^{-1}$

Proof

Exercise 9

Theorem 8

If a and b are elements of set S, then $-(a + b) = (-a) + (-b)$

Proof

(1)	$(a + b) \in S$	Closure
(2)	$-(a + b) + (a + b) = 0$	Inverse
(3)	$-a \in S$ and $-b \in S$	Inverse
(4)	$[-(a + b) + (a + b)] + (-b) = 0 + (-b)$	2, 3, Theorem 1A
(5)	$-(a + b) + [(a + b) + (-b)] = 0 + (-b)$	4, Associative
(6)	$-(a + b) + \{a + [b + (-b)]\} = 0 + (-b)$	5, Associative
(7)	$-(a + b) + [a + 0] = 0 + (-b)$	6, Inverse
(8)	$-(a + b) + [a] = (-b)$	7, Identity
(9)	$[-(a + b) + (a)] + (-a) = (-b) + (-a)$	8, 3, Theorem 1A
(10)	$-(a + b) + [(a) + (-a)] = (-b) + (-a)$	9, Associative
(11)	$-(a + b) + [0] = (-b) + (-a)$	10, Inverse
(12)	$-(a + b) = (-b) + (-a)$	11, Identity
(13)	$-(a + b) = (-a) + (-b)$	12, Commutative

Theorem 9

If a is an element of set S, then $a \times 0 = 0$ and $0 \times a = 0$

Proof

(1)	$a = a$	Reflexive
(2)	$0 = 0 + 0$	Identity
(3)	$a \times 0 = a \times (0 + 0)$	1, 2, Theorem 2B
(4)	$a \times 0 = (a \times 0) + (a \times 0)$	3, Distributive
(5)	$0 + (a \times 0) = (a \times 0)$	Identity
(6)	$0 + (a \times 0) = (a \times 0) + (a \times 0)$	Substitution, 5 in 4
(7)	$0 = a \times 0$	6, Theorem 3A
(8)	$a \times 0 = 0$	7, Symmetric
(9)	$0 \times a = 0$	8, Commutative

Theorem 10

If a is an element of set S, then $a \times (-1) = -a$

Proof

(1) $[(-1)+1]=0$ Inverse
(2) $a=a$ Reflexive
(3) $a \times [(-1)+1]=a \times 0$ 1, 2, Theorem 1B
(4) $a \times [(-1)+1]=0$ 3, Theorem 9
(5) $a \times (-1)+(a \times 1)=0$ 4, Distributive
(6) $\{[a \times (-1)]+[a \times 1]\}+[-(a \times 1)]=0+[-(a \times 1)]$ 5, Theorem 1A
(7) $[a \times (-1)]+\{[a \times 1]+[-(a \times 1)]\}=0+\{-(a \times 1)\}$ 6, Associative
(8) $[a \times (-1)]+0=0+\{-(a \times 1)\}$ 7, Inverse
(9) $a \times (-1)=-(a \times 1)$ 8, Identity
(10) $a \times (-1)=-a$ 9, Identity

Theorem 11

If a and b are elements of set S and $a+b=0$, then $a=-b$

Proof

(1) $[a+b]=0$ Given
(2) $[a+b]+(-b)=0+(-b)$ 1, Theorem 1A
(3) $a+[b+(-b)]=0+(-b)$ 2, Associative
(4) $a+0=0+(-b)$ 3, Inverse
(5) $a=-b$ 4, Identity

Theorem 12

If a and b are elements of S, then $a \times (-b)=-(a \times b)$

Proof

(1) $b+(-b)=0$ Inverse
(2) $[b+(-b)] \times a=0 \times a$ 1, Theorem 1B
(3) $a \times [b+(-b)]=a \times 0$ 2, Commutative
(4) $a \times [b+(-b)]=0$ 3, Theorem 9
(5) $[a \times b]+[a \times (-b)]=0$ 4, Distributive
(6) $[a \times (-b)]+[(a \times b)]=0$ 5, Commutative
(7) $a \times (-b)=-(a \times b)$ 6, Theorem 11

Theorem 13

If a and b are elements of set S, then $(-a) \times (-b) = a \times b$

Proof

(1) $(-a) + a = 0$ Inverse

(2) $[(-a) + a] \times (-b) = 0 \times (-b)$ 1, Theorem 1B

(3) $[(-a) \times (-b)] + [a \times (-b)] = 0 \times (-b)$ 2, Theorem 4

(4) $[(-a) \times (-b)] + [-(a \times b)] = 0 \times (-b)$ 3, Theorem 12

(5) $[(-a) \times (-b)] + [-(a \times b)] = 0$ 4, Theorem 9

(6) $\{[(-a) \times (-b)] + [-(a \times b)]\} + (a \times b) = 0 + (a \times b)$ 5, Theorem 1A

(7) $[(-a) \times (-b)] + \{-(a \times b) + (a \times b)\} = 0 + (a \times b)$ 6, Associative

(8) $[(-a) \times (-b)] + 0 = 0 + (a \times b)$ 7, Inverse

(9) $(-a) \times (-b) = (a \times b)$ 8, Identity

Theorem 14

If a and b are elements of set S and $a \times b = 0$, then $a = 0$ or $b = 0$

Proof

(1) Assume $a \neq 0$

(2) $a \times b = 0$ Given

(3) $a^{-1} \in S$ 1, Inverse

(4) $(a \times b) \times a^{-1} = 0 \times a^{-1}$ 2, Theorem 1B

(5) $(b \times a) \times a^{-1} = 0 \times a^{-1}$ 4, Commutative

(6) $(b \times a) \times a^{-1} = 0$ 5, Theorem 9

(7) $b \times (a \times a^{-1}) = 0$ 6, Associative

(8) $b \times 1 = 0$ 7, Inverse

(9) $b = 0$ 8, Identity

This shows that, if $a \neq 0$, then b must equal zero. In the same manner we can show that if $b \neq 0$, then $a = 0$. Therefore, we have proven that, if $a \times b = 0$, then either a or b must be equal to zero.

Theorem 15A

If a and b are elements of set S, then $x = b + (-a)$ is an element of set S and is a unique solution of the equation $x + a = b$

Proof

There are two distinct parts to the proof of this theorem: first, to show that there is a solution in set S and, second, to show that this solution is unique.

Part I

(1)	$x + a = b$	Given
(2)	$[x + a] + (-a) = b + (-a)$	1, Theorem 1A
(3)	$x + [a + (-a)] = b + (-a)$	2, Associative
(4)	$x + 0 = b + (-a)$	3, Inverse
(5)	$x = b + (-a)$	4, Identity
(6)	$x = b + (-a) \in S$	5, Closure

Part II

(1)	$x + a = b$	Given
(2)	$x = b + (-a)$	Given
(3)	$[b + (-a)] + a = b$	Substitution, 2 in 1
(4)	$b + [(-a) + a] = b$	3, Associative
(5)	$b + 0 = b$	4, Inverse
(6)	$b = b$	5, Identity

Therefore, $x = b + (-a) \in S$ and is a solution of $x + a = b$.

Theorem 15B

If a and b are elements of set S and $a \neq 0$ and $b \neq 0$, then $x = a \times b^{-1}$ is an element of set S and is a unique solution of the equation $x \times b = a$.

Proof

Exercise 10

We shall now summarize the postulates and some of the above theorems as they apply to the real-number system. In the descriptive comments to the right of the theorems, subtraction and division are defined as the inverse operations of addition and multiplication. That is,

$$a - b = a + (-b) \quad \text{and} \quad a \div b \quad \text{or} \quad a/b = a \times (b^{-1})$$

Given the set of real numbers, $R = \{a, b, c, \ldots\}$, with the operations of $+$ and $\times$ defined on the set, we have for any elements a, b, and c of R

P_1 Reflexive: $a = a$

P_2 Symmetric: $(a = b) \leftrightarrow (b = a)$

P_3 Transitive: $[(a = b) \wedge (b = c)] \rightarrow (a = c)$

P_4 Substitution: If $a = b$, then b can be substituted in any statement without changing the value of that statement.

P_5 Closure: $(a + b) \in S$ and $(a \times b) \in S$

P_6 Commutative: $a + b = b + a$ and $a \times b = b \times a$

P_7 Associative: $a + (b + c) = (a + b) + c$ and $a \times (b \times c) = (a \times b) \times c$

P_8 Distributive: $a \times (b + c) = (a \times b) + (a \times c)$

P_9 Identity: $a + 0 = a$ and $a \times 1 = a$

P_{10} Inverse: $a + (-a) = 0$ and $a \times a^{-1} = 1\ (a \neq 0)$

Theorem	*Description*
T1A $(a = b) \rightarrow (a + c = b + c)$	Equals added to or multiplied
T1B $(a = b) \rightarrow (a \times c = b \times c)$	by equals are still equals
T3A $(a + c = b + c) \rightarrow (a = b)$	Equals subtracted from or
T3B $(a \times c = b \times c) \rightarrow (a = b)$	divided by equals are still equals
T5A $-(-a) = a$	Minus a minus is a plus
T5B $(a^{-1})^{-1} = a$ or $1/(1/a) = a$	To divide by a fraction we invert and multiply
T9 $a \times 0 = 0 \times a = 0$	Any number times 0 equals 0
T10 $a \times (-1) = -a$	Any number times -1 equals minus that number
T12 $a \times (-b) = -(a \times b)$	Plus times a minus equals a minus
T13 $(-a) \times (-b) = (a \times b)$	Minus times a minus equals a plus
T14 $(a \times b = 0) \rightarrow [(a = 0) \vee (b = 0)]$	If a product equals 0 then at least one of the factors must equal 0
T15A $(x + a = b) \rightarrow [x = b + (-a)]$	To solve simple linear equations
T15B $[(x \times a = b) \wedge a \neq 0)] \rightarrow [x = b \times a^{-1}]$	

The rules and procedures that were used to introduce this section can now be justified by the above theorems: that is

(a) $(+2) \times (-3) = -6$ Theorem 12

(b) $(-2) \times (-3) = +6$ Theorem 13

(c) If $x + 3 = 5$, then $x = 5 - 3 = 2$ Theorem 15A

(d) To solve $x^2 - 5x + 4 = 0$ we factor the expression and set the factors equal to zero; that is

$$x^2 - 5x + 4 = 0$$
$$(x - 4)(x - 1) = 0$$
$$(x - 4) = 0 \quad \text{or} \quad (x - 1) = 0 \qquad \text{Theorem 14}$$
$$\therefore x = 4 \text{ and } x = 1 \text{ are the solutions} \qquad \text{Theorem 15A}$$

(e) To show that $(x - 4)(x - 1) = x^2 - 5x + 4$:

$[x - 4][x - 1]$ Given
$= [(x) + (-4)][(x) + (-1)]$ Definition of subtraction
$= [(x) + (-4)](x) + [(x) + (-4)](-1)$ Distributive
$= [(x)(x) + (-4)(x)] + [(x)(-1) + (-4)(-1)]$ Right distributive
$= [x^2 - 4x] + [-x + 4]$ Theorems 12 and 13
$= x^2 - 4x - x + 4$
$= x^2 - 5x + 4$

Exercises 7.3

1. If S is the set of real numbers, state the field postulates illustrated by each of the following:

(a) $4/4 = 1$
(b) $2(x + y) = 2x + 2y$
(c) $0 + 0 = 0$
(d) $0 + 1 = 1$
(e) $6 - 6 = 0$
(f) $6 + 4 \in S$
(g) $2 + 7 = 7 + 2$

2. Prove Theorem 1B.

3. State and prove Theorem 2B, using the operation of multiplication in place of addition.

4. Prove Theorem 3B.

5. Fill in the missing reasons in the proof of Theorem 4.

6. Prove Theorem 5B.

7. Prove Theorem 6B.

8. Fill in the missing reasons in the proof of Theorem 7A.

9. Prove Theorem 7B.

10. Prove Theorem 15B.

11. Prove the following statement. If a is an element of set S, then $a \times 1^{-1} = a$ (i.e., any number divided by 1 equals itself, $a/1 = a$).

12. Prove the following statement. If a and b are elements of set S, then $-(a - b)$ $= a + b$. *Hint*: $[a + (-b)] + \{-[(a) + (-b)]\} = 0$.

Review Test

1. The decimal $.\overline{41}$ expressed in fractional form is

   (a) 4/9   (b) 4/10   (c) 41/99   (d) 41/100   (e) none of these

2. The number 1/12 expressed in decimal form is

   (a) $3.\overline{3}$ (b) $8.\overline{30}$ (c) $.8\overline{3}$ (d) $.08\overline{3}$ (e) none of these

3. Determine the arithmetic mean of 1/8 and 1/9.

   (a) 17/72   (b) 1/72   (c) 1/144   (d) 17/144   (e) none of these

4. The set of negative numbers is closed under addition.

   (a) true   (b) false

5. The set of negative numbers is closed under multiplication.

   (a) true   (b) false

6. A real number is either a rational number or an irrational number but not both.

   (a) true   (b) false

7. The set of positive integers is closed under subtraction.

    (a) true   (b) false

8. The set of odd integers is closed under addition.

    (a) true   (b) false

9. Which of the following is an irrational number?

    (a) $\sqrt{4}$ (b) $\sqrt[3]{4}$ (c) -3 (d) $1.\overline{73}$

10. The set of irrational numbers is closed under the operation addition

    (a) true   (b) false

11. The sum of two integral fractions is _____ an integer.

    (a) always   (b) sometimes   (c) never

12. The product of two integral fractions is _____ an integer.

    (a) always   (b) sometimes   (c) never

13. The quotient of two rational numbers is _____ a rational number provided that the divisor is not equal to zero.

    (a) always   (b) sometimes   (c) never

14. Suppose $\triangle$ and $\perp$ represent two operations defined on a set S. Then the distributive law of $\perp$ over $\triangle$ is shown by

    (a) $(a \triangle b) \perp c = a \triangle (b \perp c)$ (b) $(a \perp b) \triangle c = a \perp (b \triangle c)$
    (c) $a \perp (b \triangle c) = (a \perp b) \triangle (a \perp c)$
    (d) $a \triangle (b \perp c) = (a \triangle b) \perp (a \triangle c)$

Problems 15–22 refer to the following information. Consider a set of elements $\{w, x, y, z\}$ together with the operation Θ, defined on the set:

Θ	w	x	y	z
w	x	z	w	y
x	z	y	x	w
y	w	x	y	z
z	y	w	z	x

15. Is the set closed for the operation Θ?

    (a) yes   (b) no

16. Is the operation $\ominus$ commutative?

    (a) yes   (b) no

17. Is the operation $\ominus$ associative?

    (a) yes   (b) no

18. The identity element for operation $\ominus$ is:

    (a) w (b) x (c) y (d) z (e) none of these

19. The inverse element for x^{-1} is:

    (a) w (b) x (c) y (d) z (e) none of these

20. The value of $(w \ominus x) \ominus (y \ominus z)$ is:

    (a) w (b) x (c) y (d) z (e) none of these

21. The value of $(w \ominus y)^{-1}$ is:

    (a) w (b) x (c) y (d) z (e) none of these

22. The value of $(w \ominus y^{-1})$ is:

    (a) w (b) x (c) y (d) z (e) none of these

Problems 23–26 refer to the following proof. In the proof, reasons for statements have been omitted. Supply the reasons (which justify each statement) from the choices following each question.

If 5 is an element of the set of integers I, then $-(-5) = 5$.

Proof

Statements	Reasons
1. $5 \in I$	1. Given
2. $(-5) \in I$	2. Refer to problem 23
3. $-(-5) + (-5) = 0$	3. Refer to problem 24
4. $[-(-5) + (-5)] + 5 = 0 + 5$	4. Refer to problem 25
5. $-(-5) + [(-5) + 5] = 0 + 5$	5. Refer to problem 26
6. $-(-5) + 0 = 0 + 5$	6. Refer to problem 27
7. $-(-5) = 5$	7. Refer to problem 28

23. (a) closure   (b) identity   (c) inverse   (d) Theorem 1B

24. (a) identity   (b) inverse   (c) Theorem 1A

    (d) Theorem 2B

25. (a) identity   (b) inverse   (c) Theorem 1A   (d) associative

26. (a) commutative   (b) Theorem 1A   (c) associative

    (d) inverse

27. (a) inverse   (b) identity   (c) associative

    (d) commutative

28. (a) inverse   (b) identity   (c) associative   (d) closure

8

STATISTICS

In our discussion of logic in Chapters 2, 3, and 4 we studied the structure of a valid argument. We were concerned there with developing our reasoning skills to enable us to express ourselves more clearly and to help us make better decisions in everyday life situations. In the chapter on probability, and again now in this chapter, we have demonstrated how to produce data that can be used in these decision making situations. We hope that this study will also enable the reader to examine more intelligently the claims and statements made by others. In newspapers and on radio and television we are bombarded with all kinds of statistical claims. If we know what the statistics mean and can evaluate them using the laws of logic, we will certainly be in a better position to make more intelligent decisions.

Example 1

A hospital has a labor dispute. Management claims that "80% of the nurses were rated as below average or average in their performance." On the other hand, the nurses' union claims that "80% of the nurses were rated as above average or average in their performance." The following table shows how *both* of these seemingly contradictory statements are *true*.

Actual Ratings

Above Average	Average	Below Average
20%	60%	20%

Without the actual data from the table, the claims made by the union and management would cause confusion or lead us to the conclusion that one side is lying.

Statistics may be defined, very loosely, as the study of numbers. A statistic may be something as simple as the sum of two numbers or as exotic as the coefficient of multiple correlation. Descriptive statistics is the branch of statistics that is concerned with the organization, summarization, and presentation of the data contained in a distribution. A distribution is a set of numerical values that represent a particular characteristic of a population. For example, the values may represent the ages of the people in the population (or a sample taken from the population); or it may represent their weights; or even the grades they achieved on an examination. Inferential statistics is the branch that deals with making projections and generalizations about a population based upon a sample. We might consider descriptive statistics as a deductive study (from the general to the specific) and inferential statistics as an inductive study (from the specific to the general). This is illustrated in the following diagram.

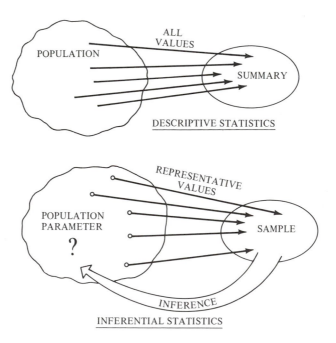

Our discussion will be confined to some of the basic and more commonly used concepts of descriptive statistics.

The main purpose of descriptive statistics is to take raw data, which are generally in an unorganized form, and to present them in an informative and meaningful manner. Suppose the test grades of 20 students were given as follows: (40, 57, 63, 94, 56, 87, 72, 48, 96, 82, 83, 70, 77, 74, 68, 94, 63, 86, 90, 80). Examining these data we can see that the unorganized form in which they are presented makes them difficult to analyze or interpret. One of the simplest methods of arranging raw data, especially when the amount of data is not too large, is that of placing the data in numerical order. A distribution which has been ordered is called an array.

Definition 1

Array: An array is a distribution of data which have been arranged in either ascending or descending order.

The array for the test scores now looks like this: (40, 48, 56, 57, 63, 63, 68, 70, 72, 74, 77, 80, 82, 83, 86, 87, 90, 94, 94, 96).
One particular statistic in which we may be interested regarding this distribution is the "average" test grade for the class.

Definition 2

Average: An average is a numerical value that is a measure of the central tendency of a distribution.

In our study we will discuss three "averages" or measures of central tendency, the mean, the median, and the mode.

Definition 3

Mean: A mean is the sum of the terms in the distribution divided by the number of terms in the distribution.

The sum of the test scores in the above distribution is 1,480. Thus the mean is 1,480 divided by 20, which is 74. The mean is the most frequently used measure of central tendency. When the word "average" is used, it is generally the mean that is being discussed.

We can now write a symbolic representation for the formula for the mean. To do this we must introduce a symbolic shorthand. We will use the capital letter X to represent the terms in the distribution. Collectively, the terms of the distribution are called the population. In this example the X can take the place of any one of our 20 test scores: $X = 82$, or $X = 94$, etc. Since X can assume the value of any of the terms in the distribution it is called a variable. Next we will use the capital Greek letter Σ (sigma) to represent the instruction "sum up" or "add up." N will stand for the number of terms in the distribution. Finally, we will use the Greek letter μ (mu) to indicate the mean of the population. Now, using the above shorthand, the symbolic representation of the formula for computing the mean is:

$$\mu = \frac{\Sigma X}{N}$$

This formula is read "the mean of the population is the sum of the terms divided by the number of terms."

Our second "average" is the median, which is simply the middle term in a distribution. It is the value that separates the lower half of the distribution from the upper half. Note that in order to locate the middle term it is necessary that the distribution be in the form of an array, that is, ordered.

Definition 4

Median: A median is the middle term of a distribution that has been ordered.

Symbolically, we use P_{50} to indicate the median of a distribution. P_{50} indicates that it is the 50% mark of the distribution. To illustrate, the median of the distribution (76, 78, 80, 80, 81) is 80 or $P_{50} = 80$.

The above distribution did have a median since there were an odd number of terms. However, when a distribution has an even number of terms, there is no middle term. What we do in this case is take the two middle terms and find their mean. There are 20 terms in the distribution of test grades example above, and the two middle terms are 74 and 77. Therefore, the median for this distribution is:

$$P_{50} = \frac{74 + 77}{2} = 75.5$$

The mode is our last "average" and the one that is used least. It is the term in the distribution that has the greatest frequency.

Definition 5

Mode: A mode is the term in the distribution that has the greatest frequency.

The mode of the distribution (1, 2, 3, 4, 4, 5) is 4, while the distribution (1, 2, 4, 5, 6, 7) does not have a mode. Sometimes a distribution may have two terms with the greatest frequency. For example, the distribution (1, 2, 2, 4, 4, 6) has two modes, 2 and 4. In this case we say the distribution is bi-modal, meaning two modes. The distribution of 20 test grades mentioned above is also bi-modal (63 and 94). In our study of statistics, if there are more than two terms with the greatest frequency, we will consider the distribution to have no mode. Therefore, a distribution will have either no mode, one mode, or two modes.

To summarize, then, we have found the following "averages" for the distribution of 20 test grades.

$$\text{Mean} = \mu = \frac{\Sigma X}{N} = 74$$

$$\text{Median} = P_{50} = 75.5$$

$$\text{Mode} = (\text{Bi-modal } 63, 94)$$

Consider a second distribution:

X
73
80
83
83
87
87
91
93
96
96
99

$\Sigma X = 968$

$$\text{Mean: } \mu \quad = \frac{\Sigma X}{N} = \frac{968}{11} = 88$$

Median: $P_{50} = 87$

Mode: None   (There is no mode since there are more than two terms that have the greatest frequency: 83, 87, 96.)

To conclude our discussion of the measures of central tendency, we will list some important facts about the mean, median, and mode.

The mean is the most frequently used measure of central tendency for several reasons. In computing the mean, the values of all the terms are used and it generally provides the most reliable "average." In addition, the mean lends itself more easily to further statistical application than either the median or the mode. Its one major weakness is that it can be affected by several extreme values. The median is also a frequently used and reliable measure of central tendency. It is not affected by extreme values, but it does not consider the values of the individual terms in its computation, only their relative positions.

The mode is the weakest and least used measure of central tendency since it considers only the term with the greatest frequency.

To illustrate the above ideas, consider the salaries of the ten employees of a certain company.

X
$ 50,000
27,000
14,000
13,500
13,000
12,500
10,000
9,000
8,000
8,000
$\Sigma X = \$165,000$

Mean: $\mu = \dfrac{\$165,000}{10} = \$16,500$

Median: $\dfrac{\$12,500 + \$13,000}{2} = \dfrac{\$25,500}{2} = \$12,750$

Mode: $8,000

During negotiations for a new contract management claimed the "average" salary of the employees was $16,500, while the labor union claimed the "average" salary was $8,000. The mediator said the "average" salary was $12,750. As we can see, these differences of opinion are due to the fact that each used a different measure of central tendency as their average. This points out that when someone uses the word "average" it is important that we know which measure of central tendency they are referring to. We can not assume that it is the mean. In the above example the median is the best "average" because it best represents the entire distribution of salaries. In addition, we can see that the mean salary was rather high due to the two extremely high salaries.

When doing a statistical report all three measures of central tendency should be given in order to present an accurate description of the distribution. Generally, the mean, median, and mode will have similar values. If, however, the mean is considerably larger than the median this indicates that there were some extremely high scores which tended to pull the mean above the median (See A). When the mean is considerably less than the median, this indicates that there are some extremely low scores which tended to pull the mean down below the median (See B). This is illustrated in the following drawings.

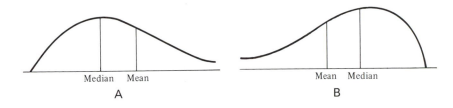

Median   Mean   Mean   Median

A   B

Exercises 8.1

1–6 For the following data sets calculate the mean, median, and mode.

1. (4, 6, 1, 2, 8, 9, 10, 7, 3, 5)

2. (7, 3, 13, 1, 19, 13, 9, 5, 11)

3. (6, 2, 4, 3, 3, 2, 7, 6, 4, 5, 5, 4)

4. (2, 7, 4, 3, 1, 8, 20, 37, 26)

5. (5, 10, 6, 16, 24, 18, 10, 12, 3, 7)

6. (10, 12, 15, 6, 7, 8, 1, 3, 2, 5, 4, 8, 27, 43, 14, 18, 26, 16, 37, 18)

Measures of Variability

We have seen that the measures of central tendency, especially the mean, are extremely useful in describing or summarizing the information contained in a distribution. However, the measures of central tendency by themselves only provide a partial description of a distribution. To illustrate, consider two students who both have a mean grade of 80 on five statistic tests.

Student I	Student II
$\mu = 80$	$\mu = 80$

By using only their respective means to compare the two students we would be forced to the conclusion that they were basically the same in their statistics achievement. Now, if we list the five examination grades for both students we can see that the conclusion we reached, based on comparing their respective means, is false.

Student I		Student II	
100		80	
70		80	
60	$\mu = 80$	80	$\mu = 80$
90		80	
80		80	

The two students were very different in their performances. Student I was very inconsistent since his exam grades varied a great deal, while Student II was extremely consistent with no variation at all.

The above example points out that the mean by itself provides only a partial description of the distribution and that we need some measure of consistency or variability if we are to present a complete and accurate summary.

In order to help us discover an appropriate measure of variability of the terms of a distribution we will examine the following exam grades of four students.

Student I	Student II	Student III	Student IV
78	76	74	74
79	78	78	75
80	80	80	80
81	82	82	85
82	84	86	76
$\Sigma X = 400$	400	400	400

Since the sum of each of the four distributions is 400, the mean grade of each student is 80, since $\mu = \dfrac{\Sigma X}{N} = \dfrac{400}{5} = 80$. Also, the median grade of each student is 80 and none of the distributions has a mode. Therefore, they all have the same measures of central tendency but an examination of the four distributions shows that they are all different. They are different in the amount of variation of the grades. Consider the table:

	74	75	76	77	78	79	80	81	82	83	84	85	86
Student I					X	X	X	X	X				
Student II			X		X		X		X		X		
Student III	X				X		X		X				X
Student IV	X	X					X					X	X

Student I's grades have the least amount of variation; Student II's grades have more variation than Student I; Student III's grades have more variation than Student II; and finally, Student IV's grades have the greatest amount of variation.

The simplest way to measure the spread of the different distributions is to compare the spread of the highest and lowest terms. This difference, which is our first measure of variability, is called the range.

Definition 6

Range: The range of a distribution is the difference between the highest and lowest terms in the distribution. (RANGE = Highest Term − Lowest Term.)

The range for the four distributions of exam grades are:

Student I	Student II	Student III	Student IV
$82 - 78 = 4$	$84 - 76 = 8$	$86 - 74 = 12$	$86 - 74 = 12$

In comparing the ranges and the variability of the four distributions, we see that as the variability of the grades increases, the value of the ranges increases. This is true with the exception of Student IV. In comparing Students III and IV, we said that the grade distribution of Student IV had more variability than the distribution of Student III. This is true since the second and fourth scores (75 and 85) of Student IV are farther apart than the second and fourth scores (78 and 82) of Student III. Their ranges, which are both 12, do not show this difference in variability since the range only considers the two extreme values of highest and lowest scores. This points out the major weakness of the range: it does not measure the internal variability of the distributions. The advantage of the range is that it is easy to compute.

Since the range does not provide an adequate measure for comparing the variability we must find a measure that takes into account the spread of all the terms. In discussing the spread or variability of a distribution it is convenient to talk about the spread from a particular point. One logical point from which to measure the spread of the terms would be the center of the distribution. The mean of the distribution was selected from among our three measures of central tendency since the mean has a unique property. The mean is the true physical center of a distribution. To illustrate this idea let us consider the scores (76, 78, 80, 80, 81). The mean of these five scores is 79. Now, if we were to represent each of the scores by a block of equal weight and were to place these weights on a balance board, we would see that the only time that the board would balance or be in equilibrium would be when the balance wedge is placed at the mean (79).

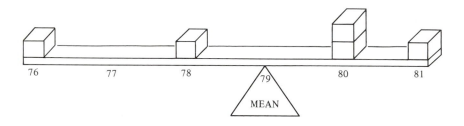

If it were placed at any other point under the board it would not balance. This balancing of the board at the mean does not happen by accident. If we examine the distance that each measurement is from the mean, we will see why the board balances at the mean.

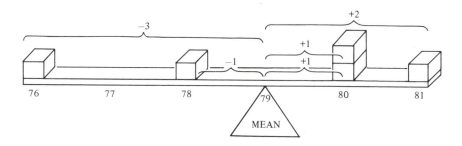

We assign the distances below the mean negative values since they are less than the mean, and positive values to the distances above the mean since they are greater than the mean. These distances that the scores are from the mean are called deviations.

Definition 7

Deviation: A deviation is the distance a term is from the mean of the distribution. A positive deviation indicates that the term is greater than the mean, and a negative deviation indicates that the term is less than the mean.

Examining the deviations $(-3, -1, +1, +1, +2)$ we see that they sum up to zero. That is, the negative weight cancels the positive weight causing the board to balance.

In symbolic form, the deviation of a term X from its mean μ is written $(X - \mu)$. The symbolic representation showing that the sum of the deviations from the mean sums up to zero is:

$$\Sigma(X - \mu) = 0$$

This is why the mean was selected as the center of the distribution and the point from which we will measure the spread of the terms.

Now we will return to the distributions of the exam grades of the four students and compute the deviations from the mean for each.

Student I		Student II		Student III		Student IV	
X	$X - 80$	X	$X - 80$	X	$X - 80$	X	$X - 80$
78	-2	76	-4	74	-6	74	-6
79	-1	78	-2	78	-2	75	-5
80	0	80	0	80	0	80	0
81	$+1$	82	$+2$	82	$+2$	85	$+5$
82	$+2$	84	$+4$	86	$+6$	86	$+6$
	0		0		0		0

We see that the sum of the deviations from the mean for each distribution equals zero or $\Sigma(X - \mu) = 0$.

Returning to the problem of trying to develop a measure of the spread of the distributions we see that as the spread of the terms in the distributions increases, the size of the deviations in the distributions also increases. That is, the deviations of Student II's grades are greater than the deviations of Student I's grades; Student III's deviations are greater than Student II's; and Student IV's deviations are greater than Student III's. Therefore, it seems that we could obtain a measure of the spread of the terms by examining the deviations. We cannot use the mean of the deviations as our measure of spread since they always sum up to zero and the means would all be zero.

This problem of the sum of the deviations always being zero was overcome by squaring the deviations. Squaring the deviations of the distributions of the grades of the four students, we have

	Student I			Student II	
X	$(X - 80)$	$(X - 80)^2$	X	$(X - 80)$	$(X - 80)^2$
78	-2	4	76	-4	16
79	-1	1	78	-2	4
80	0	0	80	0	0
81	$+1$	1	82	$+2$	4
82	$+2$	4	84	$+4$	16
		10			40

	Student III			Student IV	
X	$(X - 80)$	$(X - 80)^2$	X	$(X - 80)$	$(X - 80)^2$
74	-6	36	74	-6	36
78	-2	4	75	-5	25
80	0	0	80	0	0
82	$+2$	4	85	$+5$	25
86	$+6$	36	86	$+6$	36
		80			122

The squares of the deviations are all positive and as the spread of the terms in the distributions increases, the size of the squared deviations of the distributions also increases. That is, the greater the spread or variability of the terms in a distribution, the greater the value of the squared deviations. Therefore, it was decided that the mean or average of the squared deviations would provide a reliable measure of variability or spread. Examining the mean of the squared deviations of the four distributions of the students grades we see that as the spread of the distributions increases, the means of the squared deviations (2, 8, 16, 24.4) also increases. This mean of the squared deviations from the mean is called the variance. The symbol for the variance of a population is the Greek letter sigma squared σ^2.

Student I	Student II	Student III	Student IV
$\dfrac{10}{5} = 2$	$\dfrac{40}{5} = 8$	$\dfrac{80}{5} = 16$	$\dfrac{122}{5} = 24.4$

Definition 8

Variance: The variance of a distribution is the mean of the squared deviations from the mean of the distribution.

The formula for the variance is:

$$\sigma^2 = \frac{\Sigma(X - \mu)^2}{N}$$

The variance for the four distributions of student grades are:

Student I	Student II	Student III	Student IV
$\sigma^2 = 2$	$\sigma^2 = 8$	$\sigma^2 = 16$	$\sigma^2 = 24.4$

The variances show that Student II's grades vary more than Student I's; Student III's grades vary more than Student II's; and that Student IV's grades vary more than Student III's. The variance provides an effective measure of variability or spread that takes into consideration the spread of all the terms in the distribution.

As a final step in developing a measure of variability or spread, it was decided that since in squaring the deviations we made them numerically larger, we could reduce the variance by performing the opposite operation of taking the square root. Now the units are the same as they were in the original distribution. (If the units in the original distribution were $'s, then a variance of, say, 3.7 ^{2} would be meaningless.) The square root of the variance is called the standard deviation and is symbolically represented by σ.

Definition 9

Standard deviation: A standard deviation is the square root of the variance, that is, the square root of the mean of the squared deviations.

$$\sigma = \sqrt{\frac{\Sigma(X - \mu)^2}{N}}$$

The standard deviations for the four distributions of the student grades are:

Student I	Student II	Student III	Student IV
$\sigma = \sqrt{2} = 1.41$	$\sigma = \sqrt{8} = 2.83$	$\sigma = \sqrt{16} = 4$	$\sigma = \sqrt{24.4} = 4.94$

The standard deviations (1.41, 2.83, 4, 4.94) show that as the variability or spread of the distributions increases, the value of the standard deviations increases.

The standard deviation, which in a sense is "the average of the spread of the terms in a distribution," will be the measure of variability that we will use in the remainder of our work.

The procedures for computing the variance and standard deviation can be summarized as follows:

Procedure for Computing the Variance and Standard Deviation

1. Compute the mean. $\mu = \dfrac{\Sigma X}{N}$

2. Subtract the mean from each measurement to obtain the deviation. $(X - \mu)$

3. Square each of the deviations. $(X - \mu)^2$

4. Sum up the squares of the deviation. $\Sigma(X - \mu)^2$

5. Divide the sum of the squares by the number of terms to obtain the variance.
$$\sigma^2 = \frac{\Sigma(X - \mu)^2}{N}$$

6. Take the square root of the variance to obtain the standard deviation.
$$\sigma = \sqrt{\frac{\Sigma(X - \mu)^2}{N}}$$

To illustrate the above procedure, consider the following distributions (76, 78, 80, 80, 81) and (75, 80, 80, 83, 83, 86, 88, 91, 93, 96, 96, 99). The range, variance, and standard deviations for these distributions are computed as follows:

X	$(X - 80)^2$	$(X - 80)^2$
76	-3	9
78	-1	1
80	$+1$	1
80	$+1$	1
81	$+2$	4
395	0	16
ΣX	$\Sigma(X - \mu)$	$\Sigma(X - \mu)^2$

$$\mu = \frac{\Sigma X}{N} = \frac{395}{5} = 79$$

$$\text{Range} = H - L = 81 - 76 = \underline{\underline{5}}$$

$$\sigma^2 = \frac{\Sigma(X - \mu)^2}{N} = \frac{16}{5} = \underline{\underline{3.2}}$$

$$\sigma = \sqrt{\frac{\Sigma(X - \mu)^2}{N}} = \sqrt{3.2} = 1.79$$

For the second distribution:

X	$(X - 87.5)$	$(X - 87.5)^2$
75	-12.5	156.25
80	-7.5	56.25
80	-7.5	56.25
83	-4.5	20.25
83	-4.5	20.25
86	-1.5	2.25
88	.5	.25
91	3.5	12.25
93	5.5	30.25
96	8.5	72.25
96	8.5	72.25
99	11.5	132.25
1050	0	631.00
ΣX	$\Sigma(X - \mu)$	$\Sigma(X - \mu)^2$

$$\mu = \frac{\Sigma X}{N} = \frac{1050}{12} = \underline{87.5}$$

$$\text{Range} = H - L = 99 - 75 = \underline{24}$$

$$\sigma^2 = \frac{\Sigma(X - \mu)^2}{N} = \frac{631}{12} = \underline{52.58}$$

$$\sigma = \sqrt{\frac{\Sigma(X - \mu)^2}{N}} = \sqrt{52.58} = \underline{7.25}$$

The above examples illustrate the procedure for computing the variance and standard deviation. In the first example it was a relatively easy task to compute the deviations and their squares since the mean (79) was a whole number. However, in the second illustration the work became more tedious since the mean (87.5) was not a whole number and there were a larger number of terms in the distribution. The task of computing the deviations and their squares became more troublesome. To overcome this problem of computing the variance and standard deviation for distributions that have a large number of terms or that have a mean that is not an integer, mathematicians algebraically rearranged the formulas for the variance and standard deviations to a more workable form. They are:

Variance   Standard Deviation

$$\sigma^2 = \frac{\Sigma(X^2) - \frac{(\Sigma X)^2}{N}}{N} \qquad \sigma = \sqrt{\frac{\Sigma(X^2) - \frac{(\Sigma X)^2}{N}}{N}}$$

The above formulas for computing the variance and standard deviation are called the computational form since they are more easily used for distributions that do not have integral means, and they are especially useful when a calculator is available.

Before applying these formulas to computing the variance and standard deviation we should first examine the parts of the formulas. We see that there are three values that must be filled in before we can compute the variance and standard deviation. They are:

N—the number of terms in the distribution
ΣX—the sum of the terms in the distribution
$\Sigma(X^2)$—the sum of the squares of the terms in the distribution

It is important to note that $(\Sigma X)^2$ and $\Sigma(X^2)$ are two completely different expressions. $(\Sigma X)^2$ instructs us to sum up the terms and then square the sum, while $\Sigma(X^2)$ tells us to square the terms and then sum up the squares. Therefore, $(\Sigma X)^2$ *does not equal* $\Sigma(X^2)$. This difference will be made clear in the following examples.

Let us now compute the variance and standard deviation using the computational formulas for the distributions we did previously.

X	X^2
76	5776
78	6084
80	6400
80	6400
81	6561
395	31,221
ΣX	$\Sigma(X^2)$

$$\sigma^2 = \frac{\Sigma(X^2) - \dfrac{(\Sigma X)^2}{N}}{N} = \frac{31,221 - \dfrac{(395)^2}{5}}{5}$$

$$= \frac{31,221 - \dfrac{156,025}{5}}{5} = \frac{31,221 - 31,205}{5}$$

$$= \frac{16}{5} = \underline{3.2}$$

$$\sigma = \sqrt{\frac{\Sigma(X^2) - \dfrac{(\Sigma X)^2}{N}}{N}} = \sqrt{3.2} = \underline{1.79}$$

For the second distribution:

X	X^2
75	5625
80	6400
80	6400
83	6889
83	6889
86	7396
88	7744
91	8281
93	8649
96	9216
96	9216
99	9801
1050	92506
ΣX	$\Sigma(X^2)$

$$\sigma^2 = \frac{\Sigma(X^2) - \dfrac{(\Sigma X^2)}{N}}{N} = \frac{92506 - \dfrac{(1050)^2}{12}}{12}$$

$$= \frac{92506 - \dfrac{1102500}{12}}{12} = \frac{92506 - 91875}{12}$$

$$= \frac{631}{12} = \underline{52.58}$$

$$\sigma = \sqrt{\frac{\Sigma(X^2) - \dfrac{(\Sigma X)^2}{N}}{N}} = \sqrt{52.58} = \underline{7.25}$$

Comparing, we see that the results obtained using the computational formulas are exactly the same as those we previously obtained.

While this procedure may not seem easier than the first, the time-consuming process of obtaining ΣX and $\Sigma(X^2)$ can be rapidly and easily accomplished by the use of an electronic calculator, many of which can simultaneously compute both sums.

As a final illustration of the procedures discussed in this section we will compute the mean, median, mode, range, variance, and standard deviation for the distribution (1, 2, 2, 5, 6, 8).

$$\text{Mean: } = \frac{\Sigma X}{N} = \frac{24}{6} = \underline{4}$$

$$\text{Median: } P_{50} = \frac{2+5}{2} = \frac{7}{2} = \underline{3.5}$$

X	$(X-4)$	$(X-4)^2$
1	-3	9
2	-2	4
2	-2	4
5	$+1$	1
6	$+2$	4
8	$+4$	16
24	0	38
ΣX	$\Sigma(X-\mu)$	$\Sigma(X-\mu)^2$

$$\text{Mode: } \underline{2}$$

$$\text{Range: } H - L = 8 - 1 = \underline{7}$$

$$\text{Variance: } \sigma^2 = \frac{\Sigma(X-\mu)^2}{N}$$

$$= \frac{38}{6} = \underline{6.33}$$

$$\text{Standard deviation: } \sigma = \sqrt{\frac{\Sigma(X-\mu)^2}{N}}$$

$$= \sqrt{6.33} = \underline{2.52}$$

Using the computational formulas for the variance and standard deviation we have:

Variance:

X	X^2
1	1
2	4
2	4
5	25
6	36
8	64
24	134
ΣX	$\Sigma(X^2)$

$$\sigma^2 = \frac{\Sigma(X^2) - \frac{(\Sigma X)^2}{N}}{N}$$

$$\sigma^2 = \frac{134 - \frac{(24)^2}{6}}{6} = \frac{134 - \frac{(576)}{6}}{6}$$

$$\sigma^2 = \frac{134 - 96}{6} = \frac{38}{6} = \underline{6.33}$$

Standard Deviation:

$$\sigma = \sqrt{\frac{\Sigma(X^2) - \frac{(\Sigma X)^2}{N}}{N}} = \sqrt{6.33} = \underline{2.52}$$

Exercises 8.2

1–5 For the following data sets calculate the mean, median, mode, range, variance, and standard deviation.

1. (1, 3, 13, 11, 5, 7, 9)

2. (10, 16, 2, 4, 8, 6, 10)

3. (13, 14, 21, 5, 9, 1, 21)

4. (13, 17, 2, 1, 17, 21, 14, 3)

5. (21, 20, 4, 7, 1, 4, 20)

Percentile Rank and Z Score

So far we have been concerned with summarizing and describing a distribution by numerical values. In this section we will be concerned with locating or describing a single score's position in a distribution of scores. That is, before we were concerned with the entire distribution and now we will be concerned with describing where an individual stands within the distribution.

What is one of the first things you do when a graded examination paper is returned? Probably, after first checking it over, you look around to see how other people did on the exam. You do this in order to determine how you scored compared to the other students in the class. If you received a grade of 80 while the other students received 90's and 100's, you would most likely be disappointed because you scored lower than everyone else. On the other hand, if you received a grade of 80, while the other students received grades of 60 and 70, you probably would feel satisfied since you scored higher than everyone else. This illustrates the fact that a grade by itself does not present complete information. What is needed is the grade's position relative to the other grades.

Percentile Rank

One way to describe a measurement's position or location in a distribution of scores is to give that score's percentile rank.

Definition 10

Percentile rank: The percentile rank of a term in a distribution is the percent of terms in the distribution less than or smaller than that term.

Note: To calculate a term's percentile rank, it is necessary that the terms in the distribution be ordered.

For example, if you received a grade of 82 in a mathematics exam and you were at the 75th percentile, this would indicate that 75% of the exam scores were less than your score of 82 and also that 25% of the scores were greater than your score. Symbolically, we would write this as $82 = P_{75}$.

$\left.\uparrow\right\}$ 25% of the scores

Your grade of 82

$\left.\downarrow\right\}$ 75% of the scores

That is, a test score of 82 is the 75th percentile.

As with the other descriptive statistics, percentiles are generally used when the distributions are larger. However, for the sake of simplicity we will use small distributions in illustrating the procedure used to calculate percentile ranks.

Consider the distribution (70, 73, 75, 76, 83, 85, 86, 87, 90) of ten examination scores. What is the percentile rank of the grade of 85? That is, $85 = P_?$.

$$
\begin{array}{l}
\left.\begin{array}{l} 90 \\ 87 \\ 86 \end{array}\right\} 30\% \qquad\qquad 40\% \\[2pt]
\left.\begin{array}{l} 85 \\ 85 \end{array}\right\} 20\% \left.\begin{array}{l} 10\% \\ 10\% \end{array}\right. \\[2pt]
\left.\begin{array}{l} 83 \\ 76 \\ 75 \\ 73 \\ 70 \end{array}\right\} 50\% \qquad\qquad 60\%
\end{array}
$$

Since there are ten scores, each score represents 10% of the distribution. We see that there are five scores or 50% of the distribution less than the score of 85. However, we would not state that 85 is the 50th percentile since this would lead to the false conclusion that 50% of the scores are greater than 85. Upon examining the distribution we see that only three scores or 30% of the distribution are greater than the score of 85. This is due to the fact that there are two 85's which represent 20% of the distribution. We overcome this problem by dividing the 20%, representing the two 85's, into two equal parts of 10% each and then assigning 10% to the portion of the distribution below 85 and 10% to the portion above 85. We now state that the test score of 85 is the 60th percentile, i.e., $86 = P_{60}$. Consider another example:

$$
\begin{array}{l}
\left.\begin{array}{l}
90 \\
87 \\
86 \\
85 \\
85 \\
83 \\
76
\end{array}\right\} 70\% \\
75 \quad 10\% \\
\left.\begin{array}{l}
73 \\
70
\end{array}\right\} 20\%
\end{array}
$$

(with brackets: 70% group and 75 10%, 5%, combining to 75%; 5%, and 73/70 20% combining to 25%)

As a further illustration we will find the percentile rank of the score of 75 for this distribution. There are two scores or 20% of the distribution less than 75. There is one score of 75 or 10% of the distribution. Dividing the 10% representing the one 75 into two equal parts of 5% each and assigning 5% to portion of scores below 75, we see that a score of 75 is the 25th percentile; i.e., $75 = P_{25}$.

We can now write a formula for the above procedure for calculating the percentile rank of a score. The formula is:

$$
\text{Percentile rank of a score } X = \left[\frac{B + 1/2(E)}{N}\right] \times 100 \text{ where}
$$

B = the number of scores less than or smaller than X.

E = the number of scores with the value of X.

N = the total number of scores in the distribution.

Note: The distribution must be ordered, and the percentile rank is generally rounded off to the nearest whole number.

To illustrate the use of the formula we will use it to compute the percentile rank of the scores of 85 and 75 from the above distribution.

(a) $85 = P_?$

   $B = 5$ scores less than 85
   $E = 2$ scores of 85
   $N = 10$ scores in the distribution
   substituting in the formula we have:

$$\text{Percentile rank of } 85 = \left[\frac{5 + 1/2(2)}{10}\right] \times 100$$

$$= \left[\frac{5 + 1}{10}\right] \times 100$$

$$= \frac{6}{10} \times 100 = 60 \text{ or}$$

$$85 = P_{60}$$

(b) $75 = P_?$

   $B = 2$ scores less than 75
   $E = 1$ score of 75
   $N = 10$ scores in the distribution
   subsituting in the formula we have:

$$\text{Percentile rank of } 75 = \left[\frac{2 + 1/2(1)}{10}\right] \times 100$$

$$= \left[\frac{2 + .5}{10}\right] \times 100$$

$$= \frac{2.5}{10} \times 100 = 25 \text{ or}$$

$$75 = P_{25}$$

As a final example, consider the following problem: If 1,000 students took the math SAT exam and 685 students scored less than 550 and 305 students scored above 550, then what is the percentile of a score of 550?

Since 685 students scored below 550 and 305 students scored above 550, representing 990 of the 1,000 students, we can conclude that 10 students scored 550. Therefore, for a score of 550, we have $B = 685$,

$E = 10$, $N = 1,000$ and

$$\text{Percentile rank of a score of } 550 = \left[\frac{685 + 1/2(10)}{1,000} \right] = 100$$

$$= \left[\frac{685 + 5}{1,000} \right] \times 100$$

$$= \frac{690}{1,000} \times 100$$

$$= 69 \text{ or } 550 = P_{69}$$

Z Score

Suppose two students enrolled in two different statistics class sections with different instructors each received grades of 80 on their different final exams. Could we conclude that both students achieved about the same in statistics? The answer is no. As students know, two different instructors may have completely different standards. One of the instructors may give more difficult exams than the other or may mark the exams more critically. The only way we would compare the two students' knowledge of statistics would be by giving them both the same exam graded by the same criteria.

We cannot compare the students' knowledge of statistics based on different exam grades in different statistics sections, however it might be possible to compare them based on how each did in his respective class; that is, by comparing how each of the two students did in comparison to the other students in his class. We would do this by comparing their respective percentile ranks. If one student's exam grade of 80 represented the 90th percentile in his class while the other student's grade of 80 was the 60th percentile in his class, we would conclude that the first student did better in his statistics course relative to his class than the second student did relative to his. Now consider the case in which the grade of 80 represents the 90th percentile in both classes. Could we conclude that both students did equally as well in their respective classes? Again, the answer is no. To illustrate this conclusion suppose there were five students in each class and the grade distributions for both classes were as follows:

Class I		Class II	
$80 = P_{90}$		$80 = P_{90}$	
79		70	
79		65	
79	$\mu = 79$	60	$\mu = 66$
78		55	

We see that in both classes the grade of 80 is the 90th percentile. However, examining the two distributions we see that a grade of 80 in Class I is only one point above the mean grade of 79 while a grade of 80 in Class II is 14 points above the mean grade of 66. Therefore, it would seem logical to conclude that the student who received a grade of 80 in Class II did better with respect to his class than the student who received a grade of 80 in Class I.

Note that in making this comparison we assumed that one point on the Class I exam has the same meaning as one point on the Class II exam. This is not always the case. Consider a third statistics class with the following distribution of final exam grades.

$$\underline{\text{Class III}}$$
$$14 = P_{90}$$
$$9$$
$$9$$
$$9 \qquad \mu = 10$$
$$9$$

How can we compare the student who earned a final exam grade of 14 in Class III with the student who earned a final grade of 80 in Class II? Both the 14 and the 80 are the 90th percentile in their respective classes. The student with an 80 in Class II is 14 points above the class mean of 66, but we cannot compare this difference to the student in Class III whose grade of 14 is four points above his class mean of ten. It is obvious that one exam point in Class III has a much greater weight than one exam point in Class II.

Statisticians have overcome this problem of comparing differences that might be based on different value systems by comparing the number of standard deviations a score is different from the mean.

The standard deviation takes into account the different value systems used in grading and neutralizing them in the comparisons. That is, by comparing the number of standard deviations the different test grades are from their respective means, we are comparing the same units of measure, namely, standard deviations. We call this measurement a z score.

Definition 11

z score: The z score of a term in a distribution is the number of standard deviations that the term is either above or below the mean of the distribution. The z score of a term X, from a distribution with a mean μ and standard deviation σ, is given by

$$Z_X = \frac{X - \mu}{\sigma}$$

To illustrate, consider the final grade distribution of the three statistics classes given above. The mean and standard deviation for each class is also listed.

Class I	Class II	Class III
80	80	14
79	70	9
79	65	9
79	60	9
78	55	9
$\mu = 79$	$\mu = 66$	$\mu = 10$
$\sigma = .63$	$\sigma = 8.60$	$\sigma = 2$

We will now compute the z score for the grade of 80 in Class I, the grade of 80 in Class II, and the grade of 14 in Class III, all of which are the 90th percentile in their respective classes.

$$Z_{80} = \frac{80 - 79}{.63} \qquad Z_{80} = \frac{80 - 66}{8.60} \qquad Z_{14} = \frac{14 - 10}{2}$$

$$= \frac{+1}{.63} \qquad\qquad = \frac{14}{8.60} \qquad\qquad = \frac{4}{2}$$

$$= 1.59 \text{ standard} \qquad = 1.63 \text{ standard} \qquad = 2.00 \text{ standard}$$
deviations above the mean (Class I), deviations above the mean (Class II), deviations above the mean (Class III)

Comparing the z scores, we can conclude that the student who scored a grade of 14 in Class III did better within his class than the two students with 80's did within their classes, since he is 2 standard deviations above his class' mean while they are 1.63 and 1.59 above their class means. In addition, we see that the student with a grade of 80 in Class II did better within his class than the student with a grade of 80 in Class I.

Remember, we have not shown which of the three students has the greatest knowledge of statistics. All we have established is which student did best within his respective class.

As a further example of the usefulness of z scores consider the following problem. Suppose Ford Motor Company wants to honor their top salesman of the year. A problem arises since the company is made up of three separate divisions: Ford, Mercury, and Chrysler. They select the top salesman from each division by the number of cars sold for the year. However, they want to select one of these three division winners as Ford Motor Company's top salesman of the year. It would be unfair to select the winner by the total number of cars sold since there are more lower priced Fords sold than the medium-priced Mercury, and more medium-priced Mercurys sold than high-priced Chryslers. One way to select the top salesman from among the three division winners would be by comparing their respective z scores for the number of cars sold. To do this we need the sales statistics for each of the three divisions; that is, the mean number of cars sold per salesman and the accompanying standard deviations. The mean and standard deviation of the number of cars sold per salesman for each division as well as the number of cars sold by the division's top salesman are given below.

Cars Sold

	Ford Division	Mercury Division	Chrysler Division
Top Salesman	49	34	25
Division Mean	36	24	18
Standard Deviation	4	3	2.5

The number of standard deviations the division winners are above their divisional average is given below:

$$Z_{49} = \frac{49 - 36}{4} \qquad Z_{34} = \frac{34 - 24}{3} \qquad Z_{25} = \frac{25 - 18}{2.5}$$

$$= \frac{13}{4} \qquad\qquad = \frac{10}{3} \qquad\qquad = \frac{7}{2.5}$$

$$= 3.25 \qquad\qquad = 3.33 \qquad\qquad = 2.80$$

Comparing the z scores of the three divisional winners we see that the top salesman of the Mercury Division would be selected Ford Motor Company's Salesman of the Year.

Sometimes when we know the mean and standard deviation of a distribution we are interested in finding terms which are a specified number of standard deviations from the mean. In other words, we are looking for the raw score x of a distribution which has a specified z score. For example, if the distribution of IQ scores has a mean of 100 and a standard deviation of 15, find the IQ score that is 2 standard deviations above the mean. That is, what IQ score has a z score of $+2.00$? To simplify the solution of this problem we can algebraically rearrange the formula:

$$Z_x = \frac{x - \mu_x}{\sigma_x}$$

and solve it for x, the raw score, in terms of the z score. Doing this, we have:

$$x = \mu_x + (Z_x) \cdot (\sigma_x)$$

Now using this formula for the above problem, we have $\mu_x = 100$, $\sigma_x = 15$, $Z_x = +2.00$ and

$$X = 100 + (+2.00)(15)$$
$$X = 100 + 30$$
$$X = 130$$
$$\text{or } Z_{130} = +2.00$$

To find an IQ score one and one-half standard deviations below the mean, we have $\mu_x = 100$, $\sigma_x = 15$ and $Z_x = -1.50$.

$$X = 100 + (-1.50)(15)$$
$$X = 100 - 22.5$$
$$X = 77.5$$
$$\text{or } Z_{77.5} = -1.50$$

Note: A negative z score indicates that the raw score x is below the mean; a positive z score indicates that the raw score x is above the mean and a z score of zero indicates that the raw score is the mean.

To conclude this section we will investigate two important properties of z scores. We will demonstrate these properties by use of the following example. For the distribution (1, 4, 8, 10, 12) we will compute the mean and standard deviation.

X	$(X - 7)$	$(X - 7)^2$
1	-6	36
4	-3	9
8	$+1$	1
10	$+3$	9
12	$+5$	25
35	0	80

$$\mu_x = \frac{\Sigma X}{N} = \frac{35}{5} = \underline{7}$$

$$\sigma_x = \sqrt{\frac{\Sigma(X - \mu_x)^2}{N}} = \sqrt{\frac{80}{5}}$$

$$= \sqrt{16} = \underline{4}$$

Using the mean (7) and standard deviation (4) we will compute the z scores of the five terms in the distribution:

Z_X	$\dfrac{X - \mu_x}{\sigma_x} =$	Z_X	or
1	$\dfrac{1 - 7}{4}$	-1.50	$Z_1 = -1.50$
			$Z_4 = -\ .75$
4	$\dfrac{4 - 7}{4}$	$-\ .75$	$Z_8 = \ \ .25$
			$Z_{10} = \ \ .75$
8	$\dfrac{8 - 7}{4}$	$.25$	$Z_{12} = \ \ 1.25$
10	$\dfrac{10 - 7}{4}$	$.75$	
12	$\dfrac{12 - 7}{4}$	1.25	

Now we will calculate the mean and standard deviation for the distribution of z scores $(-1.50, -.75, +.25, +.75, +1.25)$.

Z	$Z - \mu_z$	$(Z - \mu_z)^2$	
-1.50	-1.50	2.2500	$\mu_z = \dfrac{\Sigma Z}{N} = \dfrac{0}{5} = \underline{0}$
$-\ .75$	$-\ .75$	$.5625$	
$+\ .25$	$+\ .25$	$.0625$	$\sigma_z = \sqrt{\dfrac{(Z - \mu_z)^2}{N}} = \sqrt{\dfrac{5}{5}}$
$+\ .75$	$+\ .75$	$.5625$	
$+1.25$	$+1.25$	1.5625	
$\Sigma Z = 0$	0	5.0000	$= \sqrt{1} = 1.00$

Therefore, we have the mean of the z scores equals 0 and the standard deviation of the z scores equals 1. This example illustrates the general principle that for a distribution of z scores we have: $\mu_z = 0 \qquad \sigma_z = 1$.

Frequency Table—
Frequency Polygon—Normal Curve

Suppose we have a distribution of numbers and want to determine the number of times each value appears. One way to do this is to construct a frequency table. A frequency table consists of three columns. The first column contains each of the elements of the distribution, without repetition. The second column is called the Tally. In this column we place a mark next to the value in column one each time it appears as we go through the distribution. Finally, after we have taken the tally we add up the marks for each value and place the sum in the third column, called the Frequency.

Consider the distribution 1, 3, 4, 3, 2, 3, 5, 1, 4, 2, 6, 4, 5, 6, 4 and the resulting frequency table:

Frequency Table

Value	Tally	Frequency
1	||	2
2	||	2
3	|||	3
4	||||	4
5	||	2
6	|	1

We can now construct a bar graph from this data. First, draw a scale on a vertical axis to represent the frequency and a scale on a horizontal axis to represent the individual values. Now, draw bars (vertical rectangles) above each number on the horizontal axis up to the point on the vertical scale that matches the frequency of that number.

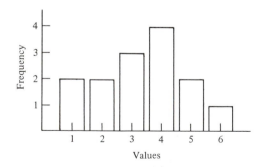

If we make the bars thick enough there will be no spaces between them (except when the frequency of a particular value is zero). When the vertical lines separating the bars are removed, we have a special kind of bar graph called a histogram.

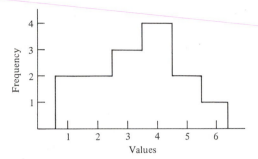

A histogram can also be constructed for a distribution with a large number of elements by letting each bar represent the frequency of elements in an interval rather than a single value. Using this method, the following frequency table and histogram have been constructed from the data set (20, 26, 11, 30, 32, 41, 40, 17, 33, 39, 43, 19, 24, 34, 36, 38, 42, 47, 30, 25, 14, 23, 20, 34, 37, 28, 16, 23, 33, 35).

Interval	Tally	Frequency
10–14	||	2
15–19	|||	3
20–24	ⅢⅢ	5
25–29	|||	3
30–34	ⅢⅢ||	7
35–39	ⅢⅢ	5
40–44	||||	4
45–49	|	1

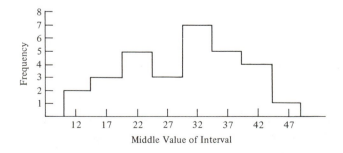

Once the histogram is constructed we may draw line segments connecting the midpoints of the top of each bar. The series of segments begins and ends on the horizontal axis. The resulting figure is called a frequency polygon.

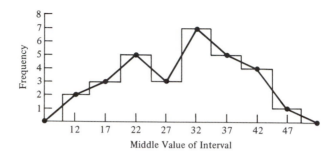

As the number of bars in the histogram increases, the segments connecting the midpoints become smaller and tend to smooth out the frequency polygon. Eventually, when the number of bars is very large, the "polygon" appears to be a smooth curve. We will find that distributions can yield many different types of "curves." For the rest of this discussion we will concentrate on one special "bell-shaped" curve which is the graph of a so-called "normal" distribution. It is given special attention because it represents many familiar distributions such as the distribution of measures of height, weight, IQ, and results on standardized tests.

The normal curve also exhibits some very special characteristics. In order to express these characteristics clearly we will place a vertical axis inside the curve where the curve reaches its maximum height. Rather than having the actual values, or even intervals of values, the horizontal axis scale will be the z scores of the elements in the distribution.

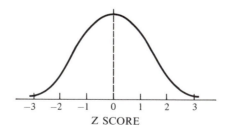

Now, if the curve is normal the following properties will be true:

1. The terms will cluster about the point $z = 0$ (i.e., the closer we are to the vertical axis the more terms we will have, or in other words, the greater the frequency).

2. The curve is symmetrical with respect to the vertical axis. We will find that there are the same number of terms a certain distance to the right of the vertical axis as there are the same distance to the left of it. In fact, we will find that each symmetrically opposite sector contains the same proportion of terms. Some examples are as follows:

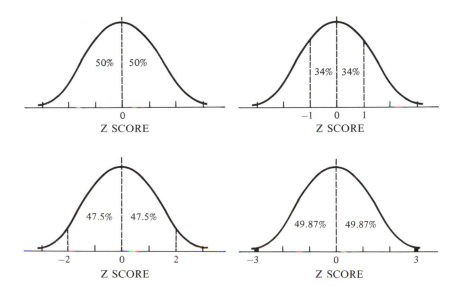

3. The normal curve is not bounded in either direction. Theoretically, z scores can have any magnitude. That is why the extreme ends of the graph do not touch the horizontal. Practically speaking, however, virtually all z scores $(99 + \%)$ are contained within the boundaries ± 3.

4. In a normal distribution, the mean, median, and mode are equal (at $z = 0$).

If a teacher assumes that the grades are normal, he might decide to mark them on a "curve." Suppose he decides that a certain percentage of the population (his class) will receive the grades A, B, C, D, and F as follows (of course, he could decide on a different arrangement as well and still be marking on a curve):

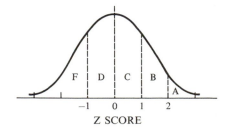

In other words, if an individual's z score is 2 or more he gets an A. If it is less than 2 and greater than or equal to 1, he gets a B; less than 1 but greater than or equal to 0, he receives a C; less than 0 and greater than or equal to -1 he gets a D; and anything below a -1 he gets an F. Using this as a guide 2.5% of the class will get A; 13.5% will get B; 34% will get C; 34% will get D; and 16% will get F.

Quite often students urge their instructors to mark them on a curve. However, what they do not realize is that under a curve a certain percentage will fail even if their average is above 90%.

Exercises 8.3

For 1–5 use the following distribution (70, 73, 75, 76, 83, 85, 85, 86, 87, 90).

1. Find the percentile rank of 76.

2. Find the percentile rank of 85.

3. Find the value whose percentile rank is 25.

4. Find the z score of 86.

5. Find the frequency of 83.

6. For the distribution (2, 7, 7, 3, 5, 1, 2, 6, 3, 7, 4, 1, 3, 3, 6, 6, 5, 4, 7, 4, 4, 2):

   (a) Prepare a frequency table
   (b) Draw a frequency polygon

7–12 Assuming that the distribution is normal find the percent of values whose z score is:

7. Greater than 0.

8. Less than 1.

9. Greater than -1 and less than 1.

10. Less than -2 or greater than 2.

11. Less than -1 or greater than 1.

12. Between -1 and 3.

13. Given $\mu = 70$, $X = 80$, $\sigma = 5$, find z_{80}.

14. If $z_{25} = 3$, $\sigma = 2$, find μ.

Chapter Review Test

1–6 For the data set (7, 4, 4, 6, 9) find the following:

1. The mean   (a) 5.5   (b) 6   (c) 4   (d) 5

2. The median   (a) 5   (b) 4   (c) 6   (d) 4.5

3. The mode   (a) 4   (b) 5   (c) 6   (d) none

4. The range   (a) 3   (b) 2   (c) 13   (d) 5

5. The variance (a) −3.6   (b) 3.6   (c) 4   (d) 4.6

6. The standard
   deviation   (a) −1.9   (b) 2   (c) 1.9   (d) 2.14

7–12 For the data set (1, 2, 2, 11, 10, 8, 9, 4, 5, 8) find the following:

7. The mean   (a) 2   (b) 6.5   (c) 7   (d) 6

8. The median   (a) 2   (b) 6.5   (c) 7   (d) 6

9. The mode   (a) 2   (b) 8   (c) none (d) 2 and 8

10. The range   (a) 7   (b) 9   (c) 10   (d) 12

11. The variance (a) 12   (b) −12   (c) 10   (d) 7.8

12. The standard (a) −3.46   (b) 2.79   (c) 3.46   (d) 3.16
    deviation

13–24 For the data set (90, 80, 89, 58, 93, 63, 94, 67, 70, 85, 100, 87, 89, 87, 80, 72, 96, 77, 80, 83) find the following:

13. The mean   (a) 82   (b) 81.5   (c) 79.2   (d) 86.1

14. The median   (a) 92.5   (b) 85   (c) 84   (d) 80

15. The mode   (a) 89   (b) 80   (c) 80 and 89 (d) none

16. The range   (a) 100   (b) 70   (c) 58   (d) 62

17. The variance   (a) 122.5   (b) 124   (c) 120   (d) 131

18. The standard
    deviation   (a) 10.95   (b) 11.13   (c) 11.01   (d) 11.45

19. The percentile
    rank of 72   (a) 20   (b) 72   (c) 25   (d) 23

20. The percentile
    rank of 80   (a) 35   (b) 30   (c) 38   (d) 40

21. The percentile
    rank of 89   (a) 78   (b) 70   (c) 65   (d) 80

22. The z score of 72 (a) $-.9083$ (b) .9083   (c) $-.8642$ (d) .8642

23. The z score of 80 (a) .1817   (b) 1.817   (c) 18.17   (d) $-.1817$

24. The z score of 89 (a) 6.358   (b) .6358   (c) 0.636   (d) 63.58

25–28 Assuming a normal distribution, find the percent of terms whose z score is:

25. Between 1 and 3   (a) 49   (b) 16   (c) 15.87   (d) 16.8

26. Between -1 and 2   (a) 15.87   (b) 81.5   (c) 49.87   (d) 84

27. Greater than 2 or
    less than 0   (a) 50   (b) 47.5   (c) 52.5   (d) 13.5

28. Greater than 0 or
    less than -2 (a) 13.5   (b) 52.5   (c) 50   (d) 47.5

Chapter 1

Exercises 1.1

1. a, b, d, f, g   2. Finite a, d; Infinite b, f, g

3. None   4. d   5. Roster, Descriptive, Graph

6. {m, i, s, p}

7. The days of the week that begin with "T."

8. (a)   (b)

9. (a) {2, 4}   (b) ∅   (c) {1, 2, 3, 4}   (d) {1, 3}

10. 63   11. T   12. F   13. T   14. F   15. T   16. T

17. Winning {a, b, c}, {a, b}, {a, c}, {b, c}
Losing {a}, {b}, {c}, ∅
Blocking: None

18. (a) Winning {a, b, c}, {a, b}, {a, c}
       Losing {b}, {c}, ∅
       Blocking {a}, {b, c}

(b)   Winning $\{a, b\}$, $\{a, b, c\}$
      Losing $\{a, c\}$, $\{b, c\}$, $\{a\}$, $\{b\}$, $\{c\}$, $\varnothing$
      Blocking: None

Exercises 1.2

1.   There are x such $x \in A$ and $x \notin B$. There are x such that $x \in A$ and $x \in B$. There are no x such that $x \in A$ and $x \in C$. There are x such that $x \in B$ and $x \notin C$. There are x such that $x \in B$ and $x \in C$. There are x such that $x \in C$ and $x \notin A$. There are x such that $x \in C$ and $x \notin B$. There are x such that $x \notin A$ and $x \notin B$ and $x \notin C$.

2.

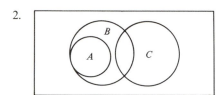

3.

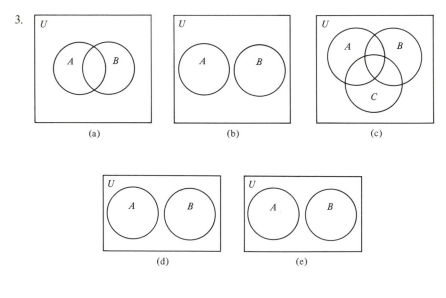

(a)   (b)   (c)

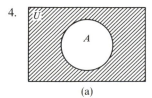

(d)

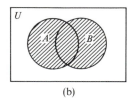

(e)

4.

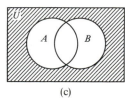

(a)   (b)   (c)

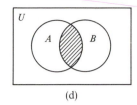

(d)

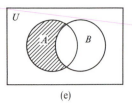

(e)

5. (a) $\{4, 5, 6, 7, 8, 9\}$ (b) $\{1, 2, 3, 4, 5, 6\}$
   (c) $\{7, 8, 9\}$ (d) $\{3\}$ (e) $\{1, 2\}$

6.   (a)   (b)   (c)

             $A \cup B'$ $A \cap B'$ $(A \cup B)'$

A	B	$A \cup B'$				$A \cap B'$				$(A \cup B)'$			
∈	∈	∈	∈	∈	∉	∈	∉	∈	∉	∈	∈	∈	∉
∈	∉	∈	∈	∉	∈	∈	∈	∉	∈	∈	∈	∉	∉
∉	∈	∉	∉	∈	∉	∉	∉	∈	∉	∉	∈	∈	∉
∉	∉	∉	∈	∉	∈	∉	∉	∉	∈	∉	∉	∉	∈
		1	3	1	2	1	3	1	2	1	2	1	3

Exercises 1.3

1.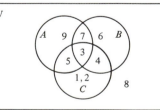

   (a) $\{3, 7\}$
   (b) $\{1, 2, 4, 5, 6, 8, 9\}$
   (c) $\{6, 7, 9\}$
   (d) $\{1, 2, 3, 4, 5, 6, 8, 9\}$

2. (a) $\{3, 5, 7, 9\} \cap \{3, 4, 6, 7\} = \{3, 7\}$
   (b) $\{3, 7\}' = \{1, 2, 4, 5, 6, 8, 9\}$
   (c) $(A \cup B) \cap C'$
       $\{3, 4, 5, 6, 7, 9\} \cap \{6, 7, 8, 9\} = \{6, 7, 9\}$

(d) $(A \cap B)' \cup C$
$\{3, 7\}' \cup \{1, 2, 3, 4, 5\}$
$\{1, 2, 4, 5, 6, 8, 9\} \cup \{1, 2, 3, 4, 5\} = \{1, 2, 3, 4, 5, 6, 8, 9\}$

3.

(a) $A \cap B$ (b) $(A \cap B)'$

A	B	$A \cap B$			$(A \cap B)'$			
∈	∈	∈	∈	∈	∈	∈	∈	∉
∈	∉	∈	∉	∉	∈	∉	∉	∈
∉	∈	∉	∉	∈	∉	∉	∈	∈
∉	∉	∉	∉	∉	∉	∉	∉	∈
		1	2	1	1	2	1	3

(c) $(A \cup B) \cap C'$ (d) $(A \cap B)' \cup C$

A	B	C	$(A \cup B) \cap C'$						$(A \cap B)' \cup C$					
∈	∈	∈	∈	∈	∈	∉	∈	∉	∈	∈	∈	∉	∈	∈
∈	∈	∉	∈	∈	∈	∈	∉	∈	∈	∈	∈	∉	∉	∉
∈	∉	∈	∈	∈	∉	∉	∈	∉	∈	∉	∉	∈	∈	∈
∈	∉	∉	∈	∈	∉	∈	∉	∈	∈	∉	∉	∈	∈	∉
∉	∈	∈	∉	∈	∈	∉	∈	∉	∉	∉	∈	∈	∈	∈
∉	∈	∉	∉	∈	∈	∈	∉	∈	∉	∉	∈	∈	∈	∉
∉	∉	∈	∉	∉	∉	∉	∈	∉	∉	∉	∉	∈	∈	∈
∉	∉	∉	∉	∉	∉	∉	∉	∈	∉	∉	∉	∈	∈	∉
			1	2	1	3	1	2	1	2	1	3	4	1

4. 4   5. (a) 9   (b) 1

6. (a) 3   (b) 16   (c) $\varnothing$ (d) $\{19, 20\}$ (e) $\{16, 17, 18, 19, 20\}$
   (f) $\{1, 2, 3, 4\}$ (g) $\cup$ (h) $\cup$ (i) $\varnothing$ (j) $\varnothing$
   (k) $\{10, 11, 12, 13\}$

7. (a) $\longleftarrow\!\!\!\!\!\bullet^{\,8}$ (b) $\overset{4}{\circ}\!\!\!\longrightarrow$
   (c) $\overset{2}{\circ}\!\!\!\!\!\longrightarrow\!\!\!\bullet^{\,5}$ (d) $\longleftarrow\!\!\!\overset{3}{\circ}\!--$
   (e) $\overset{2}{\bullet}\!\!\!\longrightarrow\!\!\!\bullet^{\,6}$

Exercises 1.4

1.

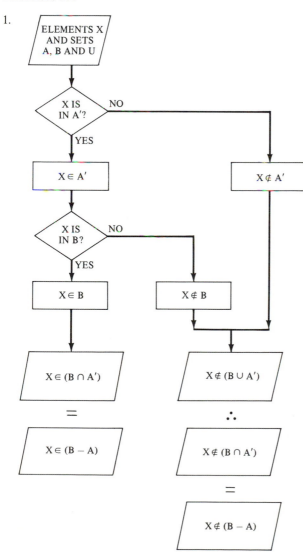

2.

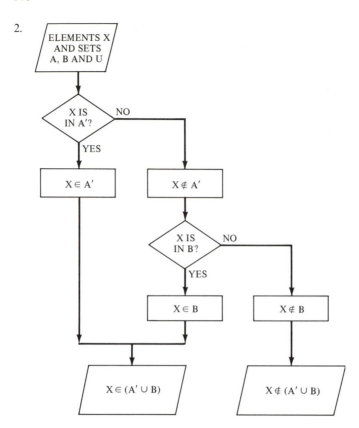

3.

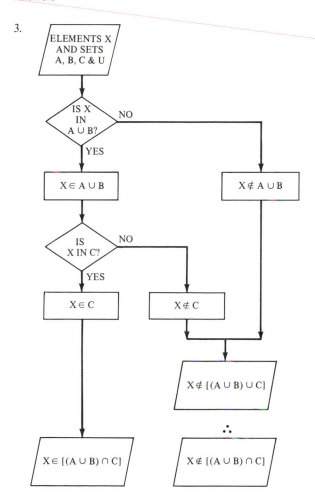

4.

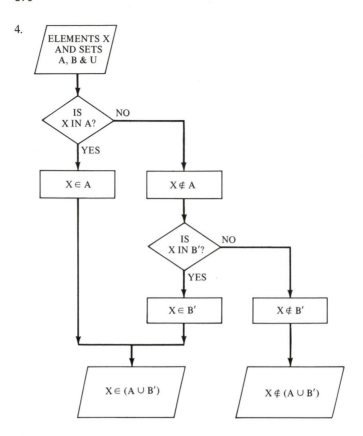

5.

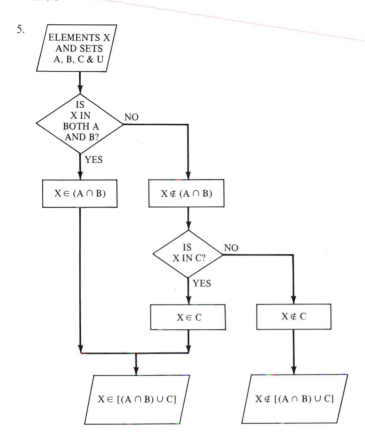

6.

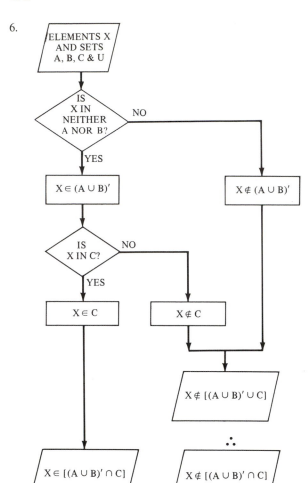

Exercises 1.5

1. F   2. T   3. T

4. $(A \cup C') \cap (A \cup C) = A$
   $A \cup (C' \cap C) \quad = A$
   $A \cup \varnothing \quad\quad\quad = A$
   $A \quad\quad\quad\quad\quad = A$

5. $A = (A \cap A) \cup \varnothing$
   $A = A \cup \varnothing$
   $A = A$

6. $[A \cup (C \cap \varnothing)] - (B \cap U) = A \cap B'$
    $(A \cup \varnothing) - B \qquad\qquad = A \cap B'$
    $A - B \qquad\qquad\qquad = A \cap B'$
    $A \cap B' \qquad\qquad\qquad = A \cap B'$

Chapter Exercises

1. (b) {1, 2, 3, 4, 5, 6}   (c) {4, 5, 6, 7, 8, 9}
    (d) $\varnothing$ (e) {12, 14, 16, 18, 20, 22}

2. (a) $\{x \mid x$ is a whole number between 0 and 10}
    (b) $\{x \mid x$ is a day of the week that begins with the letter S}
    (c) $\{x \mid x$ is a consonant}
    (d) $\{x \mid x$ is a positive odd number less than 8}
    (e) $\{x \mid x$ is a month of the year that begins with the letter J}

3. (a) {1, 3, 4}   (b) {1, 2, 3, 4, 7, 9}   (c) {1, 3, 4}
    (d) {1, 3, 4, 7, 9}   (e) {6, 8}   (f) {3, 7, 9}
    (g) $\varnothing$ (h) {7, 9}   (i) {1, 3, 4, 6, 8}
    (j) {1, 4, 7, 9}

4. (a) F   (b) F   (c) F   (d) T   (e) F   (f) F   (g) T
    (h) F

5. (a) $N = \{A, B, C\} - W$ (b) $M = \{G, H, I, J\} - W$
         $\{A, B\} - W$ $\{G, H, I\} - W$
         $\{A, C\} - W$ $\{G, H, J\} - W$
         $\{B, C\} - W$ $\{G, I, J\} - W$
         $\{A\} - L$ $\{H, I, J\} - W$
         $\{B\} - L$ $\{G, H\}$
         $\{C\} - L$ $\{G, I\}$
         $\varnothing - L$ $\{G, J\}$
                                        $\{H, I\}$
                                        $\{H, J\}$
                                        $\{I, J\}$
                                        $\{G\} - L$
                                        $\{H\} - L$
                                        $\{I\} - L$
                                        $\{J\} - L$
                                        $\varnothing - L$

6. (a) {*a, d*}   {*c, d, e*}   (b) {*a*}   {*a, b*}   {*c, e*}
        {*c, d*}   {*a, b, c, d*}   {*b*}   {*a, c*}   {*a, b, e*}
        {*a, b, d*}   {*a, c, d, e*}   {*c*}   {*a, e*}   {*b, c, e*}
        {*b, c, d*}   {*a, b, d, e*}   {*d*}   {*b, c*}   ∅
        {*a, c, d*}   {*a, b, c, e*}   {*e*}   {*b, e*}
        {*a, d, e*}   {*b, c, d, e*}
        {*b, d, e*}   *H*

    (c) {*b, d*}   (d) {*a, b, d*}   (e) None
        {*d, e*}   {*b, c, d*}
        {*a, b, c*}   {*a, c, d*}
        {*a, c, e*}   {*a, d, e*}
                          {*c, d, e*}
                          {*a, b, c, d*}
                          {*a, c, d, e*}
                          {*a, b, d, e*}
                          {*b, c, d, e*}
                          H

7. (a) $m(M) = 5$ (b) $m(M \cup N) = 10$ (c) $m(M \cup P) = 10$
    (d) $m(M \cap N) = \varnothing$ (e) $m(\varnothing) = 0$ (f) $m(M) = 5$
    (g) $m(N) = 5$ (h) $m(P) = 8$ (i) $m(N \cup P) = 8$
    (j) $m(N \cap P) = 5$ (k) $m(H) = 10$ (l) 0

8. (a)   (b)

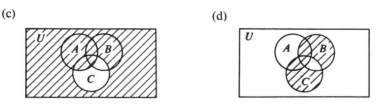

    (c)   (d)

9. (a) $A \cap B \neq \varnothing$, $A \not\subset B$ and $B \not\subset A$. $A \subset C$ and $A \neq C$. $B \subset C$ and $B \neq C$.
        $(A \cup B \cup C) \neq U$.
    (b) $A \cap B \neq \varnothing$, $A \cap C = \varnothing$, $A \not\subset B$ and $B \not\subset A$. $B \cap C \neq \varnothing$, $B \not\subset C$
        and $C \not\subset B$. $(A \cup B \cup C) \neq U$.
    (c) $C \subset B$ and $C \neq B$. $B \subset A$ and $B \neq A$. $A \neq U$.
    (d) $A \cap B \neq \varnothing$, $A \not\subset B$ and $B \not\subset A$. $A \cap C = \varnothing$. $(C \cap B) = \varnothing$.
        $(A \cup B \cup C) \neq U$.
    (e) $A \cap B \neq \varnothing$, $A \not\subset B$ and $B \not\subset A$, $C \subset B$ and $C \neq B$. $A \cap C = \varnothing$.
        $(A \cup B) = U$.

10. (a)
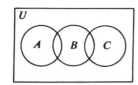

(b)
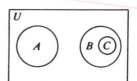

(c)
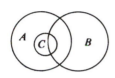

(d)
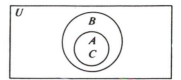

11. (a) $\{1, 2\}$ (b) $\{1, 2, 5, 6, 7, 8\}$ (c) $\{1, 2, 3, 4, 5, 6\}$
   (d) $\{1, 2, 3, 4, 5\}$ (e) $\{4, 8\}$ (f) $\{8\}$

12. (a) $\{1\} = C \cap D$
     $\{2\} = C - D = C \cap D'$
     $\{3\} = D - C = D \cap C'$
     $\{4\} = (C \cup D)' \cap B$
     $\{5\} = A - (B \cup C \cup D)$
     $\{6\} = (A \cup B \cup C \cup D)'$

   (b) $\{1\}$ $C \cap D$
      $\{2\}$ $D - C = D \cap C'$
      $\{3\}$ $(A \cap C) - D = A \cap C \cap D'$
      $\{4\}$ $(A \cap B) - (D \cup C)$
             $= A \cap B \cap D' \cap C'$
      $\{5\}$ $(C - D) - A = C \cap D' \cap A'$
      $\{6\}$ $B \cap (A \cup C)' = B \cap A' \cap C'$
      $\{7\}$ $A \cap B' \cap D' \cap C'$
      $\{8\}$ $(A \cup B \cup C \cup D)'$

13. (a)
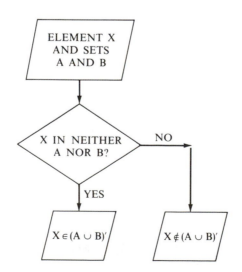

(b)

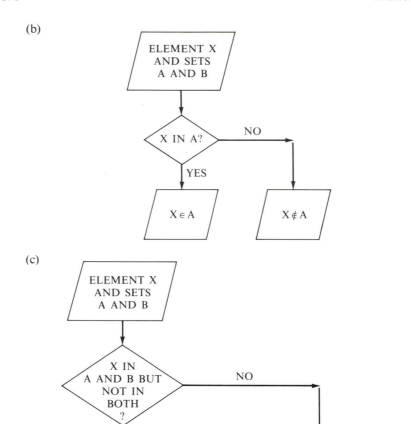

(c)

14. (a) T   (b) T   (c) F   (d) T

15. (a) Seven played only baseball
    (b) Six are taking both math and science; twenty are taking history but
        not science; and fifty-six are not taking both math and history.
    (c) 1. 7 elements   2. 3 elements
        3. 3 elements   4. 1 element
        5. 9 elements   6. 9 elements
        7. 4 elements   8. 7 elements
        9. 3 elements   10. 0 elements
    (d) 1. 39 elements   2. 11 elements
        3. 12 elements   4. 39 elements
        5. 8 elements   6. 15 elements
        7. 8 elements   8. 22 elements
        9. 30 elements   10. 0 elements

16. (a) {4}   (b) {9, 10}
    (c) {1, 2, 3, 4, 5, 6, 7}   (d) {9, 10}
    (e) {2, 3, 4}   (f) {0, 1, 2, 3, 4, 8, 9, 10}
    (g) {0, 1, 2, 9, 10}   (h) {0, 1, 2, 3, 4, 5, 6, 7, 8, 9, 10}
    (i) {2, 3, 4, 5}   (j) $\varnothing$

17. (a) $A \cap B$ (b) U (c) $A \cap B' = A - B$

18. (a) F   (b) T   (c) T   (d) T

19. (a) 1; (A_2) 2. 1; (A_6) 3. 2; (A_2) 4. 3; (A_3)
        5. 4; (thm 1)   6. 5; (A_6)
    (b) 1. Given   2. (A_5) 3. (A_6) 4. 2 & 3   5. 4; (A_4)
        6. 5; (A_2) 7. 1 & 6   8. 7; (A_5)

20. (a) *Theorem 3* U' $\varnothing$ and $\varnothing'$ U

    1 $U' = U' \cap U$ (A_5)
    2. $U' = U \cap U'$ 1; (A_2)
    3. $U' = \varnothing$ 2; (A_6)
    4. $\varnothing' = U$ Principle of duality

    (c) *Theorem 7*   For any set A of S, A' is unique. Let A_1' and A_2' be compliments of A.

    1. $A_1' = A_1' \cup \varnothing$ (A_5)
    2. $A \cap A_2' = \varnothing$ (A_6)
    3. $A_1' = A_1' \cup (A \cap A_2')$ 1 & 2
    4. $A_1' = (A_1' \cup A) \cap (A_1' \cup A_2')$ 3; (A_4)
    5. $A_1' = (A \cup A_1') \cap (A_1' \cup A_2')$ 4; (A_2)
    6. $A_1' = U \cap (A_1' \cup A_2')$ 5; (A_6)
    7. $A_1' = (A_1' \cup A_2') \cap U$ 6; (A_2)
    8. $A_1' = (A_1' \cup A_2')$ 7; (A_5)
    9. $A_2' = (A_2' \cup A_1')$ In the same manner as above.
    10. $A_1' \cup A_2' = A_2' \cup A_1'$ (A_2)
    11. $A_1' = A_2'$ 8, 9, and 10

    (e) *Theorem 10* $A \subset A$

    1. $A \cap A = A$ (Th. 1)
    2. $A \subset A$ 1; (Def. 2)

    (g) *Theorem 12*   If $A \subset B$ and $B \subset A$, then $A = B$.

    1. $A \subset B$ Given
    2. $B \subset A$ Given
    3. $A \cap B = A$ 1; (Def. 2)
    4. $B \cap A = B$ 2; (Def. 2)
    5. $(A \cap B) = (B \cap A)$ A_2
    6. $A = B$ 3, 4, and 5

Review Test

1. b	2. d	3. b	4. e	5. c	6. d	7. c	8. e
9. b	10. b	11. c	12. e	13. d	14. c	15. d	16. c
17. b	18. a	19. b	20. a	21. b	22. c	23. d	24. b
25. d	26. a	27. d	28. d	29. c	30. b		

Chapter 2

Exercises 2.1

1. (a) simple   (b) neither   (c) compound   (d) compound
   (e) compound   (f) neither   (g) simple

2. (a) conjunction   (b) disjunction   (c) conjunction
   (d) ambiguous   (e) conditional   (f) biconditional
   (g) conditional   (h) negation   (i) conjunction
   (j) biconditional   (k) conditional

3. (a) $r \to {\sim}s$ (b) $(s \land p) \lor d$ (c) $s \land p$ (d) ${\sim}(s \land h)$
   (e) $j \leftrightarrow g$ (f) $m \to h$ (g) $h \to m$

Exercises 2.2

1. (a) F   (b) T   (c) F   (d) F

2. (a) T   (b) T   (c) T   (d) F   (e) F   (f) F

3. (a) I   (b) T   (c) I   (d) F   (e) T
   (f) T   (g) T   (h) I

4.   (a)   (b)   (c)

                   $p \to {\sim}q$ ${\sim}p \lor {\sim}q$ ${\sim}(p \land {\sim}q)$

p	q	$p \to {\sim}q$				${\sim}p \lor {\sim}q$				${\sim}(p \land {\sim}q)$				
T	T	T	F	F	T	F	T	T	F	T	T	T	F	F	T
T	F	T	T	T	F	F	T	F	T	F	F	T	T	T	F
F	T	F	T	F	T	T	F	T	F	T	T	F	F	F	T
F	F	F	T	T	F	T	F	T	T	F	T	F	F	T	F
		1	3	2	1	2	1	3	2	1	4	1	3	2	1

$(p \land q) \lor \sim r$

(d)

p	q	r	$(p \land q) \lor \sim r$					
T	T	T	T	T	T	T	T	T
T	T	F	T	T	T	T	T	F
T	F	T	T	F	F	F	F	T
T	F	F	T	F	F	T	T	F
F	T	T	F	F	T	F	F	T
F	T	F	F	F	T	T	T	F
F	F	T	F	F	F	F	F	T
F	F	F	F	F	F	T	T	F
			1	3	1	4	2	1

5. (a) $p = F, q = T$ (b) $p = F, q = T$
    (c) $p = F, q = T$ (c) $p = T, q = T,$ or $p = F, q = F$

Exercises 2.3

1. (a) I   (b) T   (c) I   (d) T   (e) F   (f) I

2. (a) E   (b) C   (c) E   (d) U   (e) E
    (f) U   (g) E   (h) E   (i) C

3. Yes   (a),   (c),   (d),   (e),   (f),   (g),   (h)

Exercises 2.4

1. $p \land q$ 2. p 3. $q \lor q$ or $q \land q$ 4. $r \lor (p \to q)$

5. $p \lor (q \lor r)$ 6. $\sim q \to p$ 7. $(a \to b) \land (b \to a)$ 8. $p \to q$

9. $\sim p \lor \sim q$ 10. $(p \lor \sim q) \land (p \lor r)$ 11. $(r \land \sim p) \lor (r \land q)$

12. $\sim (p \lor \sim q)$

13. The sun is not shining or the fishing is not good.

14. If Bill does not come home early, then it is not the case that Joan is having lobster or there is a basketball game.

15. If Don and Gladys go to the beach the sun is shining, and if the sun is shining Don and Gladys go to the beach.

16. $p \to \sim q$ 17. $r \to (p \land q)$ 18. $\sim(\sim q \lor r) \to \sim p$

Chapter Exercises

1. (a) simple   (b) compound
   (c) neither   (d) compound
   (e) neither   (f) neither
   (g) neither   (h) compound
   (i) simple   (j) compound

2. (a) negation   (b) disjunction
   (c) conjunction   (d) ambiguous
   (e) disjunction   (f) biconditional
   (g) conjunction   (h) negation
   (i) conditional   (j) conditional

3. (a) If Joe is smart and studies, then he is passing math.
   (b) Joe is smart or if he does not study then he is not passing math.
   (c) It is false that Joe is smart and does not study or he is passing math.
   (d) Joe studies and is passing math if and only if he is smart.
   (e) Joe is smart and he studies, or Joe is smart and he is passing math.

4. (a) $p \to s$ conditional
   (b) $t \to p$ conditional
   (c) $(t \land p) \lor (r \land \sim p)$ disjunction
   (d) $(s \lor e) \land f$ conjunction
   (e) $\sim(\sim r \to m)$ negation
   (f) $(w \to e) \land (\sim w \to r)$ conjunction
   (g) $i \lor (\sim i \to \sim e)$ disjunction
   (h) $\sim[i \land (g \to r)]$ negation
   (i) $(m \land p) \lor (m \land c)$ disjunction
   (j) $(r \to c) \land \sim c$ conjunction

5. (a) disjunction

$\lor$	T	F
T	T	T
F	T	F

(b) conditional

$\to$	T	F
T	T	F
F	T	T

(c) biconditional $\leftrightarrow$

	T	F
T	T	F
F	F	T

6.

p	q	p	$\vee$	q
T	T	T	F	T
T	F	T	T	F
F	T	F	T	T
F	F	F	F	F
1	1	2	3	2

7. When p and q are both true and when p is true and q is false.

8. (a) false   (b) false   (c) true   (d) true   (e) true
    (f) false   (g) false   (h) true   (i) true   (j) true
    (k) false   (l) false

9. (a) false   (b) true   (c) true   (d) false   (e) false

10. (a) true   (b) true   (c) false   (d) insufficient information
    (e) true   (f) insufficient information   (g) true
    (h) insufficient information   (i) true   (j) true   (k) true
    (l) true   (m) true   (n) insufficient information   (o) true

11. (a) F T T T   (b) F F F T   (c) T F F F F F F F
    (d) T T T T F T T T   (e) F F T F T T T T
    (f) T T F T   (g) F F F F   (h) F T T F
    (i) F F T T T T T T   (j) F F T F

12. (a) $g \wedge (s \vee p)$ g is true and either s or p is true
    (b) $j \rightarrow \sim(e \wedge \sim 1)$ true for all cases except when j and e are true and l is false
    (c) $(s \wedge w) \vee \sim(\sim s \rightarrow \sim c)$ s and w are true or s is false and w and c are true or s and w are false and c is true
    (d) $\sim b \vee [b \rightarrow (\sim j \wedge \sim s)]$ b is false or b is true and j and s are false
    (e) $(a \rightarrow b) \vee \sim[\sim b \rightarrow (a \vee \sim c)]$ a is false or a and b are true

13. (a) indeterminate   (b) tautology   (c) false statement
    (d) indeterminate   (e) false statement   (f) tautology
    (g) indeterminate   (h) indeterminate

14. (a) contradictory   (b) unrelated   (c) equivalent
    (d) equivalent   (e) contradictory   (f) equivalent
    (g) unrelated   (h) contradictory

15.

	(a)	(b)	(c)	(d)	(e)
Conjunction	$\sim(p \wedge q)$	$\sim p \wedge \sim q$	$\sim(\sim p \wedge \sim q)$	$p \wedge \sim q$	$\sim(\sim p \wedge q)$
Disjunction	$\sim p \vee \sim q$	$\sim(p \vee q)$	$p \vee q$	$\sim(\sim p \vee q)$	$p \vee \sim q$
Conditional	$p \to \sim q$	$\sim(\sim p \to q)$	$\sim p \to q$	$\sim(p \to q)$	$\sim p \to \sim q$

To convert a statement containing a conjunction to a statement containing a disjunction, negate the entire statement, negate both the left and right sides of the statement, and change the connective to a disjunction (E_6). To convert a statement containing a disjunction to a statement containing a conditional, negate only the left side of the statement and change the connective to conditional (E_7).

16. (a) $q \to p$ (b) $p \to q$
    (c) $p \wedge \sim q$ (d) $\sim p \vee (q \wedge r)$
    (e) $h \leftrightarrow s$ (f) $\sim s \vee (t \wedge h)$

17. (a) T T   (b) T T T T   (c) T T T T T T T T
    (d) T T T T T T T T   (e) T T T T

18. (a) $(p \vee q); (\sim p \wedge \sim q); \sim[\sim(p \wedge q)]$
    (b) $\sim p; p \to q; p \to q$
    (c) $s \vee (n \wedge \sim d); \sim(d \to s) \wedge (l \wedge s); (l \vee n) \wedge (n \to s)$
    (d) $(p \wedge q); \sim[(\sim a \wedge \sim b) \vee (c \to d)]; \sim[\sim(c \to d) \vee a]$
    (e) $(c \vee \sim a) \vee \sim b; [a \wedge (c \to d)] \wedge (d \vee c); (a \vee b) \wedge [(a \to c) \wedge (a \vee d)]$
    (f) $(\sim s \wedge \sim a) \vee (\sim s \wedge d); [a \vee \sim(b \wedge c)] \wedge [a \vee (c \to d)];$
       $[(b \to c) \wedge (c \to a)] \vee [(b \to c) \wedge (a \vee d)]$
    (g) $(q \wedge r) \to p; r \to \sim(p \vee q); (r \to s) \to \sim(p \to q)$
    (h) $p \vee \sim(q \wedge r); \sim p \to (q \to r); \sim(p \to q)$
    (i) $\sim p \leftrightarrow \sim q; (p \wedge q) \leftrightarrow \sim r; (p \to \sim q) \wedge (\sim q \to p)$

19. (a) If prices are lower, then we save more money.
    (b) You listen to reason and are not open-minded.
    (c) Jim works hard and either he does a good job or does not have time to relax.
    (d) Taxes are higher or prices are lower.
    (e) He works hard if and only if he earns a good wage.
    (f) You listen to reason and are open-minded.

20. (a) He is a Russian leader and not a Communist.
    (b) He listens to reason and is open-minded.
    (c) He does not work hard or he does not succeed.

21. (a) $q \to \sim p$
    (b) $q \to p$

(c) $(q \wedge \sim r) \to \sim p$

(d) $\sim r \to (p \wedge q)$

(e) $\sim (q \to \sim p)$

(f) If he is a politician, he is a Senator.

(g) If he is not elected to office, then it is not the case that he is not a Supreme Court judge or not the Secretary of State.

(h) If he is not retired, he does not receive a pension.

(i) If it is not the case that his reasoning and premises are correct, the conclusion is not correct. If the conclusion is correct, his reasoning and premises are correct. If the conclusion is not correct, then it is not the case that his reasoning and premises are correct.

Review Test

1. c	2. d	3. b	4. e	5. a	6. e	7. a	8. d
9. e	10. c	11. b	12. b	13. a	14. c	15. a	
16. a	17. c	18. a	19. a	20. c	21. a	22. c	
23. b	24. b	25. c	26. a	27. e	28. d	29. d	
30. e	31. e	32. a	33. b	34. a	35. a	36. c	
37. c	38. c	39. d	40. b	41. e	42. c	43. d	
44. a	45. c	46. c	47. c	48. a			

Chapter 3

Exercises 3.1

1. (a) Invalid   (b) Valid   (c) Valid   (d) Valid
   (e) Invalid   (f) Valid

Exercises 3.2

1. Valid   2. Valid   3. Valid   4. Valid   5. Invalid

Exercises 3.3

1. Invalid   2. A_3 3. Invalid   4. A_2 5. A_6 6. Invalid

7. A_1 8. A_4 9. A_7 10. A_2 11. Invalid   12. A_5

13. A_8

Exercises 3.4

1. (a) $p \wedge q$ (b) $\sim[(p \wedge q) \rightarrow r]$

2. ... contradiction of one of the premises.

3. 1, 3, A_3 4. 3, E_6 5. 1, 4, A_1
   2, 4, A_1 4, A_6 3, 5, A_5
   4, 5, A_5 1, 5, A_3 6, E_6
                           4, A_6 2, 7, A_2
                           6, 7, A_5
                           2, 8, A_1

6. 3, 4, A_2 7. 1. $\sim(p \vee \sim q)$ premise
   2, 5, A_3 2. $\sim p \rightarrow r$ premise
   1, 6, A_1 3. $t \rightarrow \sim r$ premise
   7, A_6 4. $\sim p \wedge q$ 1, E_6
   1, 4, contradiction   5. $\sim p$ 4, A_6
                                   6. r 2, 5, A_1
                                   7. $\sim t$ 3, 6, A_2
                                   8. q 4, A_6
                                   9. $\sim t \wedge q$ 7, 8, A_5

8. $p = F, q = T$ 9. $p = F, q = F, r = F, s = T$

Chapter Exercises

1. (a) invalid   (b) valid
   (c) valid   (d) valid
   (e) valid   (f) valid
   (g) valid   (h) valid
   (i) valid   (j) valid
   (k) valid

2. (a) valid   (b) valid
   (c) valid   (d) valid
   (e) invalid   (f) invalid
   (g) valid   (h) valid
   (i) valid   (j) invalid
   (k) valid

3. (a) invalid   (b) valid
   (c) valid   (d) valid
   (e) valid

4. The argument is valid since the premises are contradictory.

5. The argument is invalid since the premises do not contain sufficient information.

6. T T F (when the premises are true and the conclusion is false, the argument is invalid)

7. (a) Modus Ponens (A_1)
    (b) Modus Tollens (A_2)
    (c) Disjunctive Syllogism (A_3)
    (d) Conjunctive Simplification (A_6)
    (e) Modus Tollens (A_2)
    (f) Conjunctive Addition (A_5)
    (g) Disjunctive Addition (A_7)
    (h) Hypothetical Syllogism (A_4)
    (i) Disjunctive Simplification (A_8)
    (j) Modus Ponens (A_1)
    (k) Disjunctive Syllogism (A_3)
    (l) Modus Tollens (A_2)
    (m) Conjunctive Simplification (A_6)
    (n) Disjunctive Syllogism (A_3)
    (o) Modus Tollens (A_2)

8. (a) (4)   (b) (4)   (c) (4)   (d) (1)   (e) (2)
    (f) (4)   (g) (4)   (h) (2)   (i) (2)   (j) (3)

9. (a) (4) 3; (E_6)   (b) (4) 2; (E_6)
         (5) 4; (A_6)   (5) $1 + 3$; (A_1)
         (6) $1 + 5$; (A_1)   (6) $4 + 5$; (A_3)
         (7) $2 + 6$; (A_2)
         (8) 4; (A_6)
         (9) $7 + 8$; (A_5)

    (c) (5) 4; (E_6)   (d) (5) $3 + 4$; (A_3)
         (6) 5; (A_6)   (6) $2 + 5$; (A_2)
         (7) $3 + 6$; (A_1)   (7) 6: (A_7)
         (8) 5; (A_6)   (8) 7; (E_6)
         (9) $1 + 8$; (A_1)   (9) $1 + 8$; (A_3)
         (10) 7; (A_6)   (10) 9; (A_6)
         (11) $9 + 10$; (A_1)
         (12) 7; (A_6)
         (13) $11 + 12$; (A_5)
         (14) $2 + 13$; (A_1)

(e) (4)   2; (E_5)
     (5)   4; (A_6)
     (6)   5; (E_7)
     (7) $1 + 6$; (A_3)
     (8) $3 + 7$; (A_1)
     (9)   4; (A_6)
    (10) $8 + 9$; (A_1)
    (11) $5 + 10$; (A_3)

(f) (4)   1; (A_6)
     (5)   4; (E_7)
     (6)   5; (E_6)
     (7)   6; (A_6)
     (8)   1; (A_6)
     (9) $7 + 8$; (A_5)
    (10)   9; (E_6)
    (11)   10; (E_7)
    (12) $2 + 11$; (A_3)
    (13) $3 + 12$; (A_3)

(g) (4)   3; (A_6)
     (5) $1 + 4$; (A_2)
     (6) $2 + 5$; (A_1)
     (7)   3; (A_6)
     (8) $6 + 7$; (A_3)
     (9) $3 + 8$; conditional
           conclusion

(h) (4)   3; (E_7)
     (5)   4; (E_6)
     (6)   5; (A_6)
     (7) $2 + 6$; (A_1)
     (8)   5; (A_6)
     (9) $1 + 8$; (A_1)
    (10)   9; (E_6)
    (11)   10; (A_6)
    (12) $7 + 11$; contradiction

(i) (4) $2 + 3$; (A_1)
     (5)   4; (A_6)
     (6) $1 + 3$; (A_2)
     (7)   6; (A_6)
     (8) $5 + 7$; contradiction

(j) (4)   2; (E_5)
     (5)   4; (A_6)
     (6) $1 + 5$; (A_1)
     (7)   3; (E_6)
     (8)   7; (A_6)
     (9) $6 + 8$; (A_1)
    (10)   4; (A_6)
    (11)   7; (A_6)
    (12) $10 + 11$; (A_3)
    (13) $9 + 12$; (A_5)
    (14) $3 + 13$; conditional
           conclusion

10. (a) *Direct*   *Indirect*

1.	$\sim r \to \sim p$	
2.	$\sim(\sim p \land q)$	
3.	$\sim r$	
4.	$\sim p$	1, 3 A_1
5.	$p \lor \sim q$	2 E_6
6.	$\sim q$	4, 5 A_3

4.	q		P.I.P.
5.	$p \lor \sim q$	2	E_6
6.	p	4, 5	A_3
7.	$\sim p$	1, 3	A_1
8.	$\therefore \sim q$	6, 7	contradict

(b) *Direct*
1. $a \rightarrow (b \wedge c)$
2. $b \rightarrow \sim a$
3. $\sim a \vee (b \wedge c)$ 1 E_7
4. $\sim b \vee \sim a$ 2 E_7
5. $(\sim a \vee b) \wedge (\sim a \vee c)$ 3 E_5
6. $\sim a \vee b$ $5a$ A_6
7. $\sim a \vee \sim b$ 4 E_3
8. $\sim a$ 6, 7 A_8

Indirect
3. a P.I.P.
4. $(b \wedge c)$ 1, 3 A_1
5. b 4 A_6
6. $\sim b$ 2, 3 A_2
7. $\therefore \sim a$ 5, 6   contradict

(c) *Direct*
1. $\sim (p \wedge \sim q)$
2. $q \rightarrow \sim s$
3. p conditional premise
4. $\sim p \vee q$ 1 E_6
5. q 3, 4 A_3
6. $\sim s$ 2, 5 A_1
7. $p \rightarrow \sim s$ conditional
             form of
             conclusion

Indirect
3. $\sim (p \rightarrow \sim s)$ P.I.P.
4. $\sim (\sim p \vee \sim s)$ 3 E_7
5. $p \wedge s$ 4 E_6
6. p 5 A_6
7. $\sim p \vee q$ 1 E_6
8. q 6, 7 A_3
9. $\sim s$ 2, 4 A_1
10. s 5 A_6
11. $\therefore (p \rightarrow \sim s)$ 9,10   contradict

(d) *Direct*
1. $\sim p \vee \sim q$
2. $(r \rightarrow p) \vee t$
3. $\sim (\sim r \vee t)$
4. $r \wedge \sim t$ 3 E_6
5. $\sim t$ 4 A_6
6. $r \rightarrow p$ 2, 5 A_3
7. r 4 A_6
8. p 6, 7 A_1
9. $\sim q$ 1, 8 A_3

Indirect
4. q P.I.P.
5. $\sim p$ 1, 4 A_3
6. $r \wedge \sim t$ 3 E_6
7. $\sim t$ 6 A_6
8. $r \rightarrow p$ 2, 7 A_3
9. r 6 A_6
10. p 8, 9 A_1
11. $\therefore \sim q$ 5, 10   contradict

(e) *Direct*
1. $p \rightarrow \sim q$
2. $p \vee q$
3. $\sim q$ conditional premise
4. p 2, 3 A_3
5. $\sim q \rightarrow p$ conditional form of conclusion
6. $(p \rightarrow \sim q) \wedge (\sim q \rightarrow p)$ 1, 5 A_5
7. $p \leftrightarrow \sim q$ 6 E_9

(f) *Direct*
1. $a \vee (b \vee \sim c)$
2. $d \rightarrow \sim a$
3. $\sim (b \wedge \sim e)$
4. $d \wedge \sim e$ conditional premise
5. d 4 A_6
6. $\sim a$ 2, 5 A_1

7. $b \lor \sim c$ 1, 6 A_3
8. $\sim b \lor e$ 3 E_6
9. $\sim e$ 4 A_6
10. $\sim b$ 8, 9 A_3
11. $\sim c$ 7, 10 A_3
12. $(d \land \sim e) \rightarrow \sim c$ 4, 11   conditional form of conclusion

Indirect

4. $\sim [(d \land \sim e) \rightarrow \sim c]$ P.I.P.
5. $\sim [\sim (d \land \sim e) \lor \sim c]$ 4 E_7
6. $(d \land \sim e) \land c$ 5 E_6
7. $d \land \sim e$ 6 A_6
8. d 7 A_6
9. $\sim a$ 2, 8 A_1
10. $b \lor \sim c$ 1, 9 A_3
11. $\sim b \lor e$ 3 E_6
12. $\sim e$ 9 A_6
13. $\sim b$ 11, 12 A_3
14. $\sim c$ 10, 13 A_3
15. c 6 A_6
16. $\therefore (b \land \sim e) \rightarrow \sim c$ 14, 15   contradict

(g) *Direct*
1. $o \rightarrow \sim t$
2. $c \lor o$
3. $\sim c$
4. o 2, 3 A_3
5. $\sim t$ 1, 4 A_1

Indirect
4. t P.I.P.
5. $\sim o$ 1, 4 A_2
6. o 2, 3 A_3
7. $\therefore \sim t$ 5, 6   contradict

(h) *Direct*
1. $\sim t \rightarrow \sim b$
2. $\sim b \rightarrow r$
3. $\sim r$
4. b 2, 3 A_2
5. t 1, 4 A_2

Indirect
4. $\sim t$ P.I.P.
5. $t \lor \sim b$ 1 E_7
6. $\sim b$ 4, 5 A_3
7. b 2, 3 A_2
8. $\therefore t$ 6, 7   contradict

(i) Invalid argument

(j) *Direct*
1. $k \rightarrow i$
2. $i \rightarrow (s \land l)$
3. $\sim l$
4. $\sim i \lor (s \land l)$ 2 E_7
5. $(\sim i \lor s) \land (\sim i \lor l)$ 4 E_5
6. $\sim i \lor l$ 5 A_6
7. $\sim i$ 3, 6 A_3
8. $\sim k$ 1, 7 A_2

Indirect
4. k P.I.P.
5. i 1, 4 A_1
6. $s \land l$ 2, 5 A_1
7. l 6 A_6
8. $\therefore \sim k$ 3, 7   contradict

(k) *Direct*

1. $t \lor \sim j$
2. $\sim p \land (t \to r)$
3. $r \to p$
4. $\sim p$ 2 A_6
5. $\sim r$ 3, 4 A_2
6. $t \to r$ 2 A_6
7. $\sim t$ 5, 6 A_2
8. $\sim j$ 1, 7 A_3

Indirect

4. j P.I.P.
5. t 1, 4 A_3
6. $\sim p$ 2 A_6
7. $\sim r$ 3, 6 A_2
8. $t \to r$ 2 A_6
9. $\sim t$ 7, 8 A_2
10. $\therefore \sim j$ 5, 9   contradict

(l) *Direct*

1. $r \to (c \land i)$
2. $\sim(i \lor \sim r)$
3. $\sim i \land r$ 2 E_6
4. r 3 A_6
5. $c \land i$ 1, 4 A_1
6. c 5 A_6
7. $c \lor o$ 6 A_7

Indirect

4. $\sim(c \lor o)$ P.I.P.
5. $\sim c \land \sim o$ 4 E_6
6. $\sim c$ 5 A_6
7. $\sim i \land r$ 2 E_6
8. r 7 A_6
9. $c \land i$ 1, 8 A_1
10. c 9 A_6
11. $\therefore c \lor o$ 6, 10   contradict

(m) *Direct*

1. $(c \land d) \to (s \lor h)$
2. $c \land \sim h$
3. d conditional premise
4. c 2 A_6
5. $c \land d$ 3, 4 A_5
6. $s \lor h$ 1, 5 A_1
7. $\sim h$ 2 A_6
8. s 6, 7 A_3
9. $d \to s$ 3, 8   conditional form of conclusion

Indirect

3. $\sim(d \to s)$ P.I.P.
4. $\sim(\sim d \lor s)$ 3 E_7
5. $d \land \sim s$ 4 E_6
6. $\sim s$ 5 A_6
7. c 2 A_6
8. d 5 A_6
9. $c \land d$ 7, 8 A_5
10. $s \lor h$ 1, 9 A_1
11. $\sim h$ 2 A_6
12. s 10, 11 A_3
13. $\therefore d \to s$ 6, 12   contradict

(n) *Direct*

1. $h \to \sim(r \land t)$
2. $t \land (r \lor f)$
3. $\sim f$
4. $r \lor f$ 2 A_6
5. r 3, 4 A_3
6. t 2 A_6
7. $r \land t$ 5, 6 A_5
8. $\sim h$ 1, 7 A_2

Indirect

4. h P.I.P.
5. $\sim(r \land t)$ 1, 4 A_1
6. $\sim r \lor \sim t$ 5 E_6
7. t 2 A_6
8. $\sim r$ 6, 7 A_3
9. $r \lor f$ 2 A_6
10. r 3, 9 A_3
11. $\therefore \sim h$ contradict

(o) *Direct*

1. $o \to (b \land a)$
2. $\sim(\sim g \land h)$
3. $o \land \sim g$ conditional premise

Indirect

3. $\sim[(o \land \sim g) \to a]$ P.I.P.
4. $\sim[\sim(o \land \sim g) \lor a]$ 3 E_7
5. $(o \land \sim g) \land \sim a$ 4 E_6
6. $\sim a$ 5 A_6

4.	o		3	A_6
5.	$b \wedge a$	1, 4	A_1
6.	a	5	A_6
7.	$(o \wedge \sim g) \rightarrow a$	conditional form of conclusion

7.	$o \wedge \sim g$		5	A_6
8.	o		5	A_6
9.	$b \wedge a$	1, 8	A_1
10.	a		9	A_6
11.	$\therefore (o \wedge \sim g) \rightarrow a$		6, 10	contradict

(p) *Direct*

1.	$p \vee (w \wedge s)$		
2.	$w \rightarrow (r \wedge o)$		
3.	$\sim r$		conditional premise
4.	$\sim w \vee (r \wedge o)$	2	E_7
5.	$(\sim w \vee r) \wedge (\sim w \vee o)$	4	E_5
6.	$\sim w \vee r$	5	A_6
7.	$\sim w$	3, 6	A_3
8.	$(p \vee w) \wedge (p \vee s)$	1	E_5
9.	$p \vee w$	8	A_6
10.	p	7, 9	A_3
11.	$\sim r \rightarrow p$	3, 10	conditional form of conclusion

Indirect

3.	$\sim(\sim r \rightarrow p)$		P.I.P.
4.	$\sim(r \vee p)$	3	E_7
5.	$\sim r \wedge \sim p$	4	E_6
6.	$\sim p$	5	A_6
7.	$\sim r$	5	A_6
8.	$\sim w \vee (r \wedge o)$	2	E_7
9.	$(\sim w \vee r) \wedge (\sim w \wedge o)$	8	E_5
10.	$\sim w \vee r$	9	A_6
11.	$\sim w$	7, 10	A_3
12.	$(p \vee w) \wedge (p \vee s)$	1	E_5
13.	$p \vee w$	12	A_6
14.	p	11, 13	A_3
15.	$\therefore r \rightarrow p$	6, 14	contradict

(q) *Direct*

1.	$(s \wedge d) \rightarrow a$		
2.	$s \wedge (p \vee \sim a)$		
3.	d		
4.	s	2	A_6
5.	$s \wedge d$	3, 4	A_5
6.	a	1, 5	A_1
7.	$p \vee \sim a$	2	A_6
8.	p	6, 7	A_3

Indirect

4.	$\sim p$		P.I.P.
5.	$p \vee \sim a$	2	A_6
6.	$\sim a$	4, 5	A_3
7.	$\sim(s \wedge d)$	1, 6	A_2
8.	$\sim s \vee \sim d$	7	E_6
9.	$\sim s$	3, 8	A_3
10.	s	2	A_6
11.	$\therefore p$	9, 10	contradict

(r) *Direct*

1.	$a \rightarrow (e \vee h)$		
2.	$\sim e \vee p$		
3.	$x \rightarrow \sim h$		
4.	$x \wedge \sim p$		
5.	$\sim p$	4	A_5
6.	$\sim e$	2, 5	A_3
7.	x	4	A_6
8.	$\sim h$	3, 7	A_1
9.	$\sim e \wedge \sim h$	6, 8	A_5
10.	$\sim(e \vee h)$	9	E_6
11.	$\sim a$	1, 10	A_2

Indirect

5.	a		P.I.P.
6.	$e \vee h$	1, 5	A_1
7.	$\sim p$	4	A_6
8.	$\sim e$	2, 7	A_3
9.	h	6, 8	A_3
10.	x	4	A_6
11.	$\sim h$	3, 10	A_1
12.	$\sim a$	9, 11	contradict

(s) *Direct*

1.	$r \to \sim l$		
2.	$l \to \sim t$		
3.	$r \lor t$		
4.	$t \lor r$	3	E_3
5.	$\sim t \to r$	4	E_7
6.	$\sim t \to \sim l$	5, 1	A_4
7.	$\sim l \lor \sim t$	2	E_7
8.	$t \lor \sim l$	6	E_7
9.	$\sim l \lor t$	8	E_3
10.	$\sim l$	7, 9	A_8

Indirect

4.	l		P.I.P.
5.	$\sim r$	1, 4	A_2
6.	$\sim t$	2, 4	A_1
7.	r	3, 6	A_3
8.	$\therefore \sim l$	5, 7	contradict

(t) *Direct*

1.	$(r \lor d) \to t$		
2.	$\sim r \to t$		
3.	$\sim (r \lor d) \lor t$	1	E_7
4.	$(\sim r \land \sim d) \lor t$	3	E_6
5.	$(\sim r \lor t) \land$ $(\sim d \lor t)$	4	E_5
6.	$\sim r \lor t$	5	A_6
7.	$r \lor t$	2	E_7
8.	t	6, 7	A_8

Indirect

3.	$\sim t$		P.I.P.
4.	$\sim (r \lor d)$	1, 3	A_2
5.	$\sim r \land \sim d$	4	E_6
6.	$\sim r$	5	A_6
7.	r	2, 3	A_2
8.	$\therefore t$	6, 7	contradict

(u) *Direct*

1.	$p \to (m \lor w)$		
2.	$m \to (t \lor i)$		
3.	$\sim (\sim t \to \sim p)$		
4.	$(f \lor \sim w) \land g$		
5.	$\sim g \lor \sim f$		conditional premise
6.	$\sim (t \lor \sim p)$	3	E_7
7.	$\sim t \land p$	6	E_6
8.	p	7	A_6
9.	$m \lor w$	1, 8	A_1
10.	g	4	A_6
11.	$\sim f$	5, 10	A_3
12.	$f \lor \sim w$	4	A_6
13.	$\sim w$	11, 12	A_3
14.	m	9, 13	A_3
15.	$t \lor i$	2, 14	A_1
16.	$\sim t$	7	A_6
17.	i	15, 16	A_3
18.	$(\sim g \lor \sim f) \to i$		conditonal form of conclusion

Indirect

5.	$\sim [(\sim g \lor \sim f) \to i]$		P.I.P.
6.	$\sim [\sim (\sim g \lor \sim f) \lor i]$	5	E_7
7.	$(\sim g \lor \sim f) \land \sim i$	6	E_6
8.	$\sim i$	7	A_6
9.	$\sim g \lor \sim f$	7	A_6
10.	g	4	A_6
11.	$\sim f$	9, 10	A_3
12.	$(f \lor \sim w)$	4	A_6
13.	$\sim w$	11, 12	A_3
14.	$\sim (t \lor \sim p)$	3	E_7
15.	$\sim t \land p$	14	E_6
16.	p	15	A_6
17.	$m \lor w$	1, 17	A_1
18.	m	13, 17	A_3
19.	$t \lor i$	2, 18	A_1
20.	$\sim t$	15	A_6
21.	i	19, 20	A_3
22.	$\therefore (\sim g \lor \sim f) \to i$	8, 21	contradict

Chapter 4

Exercises 4.1

1. (a)   (b)

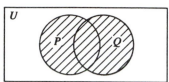   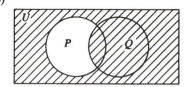

(c)   (d)

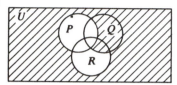   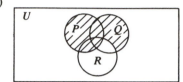

2. (a) invalid, since $\{(P \cup Q') \cap Q'\} \not\subseteq P$
   (b) valid, since $\{(P \cup Q) \cap (P' \cup R) \cap R'\} \subseteq Q$

3. (a) Premises   conclusion

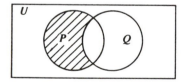

$(q \rightarrow \sim p) \wedge (\sim(\sim p))$ $\sim q$
$(\sim q \vee \sim p) \wedge p$ Q'
$(Q' \cup P') \cap P$

   valid, since $\{(Q' \cup P') \cap P\} \subseteq Q'$
   (b) invalid, since $\{(P' \cup Q) \cap P'\} \not\subseteq Q$
   (c) valid, since $[(P' \cup R) \cap (P \cup Q') \cap R'] \subseteq Q'$
   (d) Premises   Conclusion

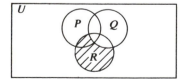   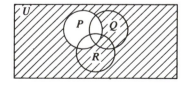

$(p \rightarrow \sim q) \wedge (q \rightarrow \sim r) \wedge r$ $\sim p$
$(P' \cup Q') \cap (Q' \cup R') \cap R$ P'

Invalid, since $[(P' \cup Q') \cap (Q' \cup R') \cap R] \not\subseteq P$

Exercises 4.2

1.

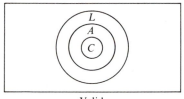

Valid

2.

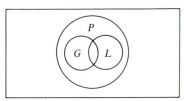

Invalid

3.

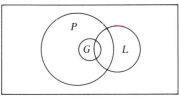

Valid

4.

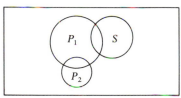

Invalid

5.

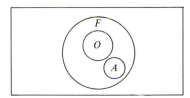

Invalid

6.

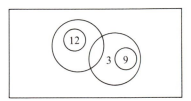

Invalid

7.

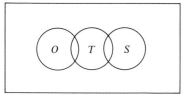

Invalid

8.

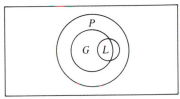

Invalid

9.

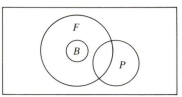

Invalid

10. $(x)(C_x \to A_x)$ 11. $(\exists x)(G_x \land \sim L_x)$ 12. $(x)(G_x \to P_x)$
     $(x)(A_x \to L_x)$ $(x)(L_x \to P_x)$ $(\exists x)(L_x \land \sim P_x)$
     $\therefore (x)(C_x \to L_x)$ $\therefore (\exists x)(G_x \land P_x)$ $\therefore (\exists x)(L_x \land \sim G_x)$

13. $(x)(G_x \to P_x)$ 14. $(\exists x)(P_x \land S_x)$ 15. $(x)(O_x \to F_x)$
     $(\exists x)(G_x \land \sim L_x)$ $(\exists x)(P_x \land T_x)$ $(x)(A_x \to F_x)$
     $\therefore (\exists x)(L_x \land \sim P_x)$ $\therefore (\exists x)(S_x \land T_x)$ $\therefore (\exists x)(A_x \land O_x)$

16. $(\exists x)(T_x \land S_x)$ 17. $(\exists x)(L_x \land D_x)$ 18. $(x)(B_x \to F_x)$
     $(\exists x)(T_x \land O_x)$ $(9)(9 \to D_x)$ $(\exists x)(F_x \land P_x)$
     $\therefore (\exists x)(O_x \land S_x)$ $\therefore (9)(9 \to L_x)$ $\therefore (\exists x)(B_x \land P_x)$

Exercises 4.3

1. (a)   1. Given   (b) 1a. Conditional Premise
          2. 1; (S_1) 2a. 1a; (S_4)
          3. 2; (E_7) 3a. 2a; (A_6)
          4. 3; (S_7) and (S_8) 4a. 1a, 3a; Conditional form of the conclusion
          5. 4; (S_1)

                                             1b. Conditional Premise
                                             2b. 1b; (A_7)
                                             3b. 2b; (S_3)
                                             4b. 1b, 3b; (A_5)
                                             5b. 4b; (S_4)
                                             6b. 1b, 5b; Conditional form of the conclusion
                                             ----
                                             7. 4a, 6b; (A_5)
                                             8. 7; (E_9)
                                             9. 8: (S_2)

2. (a) *Theorem 2* If $A \subset B$ and $B \subset C$, then $A \subset C$.

         1. $A \subset B$ Given
         2. $B \subset C$ Given
         3. $(x \in A) \to (x \in B)$ 1; (S_1)
         4. $(x \in B) \to (x \in C)$ 2; (S_1)
         5. $(x \in A) \to (x \in C)$ 3, 4; (A_4)
         6. $A \subset C$ 5; (S_1)

    (c) *Theorem 6* $(A')' = A$

         1. $(x \in A) \leftrightarrow (x \notin A')$ S_7
         2. $(x \notin A') \leftrightarrow [x \in (A')']$ S_7
         3. $(x \in A) \leftrightarrow (x \in (A')')$ 1, 2,; (A_4)
         4. $A = (A')'$ 3; (S_2)

    (d) *Theorem 8* $(A \cup B) = (B \cup A)$ and $(A \cap B) = (B \cap A)$

         We will prove the first of these two statements. The proof of the second
         is similar. To show $(A \cup B) = (B \cup A)$ we must show that $[x \in (A \cup B)]$
         $\leftrightarrow [x \in (B \cup A)]$.

1a. $[x \in (A \cup B)]$ Conditional Premise
2a. $(x \in A) \vee (x \in B)$ 1a; (S_3)
3a. $(x \in B) \vee (x \in A)$ 2a; (E_3)
4a. $[x \in (B \cup A)]$ 3a; (S_3)
5a. $[x \in (A \cup B)] \to [x \in (B \cup A)]$ 1a, 4a; Conditional form of
                                                                 the conclusion

---

1b. $[x \in (B \cup A)]$ Conditional premise
2b. $(x \in B) \vee (x \in A)$ 1b; (S_3)
3b. $(x \in A) \vee (x \in B)$ 2b; (E_3)
4b. $[x \in (A \cup B)]$ 3b; (S_3)
5b. $[x \in (B \cup A)] \to [x \in (A \cup B)]$ 1b, 4b; Conditional form of
                                                                 the conclusion

---

6. $\{[x \in (A \cup B)] \to [x \in (B \cup A)]\}$
   $\wedge \{[x \in (B \cup A)] \to [x \in (A \cup B)]\}$ 5a, 5b; (A_5)
7. $[x \in (A \cup B)] \leftrightarrow [x \in (B \cup A)]$ 6; (E_9)
8. $(A \cup B) = (B \cup A)$ 7; (S_2)

(f) *Theorem 11* $\varnothing' = U$ and $U' = \varnothing$

We will prove the first of these two statements. The proof of the second
is similar.

1. $x \in (\varnothing' \cup \varnothing) \leftrightarrow (x \in \varnothing')$ (S_6)
2. $x \in (\varnothing \cup \varnothing') \leftrightarrow (x \in U)$ (S_7)
3. $\{[x \in (\varnothing' \cup \varnothing) \to (x \in \varnothing')] \wedge$
   $\{(x \in \varnothing') \to [x \in (\varnothing' \cup \varnothing)]\}$ 1; (E_9)
4. $(x \in \varnothing') \to [x \in (\varnothing' \cup \varnothing]$ 3; (A_6)
5. $\{[x \in (\varnothing \cup \varnothing')] \to (x \in U)\}$
   $\wedge \{(x \in U) \to [x \in (\varnothing \cup \varnothing')]\}$ 2; (E_9)
6. $[x \in (\varnothing \cup \varnothing')] \to (x \in U)$ 5; (A_6)
7. $[x \in (\varnothing' \cup \varnothing)] \to (x \in U)$ 6; (Th. 8)
8. $(x \in \varnothing') \to (x \in U)$ 4, 7; (A_4)
9. $(x \in U) \to x \in (\varnothing \cup \varnothing')$ 5; (A_6)
10. $x \in (\varnothing' \cup \varnothing) \to (x \in \varnothing')$ 3; (A_6)
11. $x \in (\varnothing \cup \varnothing') \to (x \in \varnothing')$ 10; (Th. 8)
12. $(x \in U) \to (x \in \varnothing')$ 9, 11; (A_4)
13. $[(x \in \varnothing') \to (x \in U)]$
    $\wedge [(x \in U) \to (x \in \varnothing')]$ 8, 12; (A_5)
14. $(x \in \varnothing') \leftrightarrow (x \in U)$ 13; (E_9)
15. $\varnothing' = U$ 14; (S_2)

Switching Circuits

Exercises 4.4

1. (a) $(A \vee B) \wedge (\sim A \vee C)$
   (c) $A \wedge \{[B \wedge (C \vee \sim C)] \vee [(\sim A \vee \sim B) \wedge (C \vee B)]\}$

2. (a) $(A \lor B) \land (\sim A \lor C)$
(0 0 0) 0 (1 0 1 1)
1 4 1 <u>5</u> 2 1 3 1

(c) $A \land \{[B \land (C \lor \sim C)] \lor [(\sim A \lor \sim B) \land (C \lor B)]\}$
0 0 {[0 0 (1 1 0 1)] 1 [(1 0 1 1 0) 1 (1 1 0)]}
1 <u>11</u> 1 6 1 5 2 1 10 3 1 7 4 1 9 1 8 1

3. (a)

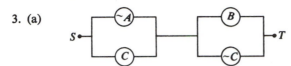

(c)

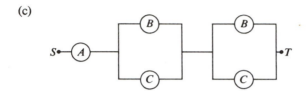

4. (a) $(\sim A \lor C) \land (B \lor \sim C)$
(1 0 1 0) 1 (1 1 1 0)
3 1 4 1 <u>6</u> 1 5 2 1

(c) $\{A \land [B \lor \overline{C}]\} \land (B \lor C)$
{0 0 1 1 0} 0 1 1 0
1 4 1 2 1 <u>5</u> 1 3 1

5. (a)

A	B	$[A \lor (B \land \sim B)] \land [A \land (A \lor B)]$
1	1	1 1 1 0 0 1 1 1 1 1 1 1 1
1	0	1 1 0 0 1 0 1 1 1 1 1 1 0
0	1	0 0 1 0 0 1 0 0 0 0 1 1
0	0	0 0 0 0 1 0 0 0 0 0 0 0
Ans.		1 4 1 3 2 1 <u>7</u> 1 6 1 5 1

(c) $A \land (B \lor C)$

Review Test

1. (a) Premise   (b) Conclusion

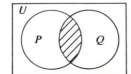

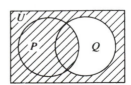

(c)   Yes, since $(Q \cap (P \cup Q')] \subset (P \cup Q')$

2. (a)   Premise   (b)   Conclusion

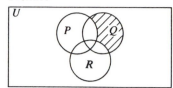   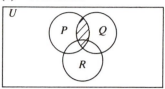

(c)   No, since $[(P \cup Q) \cap (P' \cup R) \cap R'] \not\subset (P \cap Q)$

3. $[A \wedge (B \vee C)] \vee [(\sim A \vee D) \wedge E]$

4. a   5. d   6. e   7. b   8. a   9. b   10. c   11. d

Chapter 5

Exercises 5.1

1. (a) $S = \{h, t\}$ (b) $P(h) = 1/2$

2. $P(t) = .3$

3. (a) $S = \{(h, t), (h, h), (t, t), (t, h)\}$
   (b)   1. $P(h, h) = 1/4$
         2. $P(h, t) = 1/4$
         3. $P(h, t) + P(t, h) = 1/4 + 1/4 = 1/2$
         4. $P(t, t) = 1/4$

4. (a) $S = \{(h, h, h), (h, h, t), (h, t, h), (h, t, t), (t, h, h), (t, h, t), (t, t, h), (t, t, t)\}$
   (b)   1. $P(h, h, h) = 1/8$
         2. $P(h, h, t) + P(h, t, h) + P(t, h, h) = 1/8 + 1/8 + 1/8 = 3/8$
         3. $P(t, t, t) = 1/8$
         4. $P(h, t, t) + P(t, h, t) + P(t, t, h) = 1/8 + 1/8 + 1/8 = 3/8$
         5. $P(h, h, h) + P(h, h, t) + P(h, t, h) + P(h, t, t) + P(t, h, h) + P(t, h, t)$
            $+ P(t, t, h) = 1/8 + 1/8 + 1/8 + 1/8 + 1/8 + 1/8 + 1/8 = 7/8$
         6. $P(h, t, t) + P(t, h, t) + P(t, t, h) + P(t, t, t) = 1/8 + 1/8 + 1/8 + 1/8 = 1/2$
         7. $P(t, h, t) = 1/8$

5. (a) $P(\text{ace of spades}) = 1/52$
   (b) $P(\text{ace of spades}) + P(\text{ace of clubs}) = 1/52 + 1/52 = 2/52 = 1/26$
   (c) $P(\text{any of the 4 aces}) = 4/52 = 1/13$
   (d) $P(\text{any one of 13 spades}) = 13/52 = 1/4$
   (e) $P(\text{any one of 12 face cards}) = 12/52 = 3/13$
   (f) $P(\text{any 7 or 8}) = 8/52 = 2/13$
   (g) $P(\text{any card}) = 52/52 = 1$

6. (a) P(ace of spades) $= 1/13$
   (b) P(any spade) $= 13/13 = 1$
   (c) P(any diamond) $= 0$
   (d) $P(3, 6, 9) = 3/13$
   (e) P(Jack, Queen, King) $= 3/13$

7. (a) $P(a) = 1/9, P(b) = 1/9, P(c) = 2/9, P(d) = 2/9, P(e) = 3/9 = 1/3$
   (b) 1. $P(a) + P(e) = 1/9 + 3/9 = 4/9$
       2. $P(b) + P(c) + P(d) = 1/9 + 2/9 + 2/9 = 5/9$
       3. $P(a) + P(c) + P(e) = 1/9 + 2/9 + 3/9 = 6/9 = 2/3$

8. (a) $P(1) = P(2) = P(3) = 1/10, P(4) = P(5) = 2/10, P(6) = 3/10$
   (b) 1. $P(2) + P(4) + P(6) = 1/10 + 2/10 + 3/10 = 6/10$
       2. $P(1) + P(3) + P(5) = 1/10 + 1/10 + 2/10 = 4/10$
       3. 0
       4. $P(1) + P(2) + P(3) + P(4) + P(5) + P(6) = 1/10 + 1/10 + 1/10$
          $+ 2/10 + 2/10 + 3/10 = 10/10 = 1$
       5. $P(2) + P(3) + P(4) = 1/10 + 1/10 + 2/10 = 4/10$

9. (a) $S = \{$(Alice, Bruce, Carl), (Alice, Bruce, Donna),
       (Alice, Donna, Ed), (Alice, Bruce, Ed), (Alice, Carl, Donna),
       (Alice, Carl, Ed), (Bruce, Carl, Donna), (Bruce, Carl, Ed),
       (Bruce, Donna, Ed), (Carl, Donna, Ed)$\}$

   (b) 1. P(Bruce, Carl, Ed) $= 1/10$
       2. P(Alice, Bruce, Donna) $+ P$(Alice, Donna, Ed) $+$
          P(Alice, Carl, Donna) $= 1/10 + 1/10 + 1/10 = 3/10$
       3. $3/10$
       4. P(Alice, Bruce, Carl) $+ P$(Alice, Bruce, Ed) $+$
          P(Alice, Carl, Ed) $+ P$(Bruce, Carl, Donna) $+$
          P(Bruce, Carl, Ed) $+ P$(Bruce, Donna, Ed) $+$
          P(Carl, Donna, Ed) $= 1/10 + 1/10 + 1/10 + 1/10 + 1/10 + 1/10$
          $+ 1/10 = 7/10$
       5. $6/10$
       6. $4/10$

10. (a)

Political affiliation Sex	Democrat	Republican	Total
Male	40	20	60
Female	25	15	40
Total	65	35	100

   (b) $P(C_1) = .4,$ $P(C_2) = .2,$ $P(C_3) = .25,$ $P(C_4) = .15$
   (c) P(Male) $= .6$
   (d) P(Republican) $= .35$

Exercises 5.2

1. (a) 1/4   (b) 3/4   (c) 3/4   (d) 1/2

2. (a) $S = \{(1, 1), (1, 2), (1, 3), (2, 1), (2, 2), (2, 3), (3, 1), (3, 2), (3, 3)\}$
   (b) 8/9   (c/) 4/9   (d) 1   (e) 1/3

3. (a) 51/52   (b) 3/52   (c) 22/52   (d) 15/52   (e) 19/52

4. (a) 1/3   (b) 1/3   (c) 8/9   (d) 2/3   (e) 1/9

5. (a)

	G	G'	Total
F	10	5	15
F'	15	20	35
Total	25	25	50

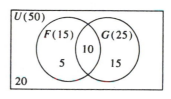

   (b) 1/2   (c) 30/50   (d) 20/50   (e) 15/50   (f) 45/50
   (g) 30/50

6. (b) 30/40   (c) 20/40   (d) 20/40   (e) 5/40   (f) 30/40

7. (b) 15/50   (c) 20/50   (d) 30/50   (e) 20/50   (f) 35/50

8. 4/13

Exercises 5.3

1. (a) 1/78   (b) 1/13   (c) 1/12   (d) 1/4

2. (a) $S = \{(h, h) (h, t), (t, h), t, t)\}$
   (b) 3/4   (c) 1/2   (d) 1/2   (e) 1/4   (f) 1/2

3. (a) $S = \{(h,\ h), (h, t), (t, h), (t, t)\}$
   (b) 68/100   (c) .4   (d) .4   (e) .32   (f) 32/44

4. (a) $S = \{(1, 1), (1, 2), (1, 3), (1, 4), (2, 1), (2, 2), (2, 3), (2, 4), (3, 1), (3, 2),$
          $(3, 3) (3, 4), (4, 1), (4, 2), (4, 3), (4, 4)\}$
   (b) 7/16   (c) 9/16   (d) 11/16   (e) 3/16   (f) 3/10
   (g) 1/2   (h) 1/2   (i) 1/2

5. (a)

	Failed Science	Passed Science	Total
Failed Math	3	2	5
Passed Math	7	28	35
Total	10	30	40

(b) 28/40   (c) 7/10   (d) 1/15

6. (b) 2/3   (c) 1/3   (d) 7/9   (e) 7/8   (f) The two events are not independent.

7. 7

8. .8 (assuming independence)

9. (a) 1/1296   (b) 1/16   (c) 671/1296   (d) 1/16   (e) 1/4

10. (a) 26/100   (b) 7/100   (c) The events are not independent.
    (d) 3/10

Exercises 5.4
1.

Event (E)	$P(E)$	x	Number on top of die	$x \cdot P(E)$
E_1	1/6	x_1	1	1/6
E_2	1/6	x_2	2	2/6
E_3	1/6	x_3	3	3/6
E_4	1/6	x_4	4	4/6
E_5	1/6	x_5	5	5/6
E_6	1/6	x_6	6	6/6

$$E(x) = \Sigma\, x\, P(x) = 1/6 + 2/6 + 3/6 + 4/6 + 5/6 + 6/6 = 21/6 = 3.5$$

2. $E(x) = 1$

3. $P(1) + P(2) + P(3) + P(4) + P(5) + P(6) = 1$
   $P(1) = P(2) = P(4) = P(5) = x$ $P(3) = P(6) = 2x$
   $x + x + 2x + x + x + 2x = 1$
   $8x = 1$
   $x = 1/8$

Event (E)	$P(E)$	x	Number on top of die	$x \cdot P(E)$
E_1	1/8	x_1	1	1/8
E_2	1/8	x_2	2	2/8
E_3	2/8	x_3	3	6/8
E_4	1/8	x_4	4	4/8
E_5	1/8	x_5	5	5/8
E_6	2/8	x_6	6	12/8

$$E(x) = \Sigma x P(E) = 1/8 + 2/8 + 6/8 + 4/8 + 5/8 + 12/8 = 30/8 = 3.75$$

4. $E(x) = 1.2$ males   5. $E(y) = -\$38/36$ 6. $E(y) = \$1.62$ 7. $\$20.20$

Review Test

1. b	2. c	3. b	4. a	5. c
6. b	7. a	8. b	9. b	10. c
11. d	12. b	13. c	14. b	15. e
16. a	17. c	18. c	19. c	20. c

Chapter 6

Exercises 6.1

1. (a) 110000000_2 (b) 123_8 (c) 11001_2 (d) 614_8
   (e) 10011111_2 (f) 274_8 (g) 1101.1_2 (h) 15.4_8
   (i) 11001.101_2 (j) 31.5_8

2. (a) 43   (b) 215   (c) 11   (d) 440
   (e) 5.75   (f) 7.6875   (g) 13.25   (h) 25.25
   (i) 7.5   (j) 35.75

3. (a) 27.6_8 (b) 5.2_8 (c) 6.4_8 (d) 25.34_8

4. (a) 10100001.011_2 (b) 10.1011_2
   (c) 10110.001111_2 (d) 111110.101010_2

5. (a) 110110.10_2 (b) 1410.76_8 (c) 59.25 (d) 10001_2
   (e) 101.1_2 (f) 302.63_8 (g) 22.0_8 (h) 1000.0111_2
   (i) 1111.010101_2 (j) 2272.334_8 (k) 1101011.1111_2 (l) 111_2
   (m) 101_2 (n) 4_8 (o) 11.2_8 (p) 100_2

7. (a) $b0_{16}$ (b) ff_{16} (c) $c.8_{16}$ (d) $1b.4_{16}$

8. (a) 987 (b) 3567 (c) 302 (d) 175.5

9. (a) 111110_2 (b) 59_{16} (c) 1000_{16} (d) 575_{16}
   (e) 750_{16} (f) $f10_{16}$ (g) $32a_{16}$ (h) $e49a_{16}$
   (i) 4_{16} (j) 8_{16}

Exercises 6.2
1.

A	B	Difference	Borrow
0	0	0	0
1	0	1	0
0	1	1	1
1	1	0	0

Difference   Borrow

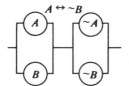

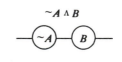

Exercises 6.3

1.

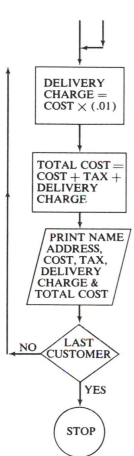

3.

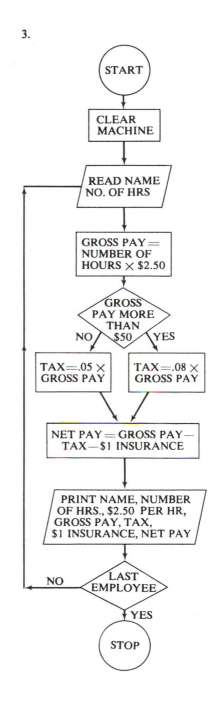

5.

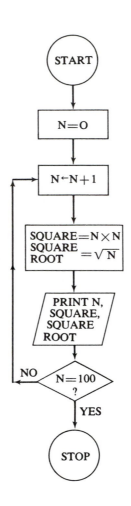

START

N=0

N←N+1

SQUARE=N×N
SQUARE
ROOT   =√N̄

PRINT N,
SQUARE,
SQUARE
ROOT

NO   N=100
?

YES

STOP

Review Test

1. c	2. e	3. b	4. a	5. c	6. e
7. b	8. c	9. d	10. a	11. b	12. e
13. d	14. b	15. a	16. a	17. a	18. c
19. b	20. b	21. b	22. b	23. d	24. c
25. c	26. c	27. c	28. e	29. b	30. c

Chapter 7

Exercises 7.1

1. (a) 413.65   (b) 709.02   (c) 2001
   (d) .52   (e) 6.87

2. (a) $(4 \times 10^2) + (0 \times 10^1) + (2 \times 10^0) + (5 \times 10^{-1}) + (0 \times 10^{-2})$
       $+ (8 \times 10^{-3})$
   (b) $(3 \times 10^1) + (0 \times 10^0) + (0 \times 10^{-1}) + (0 \times 10^{-2}) + (3 \times 10^{-3})$
       $+ (4 \times 10^{-4})$
   (c) $-[(4 \times 10^0) + (5 \times 10^{-1}) + (2 \times 10^{-2})]$
   (d) $(0 \times 10^0) + (0 \times 10^{-1}) + (0 \times 10^{-2}) + (1 \times 10^{-3}) + (7 \times 10^{-4})$
   (e) $(6 \times 10^2) + (9 \times 10^1) + (0 \times 10^0)$

3. (a) $.8\overline{3}$ (b) $.\overline{3}$ (c) $.\overline{2}$
   (d) $.\overline{142857}$ (e) $.0\overline{9}$

4. (a) $N = 2/3$ (b) $N = 9/11$ (c) $N = 136/33$
   (d) $N = -664/55$ (e) $N = -76/15$

5. (a) irrational   (b) rational   (c) irrational
   (d) rational   (e) irrational   (f) irrational
   (g) rational   (h) irrational   (i) rational

6. (a) $R^1 = \{x | x \text{ is an irrational number}\}$
   (b)

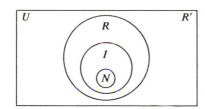

   (c) (a) True   (b) True   (c) False
       (d) False   (e) False   (f) False

7. (a) real, rational, integer, nonnegative, natural
   (b) real, rational, integer
   (c) real, irrational
   (d) real, rational, integer, nonnegative
   (e) real, irrational
   (f) real, irrational
   (g) real, rational
   (h) real, rational
   (i) real, rational
   (j) real, rational, integer
   (k) real, rational
   (l) real, rational, integer, nonnegative, natural
   (m) real, rational, integer, nonnegative, natural
   (n) real, rational

8. (a) 9/40   (b) 7/10   (c) 71/90
   (d) The set of integers is not dense.

Exercises 7.2
1. (a) 3   (b) 2   (c) 3   (d) 3   (e) 2
   (f) 2   (g) 4   (h) 2   (i) 3   (j) 0

2. (a) 0   (b) 0   (c) 1   (d) 0   (e) 0
   (f) 3   (g) 4   (h) cannot be solved
   (i) 3   (j) cannot be solved

3.

+	0	1	2		×	0	1	2	{0, 1, 2}
0	0	1	2		0	0	0	0	
1	1	2	0		1	0	1	2	
2	2	0	1		2	0	2	1	

   (a) yes   (b) yes   (c) yes   (d) yes   (e) yes
   (f) yes   (g) 0   (h) 1   (i) $-0 = 0$, $-1 = 2$ & $-2 = 1$
   (j) 0^{-1} does not exist, $1^{-1} = 1$ and $2^{-1} = 2$ (k) yes
   (l) 1   (m) 0   (n) 0   (o) 2
   (p) cannot be solved since 0^{-1} does not exist.

4.

+	0	1	2	3		×	0	1	2	3	{0, 1, 2, 3}
0	0	1	2	3		0	0	0	0	0	
1	1	2	3	0		1	0	1	2	3	
2	2	3	0	1		2	0	2	0	2	
3	3	0	1	2		3	0	3	2	1	

   (a) yes   (b) yes   (c) yes   (d) yes   (e) yes
   (f) yes   (g) 0   (h) 1
   (i) $-0 = 0$, $-1 = 3$, $-2 = 2$ & $-3 = 1$
   (j) 0^{-1} does not exist, $1^{-1} = 1$, 2^{-1} does not exist and $3^{-1} = 3$
   (k) yes   (l) 1   (m) 1   (n) 1   (o) 1
   (p) cannot be solved

5. (a) yes   (b) yes   (c) yes   (d) e
   (e) $s^{-1} = t$, $e^{-1} = e$, $t^{-1} = s$
   (f) We cannot test for a distributive property with only one operation (*).

6. (a) yes   (b) yes   (c) yes   (d) yes   (e) yes
   (f) yes   (g) T (h) F
   (i) The conjunctive inverse of T is T. There is no conjunctive inverse for F
   (j) The disjunctive inverse of F is F. There is no disjunctive inverse for T
   (k) yes   (l) yes

7.

	Real Numbers	Irrational Numbers	Rational Numbers	Integers	Nonnegative Integers	Natural Numbers
(a)	T	F	T	T	T	T
(b)	T	T	T	T	T	T
(c)	T	T	T	T	T	T
(d)	T	F	T	T	T	F
(e)	T	T	T	T	F	F
(f)	T	F	T	T	T	T
(g)	T	T	T	T	T	T
(h)	T	T	T	T	T	T
(i)	T	F	T	T	T	T
(j)	T	T	T	F	F	F
(k)	T	T	T	T	T	T

Exercises 7.3

1. (a) inverse postulate
   (b) distributive postulate
   (c) identity postulate
   (d) identity postulate
   (e) inverse postulate
   (f) closure postulate
   (g) commutative postulate

2. *Theorem 1B* If a, b, and c are elements of set S, and $a = b$, then $a \times c = b \times c$.

Proof

(1) $(a \times c) \in S$ — Closure P_5
(2) $(a \times c) = (a \times c)$ — Reflexive P_1
(3) $a = b$ — Given
(4) $(a \times c) = (b \times c)$ — Substitution (3) in (2) — P_4

4. If a, b, and c are elements of set S, and $a \times c = b \times c$ and $c \neq 0$, then $a = b$.

Proof

(1) $(a \times c) \in S$ and $(b \times c) \in S$ — Closure — P_5
(2) $(a \times c) = (b \times c)$ — Given
(3) $c^{-1} \in S$ — Inverse — P_{10}
(4) $(a \times c) \times c^{-1} = (b \times c) \times c^{-1}$ — (2) & (3) Th — $1B$
(5) $a \times (c \times c^{-1}) = b \times (c \times c^{-1})$ — (4) Associative — P_7
(6) $(c \times c^{-1}) = 1$ — Inverse — P_{10}
(7) $a \times 1 = b \times 1$ — Substitution (6) in (5) — P_4
(8) $a \times 1 = a$ and $b \times 1 = b$ — Identity — P_9
(9) $a = b$ — Substitution (8) in (7) — P_4

5. (1) Closure — P_5
   (4) Commutative — P_6
   (3) Distributive — P_8
   (5) Commutative — P_6

11. If a is an element of set S, then $a \times 1^{-1} = a$.

Proof

(1)	$a \times 1 = a$	Identity $- P_9$
(2)	$1 = 1^{-1}$	Th.$- 6B$
(3)	$a \times 1^{-1} = a$	Substitution (2) in (1) $- P_4$

Review Test

1. c	2. d	3. a	4. a	5. b	6. a	7. b
8. b	9. b	10. b	11. b	12. b	13. a	14. c
15. a	16. a	17. a	18. c	19. b	20. b	21. d
22. a	23. c	24. b	25. c	26. c	27. a	28. b

Chapter 8

Exercises 8.1

(1–6)

	(1)	(2)	(3)	(4)	(5)	(6)
Mean	5.5	9	4.25	12	11.1	14
Median	5.5	9	4	7	10 m	11
Mode	none	13	4	none	10	bimodal (8 & 18)

Exercises 8.2

(1–5)

	(1)	(2)	(3)	(4)	(5)
Mean	7	8	12	11	11
Median	7	8	13	13.5	7
Mode	none	10	21	17	bimodal (4 & 20)
Range	12	14	20	20	20
Variance	16	18.29	49.43	53.75	68
Standard Deviation	4	4.28	7.03	7.33	8.25

Exercises 8.3

1. $76 = P_{45}$ 2. $85 = P_{60}$ 3. 75   4. .7680   5. 1

6. (a)

Value	Tally	Frequency
1	||	2
2	|||	3
3	||||	4
4	||||	4
5	||	2
6	|||	3
7	||||	4

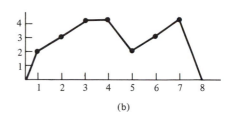

(b)

7. 50%   8. 84%   9. 68%   10. 5%   11. 32

12. 83.8%   13. 2   14. $u = 19$

Chapter Exercises

1. (b)   2. (c)   3. (a)   4. (d)   5. (b)   6. (c)   7. (d)

8. (b)   9. (d)   10. (c)   11. (a)   12. (c)   13. (a)   14. (c)

15. (b)   16. (d)   17. (a)   18. (c)   19. (d)   20. (c)   21. (b)

22. (a)   23. (d)   24. (b)   25. (c)   26. (b)   27. (c)   28. (b)

INDEX